Separation Hydrometallurgy of Rare Earth Elements

Jack Zhang • Baodong Zhao • Bryan Schreiner

Separation Hydrometallurgy of Rare Earth Elements

Jack Zhang
Saskatchewan Research Council
Saskatoon, SK, Canada

Baodong Zhao
Saskatchewan Research Council
Saskatoon, SK, Canada

Bryan Schreiner
Saskatchewan Research Council
Saskatoon, SK, Canada

ISBN 978-3-319-80302-9 ISBN 978-3-319-28235-0 (eBook)
DOI 10.1007/978-3-319-28235-0

Springer Cham Heidelberg New York Dordrecht London

Softcover reprint of the hardcover 1st edition 2016

Printed on acid-free paper

Springer International Publishing AG Switzerland is part of Springer Science+Business Media (www.springer.com)

This book is dedicated to all the researchers that contributed to the development of rare earth element separation technology. In particular we recognize Professors Guangxian Xu and Deqian Li for their extensive contributions to the separation hydrometallurgy of rare earth elements. Professors Xu and Li were engaged in the study of separation chemistry and clean hydrometallurgy of rare earths each for over 50 years and received numerous awards. Professors Xu's and Li's achievements were critical to rare earth separation hydrometallurgy. Professor Li's initiative and encouragement convinced the authors to undertake the development of rare earth separation hydrometallurgy at the Saskatchewan Research Council and inspired the writing of this book for all of which we are very grateful.

Preface

The expanding population, economic growth, and desire for new technologies are driving the demand for rare earth elements which are critical. This growth requires new production of rare earths that is secure and environmentally sustainable. There are hundreds of rare earth deposits in the world at different stages of development, from exploration to proven mineral reserves, and detailed engineering.

Canada has a significant proportion of these resources. Canada has significant, high-quality, and unique rare earth deposits across the country and has a long history of mining and metallurgical processing, and mineral development and production. The rare earth sector is just emerging in Canada; therefore we need to ensure that Canada has the capability to develop these resources to their highest value. High-purity rare earth element production is a critical component in commercial defence and other high-value applications.

The rare earth needs for the economy are wide and complex. Customers include manufacturers of rare earth functional materials, such as magnets, catalysts, metallurgical additives, batteries, polishing powders, phosphors, glass additives, and ceramics. These customers supply products that are used in a wide variety of industries including health care; hybrid, electric, and other vehicles; defense; wind power; communications; lighting; fiber optic; and other applications. These products and services are enabled by rare earth materials to provide solutions to manufacturing needs.

Mineral processing and hydrometallurgy, including separation, of rare earths are critical for Canada to advance the development of resources to contribute to the supply chain and acquire greater economic benefits from the emerging rare earth sector. Canada has more than 50 % of the world's demonstrated rare earth resources, with a number of projects at advanced stages compared to other jurisdiction in the world. This is an opportunity for Canada to enhance its expertise in exploration, development, mining, and metallurgy to develop the rare earth industry that will generate substantial value.

The Saskatchewan Research Council has taken the initiative to develop expertise, capabilities, and facilities to enable the rare earth industry to succeed with the development and delivery of timely solutions needed for Canada's emerging rare earth sector. This includes the entire rare earth production cycle from exploration, processing, and separation of the rare earth elements. The purpose of this book is to provide the background and basis for the advancement of separation hydrometallurgy in Canada. From this basis we can develop and advance our capabilities to enable Canada and other parts of the world to successfully build supplies of these critical rare earth elements for the future.

Saskatchewan Research Council
Saskatoon, SK, Canada

Jack Zhang
Baodong Zhao
Bryan Schreiner

Acknowledgements

Dr. Bryan Schreiner, Chief Geoscientist and Manager Minerals at the Saskatchewan Research Council (SRC), has provided reviews throughout the authorship process of this manuscript. He worked closely with the authors Jack Zhang and Baodong Zhao to enable the production of this publication.

Contents

Chapter 1
Rare Earth Elements and Minerals

1.1 Rare Earth Elements (REEs)

The rare earth elements (REEs) include the 15 lanthanide elements, plus yttrium and scandium. They are Group 3 transition metals of Period 4, Period 5, and Period 6 in the d-block of the periodic table. The lanthanide elements are generally categorized into two groups based on their double-salt solubility: the light rare earth elements (LREEs), also called cerium group (Ce-group), i.e., lanthanum to europium (atomic number $Z=57$ to 63); and the heavy rare earth elements (HREEs), i.e., gadolinium to lutetium ($Z=64$ to 71). Yttrium is usually grouped with the HREEs due to their similarity in physical and chemical properties, as well as their co-occurrence in nature. That's why the HREEs group is also called the yttrium group (Y-group). The REEs are also categorized into three groups based on their extractability with acidic extractants: the REEs from lanthanum to neodymium are called LREEs; from samarium to gadolinium, the medium rare earth elements (MREEs) and from terbium to lutetium including yttrium, the HREEs. Scandium is included in neither the LREEs group nor the HREEs group because scandium does not occur in rare earth minerals due to its significantly different chemistry determined by its smaller atomic and trivalent ionic radii than those of the other rare earth elements.

Table 1.1 lists the REEs, atomic numbers, and abundances in the upper earth's crust (Taylor and McLennan 1985). REEs with low atomic numbers are more abundant than those with high atomic numbers. More interestingly, REEs with even atomic numbers are much more abundant than adjacent lanthanide elements with odd atomic numbers (Table 1.1).

J. Zhang et al., *Separation Hydrometallurgy of Rare Earth Elements*,
DOI 10.1007/978-3-319-28235-0_1

Table 1.1 REEs, atomic numbers, and abundances in the upper earth crust

Element	Symbol	Atomic number	Abundance (ppm)
Lanthanum	La	57	30
Cerium	Ce	58	64
Praseodymium	Pr	59	7.1
Neodymium	Nd	60	26
Promethium	Pm	61	N/A
Samarium	Sm	62	4.5
Europium	Eu	63	0.88
Gadolinium	Gd	64	3.8
Terbium	Tb	65	0.64
Dysprosium	Dy	66	3.5
Holmium	Ho	67	0.80
Erbium	Er	68	2.3
Thulium	Tm	69	0.33
Ytterbium	Yb	70	2.2
Lutetium	Lu	71	0.32
Scandium	Sc	21	13.6
Yttrium	Y	39	22

1.1.1 Scandium (Sc)

Scandium is the first element of Period 4 with the atomic number of 21 in Group 3. Its electronic configuration is $3d^1 4s^2$. Sc is a silvery-white non-lanthanide rare earth element with many applications in aluminum alloys for aerospace and guns, and consumer products such as phosphors and fluorescent and energy-saving lamps.

1.1.2 Yttrium (Y)

Yttrium is the first element of Period 5 with the atomic number of 39 in Group 3. Its electronic configuration is $4d^1 5s^2$. Y is a silvery-metallic non-lanthanide rare earth element. Y is widely used in consumer products, garnets, lasers, phosphors, alloys, medical devices, and superconductors.

1.1.3 Lanthanum (La)

Lanthanum is the first lanthanide rare earth element of Period 6 with the atomic number of 57 in Group 3. Its electronic configuration is $5d^1 6s^2$. La is a silvery-white metallic rare earth element. La is used in optical glasses, hybrid batteries and engines, alloys, and catalysts for petroleum cracking.

1.1.4 Cerium (Ce)

Cerium is the second lanthanide rare earth element of Period 6 with the atomic number of 58. It is the most abundant rare earth element. The electronic configuration of Ce is $4f^1 5d^1 6s^2$. Ce metal has a silvery-white luster. Ce is widely used in glass polishing, solar panels, light-emitting diodes (LEDs), catalysts, alloys, pigments, and permanent magnets.

1.1.5 Praseodymium (Pr)

Praseodymium is the third lanthanide rare earth element of Period 6 with the atomic number of 59. Its electronic configuration is $4f^3 6s^2$. Pr is a soft, silvery, malleable, and ductile metal with major application in permanent magnets, aircraft engine alloys, computerized axial tomography (CAT) scan machines, and fiber optics.

1.1.6 Neodymium (Nd)

Neodymium is the fourth lanthanide rare earth element of Period 6 with the atomic number of 60. Its electronic configuration is $4f^4 6s^2$. Nd is a soft and silvery metal with major applications in permanent magnets, catalysts, and hybrid engines.

1.1.7 Promethium (Pm)

Promethium is the rarest lanthanide element due to the lack of stable isotopes. It is in Period 6 with the atomic number of 61. The electronic configuration of Pm is $4f^5 6s^2$. Due to the lack of stable isotopes, the major application of Pm is for research purposes.

1.1.8 Samarium (Sm)

Samarium is the sixth lanthanide rare earth element of Period 6 with the atomic number of 62. The electronic configuration of Sm is $4f^6 6s^2$. Sm is a silvery metal and has important applications in permanent magnets for defense applications, medical devices, and cancer drugs.

1.1.9 Europium (Eu)

Europium is the seventh lanthanide rare earth element of Period 6 with the atomic number of 63. The electronic configuration of Eu is $4f^7 6s^2$. Eu is a moderately hard, silvery metal with major applications in phosphors and panel displays such as red and green colors in TV sets, control rods for nuclear reactors and alloys.

1.1.10 Gadolinium (Gd)

Gadolinium is the eighth lanthanide rare earth element of Period 6 with the atomic number of 64. The rare earth elements from Gd are normally categorized into heavy rare earth elements. Its electronic configuration is $4f^7 5d^1 6s^2$. Gd is silvery-white, malleable and ductile metal with a variety of specialized uses such as in nuclear reactors, magnets, high strength alloys, phosphors, and garnets.

1.1.11 Terbium (Tb)

Terbium is the ninth lanthanide rare earth element of Period 6 with the atomic number of 65. Its electronic configuration is $4f^9 6s^2$. Tb is a hard, silvery-white rare earth metal with major applications in magnets and phosphors.

1.1.12 Dysprosium (Dy)

Dysprosium is the tenth lanthanide rare earth element of Period 6. Its atomic number is 66 and its electronic configuration is $4f^{10} 6s^2$. Dy has a metallic silvery luster. Its major applications are found in magnets, alloys, nuclear reactors, and hybrid cars.

1.1.13 Holmium (Ho)

Holmium is the eleventh lanthanide rare earth element of Period 6. Its atomic number is 67 with an electronic configuration of $4f^{11} 6s^2$. It is a soft and malleable, silvery-white metal. Ho is mainly used in magnets, nuclear reactors, medical devices, and fiber optics.

1.1.14 Erbium (Er)

Erbium is the twelfth lanthanide rare earth element of Period 6. Its atomic number is 68 with an electronic configuration of $4f^{12}6s^2$. It is silvery-white metal with major applications found in nuclear reactors, optical products, and medical devices.

1.1.15 Thulium (Tm)

Thulium is the thirteenth lanthanide rare earth element of Period 6. It is one of least abundant rare earth elements. Its atomic number is 69 and its electronic configuration is $4f^{13}6s^2$. It has a bright silvery-gray luster. Its major applications are found in lasers and X-ray sources.

1.1.16 Ytterbium (Yb)

Ytterbium is the fourteenth lanthanide rare earth element of Period 6. Its atomic number is 70 with electronic configuration of $4f^{14}6s^2$. It is a soft and malleable ductile metal with a silvery luster. Yb is mainly used in lighting and health care products as well as alloys.

1.1.17 Lutetium (Lu)

Lutetium is the fifteenth lanthanide rare earth element of Period 6. It is one of the least abundant rare earth elements. Lu has an atomic number of 71 with electronic configuration of $4f^{14}5d^1 6s^2$. Lutetium is the smallest among the lanthanide atoms due to lanthanide contraction. It is a silvery metal with major applications in catalysts.

1.2 REE Electronic Configurations and Lanthanide Contraction

1.2.1 REE Electronic Configurations

Table 1.2 lists the electronic configurations, atomic and trivalent ionic radii as well as the valence state of REEs. There are two types of atomic electronic configurations in lanthanide elements based on the Lowest Energy Principle, $[Xe]4f^n6s^2$ or $[Xe]4f^{n-1}5d^16s^2$, here [Xe] is the electronic configuration of krypton and $n = 1 - 14$.

Table 1.2 Electronic configuration of REEs

REEs	Atomic number	Atomic electronic configuration						RE^{3+} electronic configuration	Atomic radii Å	RE^{3+} radii Å	Valence state
		K, *L*, *M*, N		*O*			*P*				
		Inner	4*f*	5 *s*	5*p*	5*d*	6 *s*				
La	57	Full 46	0	2	6	1	2	[Xe]$4f^0$	1.877	1.061	+3
Ce	58		1	2	6	1	2	[Xe]$4f^1$	1.825	1.034	+3, +4
Pr	59		3	2	6		2	[Xe]$4f^2$	1.828	1.013	+3, +4
Nd	60		4	2	6		2	[Xe]$4f^3$	1.821	0.995	+3
Pm	61		5	2	6		2	[Xe]$4f^4$	(1.810)	0.979	+3
Sm	62		6	2	6		2	[Xe]$4f^5$	1.802	0.964	+2, +3
Eu	63		7	2	6		2	[Xe]$4f^6$	2.042	0.950	+2, +3
Gd	64		7	2	6	1	2	[Xe]$4f^7$	1.802	0.938	+3
Tb	65		9	2	6		2	[Xe]$4f^8$	1.782	0.923	+3, +4
Dy	66		10	2	6		2	[Xe]$4f^9$	1.773	0.908	+3
Ho	67		11	2	6		2	[Xe]$4f^{10}$	1.766	0.894	+3
Er	68		12	2	6		2	[Xe]$4f^{11}$	1.757	0.881	+3
Tm	69		13	2	6		2	[Xe]$4f^{12}$	1.746	0.869	+3
Yb	70		14	2	6		2	[Xe]$4f^{13}$	1.94	0.859	+2, +3
Lu	71		14	2	6	1	2	[Xe]$4f^{14}$	1.734	0.848	+3
			3*d*	4 *s*	4*p*	4*d*	5 *s*				
Sc	21	Full 18	1	2				[Ar]	1.801	0.680	+3
Y	39		10	2	6	1	2	[Kr]	1.641	0.880	+3

The atomic electronic configurations of lanthanum, cerium, and gadolinium belong to the [Xe]$4f^{n-1}5d^16s^2$ type; lutetium is of the [Xe]$4f^{14}5d^16s^1$ type (a special case of [Xe]$4f^{n-1}5d^16s^2$ type); all of the other lanthanide elements are of the [Xe]$4f^n6s^2$ type. There is no 4*f* electron in scandium and yttrium atoms; however, the electrons in their outermost orbitals have the configuration of $(n-1)d^1ns^2$, that is why their chemistry is similar to that of lanthanide elements and are called rare earth elements (Xu 2005; Li 2011; Chen 2005; Wu 2005; Li 1990; Gupta and Krishnamurthy 2005).

The similarity of the physical and chemical properties of REEs is dependent on their electronic configurations. It is very difficult to separate them using common chemical methods. With the increase of the atomic number of the lanthanide elements, the electronic configurations in the outermost two shells (*O* and *P*) are the same ($5s^25p^65d^{0,1}6s^2$); the increased electrons fill in the 4*f* orbital that has a capacity of 14 electrons, which determines the number of lanthanide elements, i.e., the elements from cerium to lutetium or *f* elements.

REEs easily lose their electrons in $(n-1)d^1ns^2$ or $4f^1$ and form RE^{3+} ions. Due to the special stability of the inert gas configuration and the f^0, f^7, and f^{14} configurations, scandium, yttrium, lanthanum, gadolinium, and lutetium only form the RE^{3+} ions. Neodymium, Dysprosium, Holmium, Erbium, Thulium, and Ytterbium only form the RE^{3+} ions as well. Besides the RE^{3+} ion, cerium, praseodymium, and terbium can also form RE^{4+} ions, and samarium, europium, and terbium have RE^{2+} ions.

1.2.2 Lanthanide Contraction

Both the atomic radii and trivalent ionic radii of the lanthanide elements decrease with the increase in atomic number (Table 1.2), this phenomenon is called lanthanide contraction. The reason for this contraction is due to the weak shielding effect of the 4*f* electrons to the nuclear charge, causing the increase of the effective nuclear charge with the increase in atomic number, thus increasing the attraction to the electrons in the outer shell.

There are only two conductive electrons ($6s^2$) for europium and ytterbium, while the other lanthanide elements have three conductive electrons ($6s^24f^1$). This explains why the atomic radius of europium and ytterbium are far greater than those of the other lanthanide elements.

As a result of the lanthanide contraction, the radii of RE^{3+} decreases from 1.061 Å (La^{3+}) to 0.848 Å (Lu^{3+}). The average of the radii change between two neighbor RE^{3+} ions is 0.0152 Å. The tiny radii discrepancy makes the lanthanide elements able to substitute each other in the crystal lattice of their minerals, i.e., the isomorphism. The radii of Y^{3+} is close to that of Er^{3+}, therefore yttrium is found in coexistence with HREE minerals in REE deposits. Because the radius of Sc^{3+} is far smaller than those of the other RE^{3+}, scandium generally does not occur in rare earth deposits.

It is the lanthanide contraction that makes their separation possible. The lanthanide contraction results in the regular property changes of the lanthanides with their atomic number. With the increase of their atomic number, their radii decrease, making the coordination ability of some of the ligands with lanthanide cations increase. Taking these subtle differences of the coordination ability, SX and IX technologies are able to separate REEs to produce high purity single rare earth compounds.

1.2.3 Physicochemical Properties of Rare Earth Metals

Rare earth metals are typical metallic elements; most of them are silvery-white and silvery-gray luster, with the exception of Pr and Nd that are pale yellow. The physical properties of rare earth metals are listed in Table 1.3 (Chen 2005; Wu 2005).

Except for Yb, the melting points of the HREE metals (1312–1652 °C) are higher than those of the LREE metals, while the boiling points of the LREE metals (with exceptions of Sm and Eu) are higher than those of the HREE metals (except for Tb, Lu, and Y).

Table 1.3 Physical properties of rare earth metals

REEs	Density (g/cm^3)	Melting point (°C)	Boiling point (°C)	Resistivity (25 °C) ($\times 10^{-4} \Omega$ cm)	Thermal neutron absorption capture cross section (b)
Sc	2.985	1538	2730	66	24.0
Y	4.472	1502	2630	53	1.31
La	6.166	920	3470	57	9.3
Ce	6.773	793	3468	75	0.73
Pr	6.475	935	3017	68	11.6
Nd	7.003	1024	3210	64	46
Sm	7.536	1072	1670	92	5600
Eu	5.245	826	1430	81	4300
Gd	7.886	1312	2800	134	46000
Tb	8.253	1356	2480	116	46
Dy	8.559	1407	2330	91	950
Ho	8.78	1461	2490	94	65
Er	9.054	1497	2420	86	173
Tm	9.318	1545	1720	90	127
Yb	6.972	824	1320	28	37
Lu	9.84	1652	3330	68	112

Table 1.4 Colors of rare earth cations (RE^{3+})

RE^{3+}	Uncoupled electrons	Color	Uncoupled electrons	RE^{3+}
La^{3+}	$0(4f^{0})$	Colorless	$0(4f^{14})$	Lu^{3+}
Ce^{3+}	$1(4f^{1})$	Colorless	$1(4f^{13})$	Yb^{3+}
Pr^{3+}	$2(4f^{2})$	Yellow green	$2(4f^{12})$	Tm^{3+}
Nd^{3+}	$3(4f^{3})$	Red	$3(4f^{11})$	Er^{3+}
Pm^{3+}	$4(4f^{4})$	Pink/pale yellow	$4(4f^{10})$	Ho^{3+}
Sm^{3+}	$5(4f^{5})$	Pale yellow	$5(4f^{9})$	Dy^{3+}
Eu^{3+}	$6(4f^{6})$	Pale red	$6(4f^{8})$	Tb^{3+}
Gd^{3+}	$7(4f^{7})$	Colorless	$7(4f^{7})$	Gd^{3+}

Sm, Eu, and Gd have very high thermal neutron absorption capture cross sections and are used as thermal neutron absorption capture materials in nuclear reactors.

Rare earth metals have low conductivities. However, La shows superconductivity at 47 K.

There is no uncoupled electron on the 3*d* orbital of Sc, the 4*d* orbital of Y, and the 4*f* orbital of La and Lu, therefore, Sc, Y, La, and Lu show diamagnetism. All the other rare earth metals show paramagnetism, with Gd, Tb, Dy, and Tm having ferromagnetism.

The colors of the rare earth cations (RE^{3+}) are determined by their 4*f* configurations, as shown in Table 1.4 (Chen 2005).

The average life of some excited states of the REE cations is very long and up to 10^{-6} to 10^{-2}s, which is much longer than that of the general cations (10^{-10} to 10^{-8}s). The REEs with this property are used to produce the long afterglow luminescent materials.

Rare earth elements are very reactive metals and have to be kept in kerosene to prevent oxidation from air and moisture. The reactivity of rare earth elements increases in the order of Sc, Y, and La, but decreases from La to Lu.

The ignition point of rare earth metals is very low. It is 165, 290, and 270 °C for Ce, Pr, and Nd, respectively. Therefore, the Ce-based mixed light rare earth metals are used as the pyrophoric alloy, such as flint.

Rare earth metals readily react with oxygen, hydrogen, halogen, sulfur, nitrogen, and carbon, forming stable compounds. They react with water and inorganic acids as well.

Rare earth metals form various intermetallic compounds with many metals. The rare earth permanent magnets (Sm-Co alloys and Nd-Fe-B alloys) and the rare earth hydrogen storage material ($RENi_5$) are two classes of the rare earth intermetallic compounds.

1.3 Applications of Rare Earth Elements

REEs have wide applications in various areas due to their special physical and chemical properties. The applications of REEs can be categorized into the traditional and the high-tech areas. The traditional areas include the industries of metallurgy and machinery, glass and ceramics, and petroleum and chemical industry. The consumption in these areas accounts for about 85 % of the total REE consumption. The applications of REEs in the traditional areas are in the forms of mixed rare earth metals and alloys as well as rare earth oxides/salts. The high-tech areas, which have been developed in the last 40 years, include the industries of phosphors, permanent magnets, batteries (hydrogen storage material, NiMH) and nuclear industry, as well as superconductive materials and magnetostrictive materials. The applications of REEs in the high-tech areas are in the form of high purity single rare earth oxides.

1.3.1 Metallurgy and Machinery

Due to the high chemical activity of the rare earth metals, rare earth metals and alloys can be used to purify the molten metal or alloy by removal of the harmful impurities such as oxygen and sulfur, modify the configuration and distribution of the impurities, reduce the size and amount of inclusions, and make the grains finer. All these functions improve the mechanical properties, the corrosion resistance, and the oxidization resistance of the materials.

REEs in the cast iron change the graphite into spherical particles, thus forming the so-called nodular cast iron. The nodular cast iron has excellent mechanical properties and machinability. Cerium is the most effective rare earth element for this function.

REEs in non-ferrous metals significantly improve their machinability and physical properties. Containing 0.15–0.25 % REEs, the RE–Al–Zr conducting wire has been widely used in China as the high-current conducting wire. Its ampacity is 1.6–2.0 times that of the pure aluminum wire at high temperature (less than 150 °C). Another example is the Al–Mg–Si–RE conducting wire; it is widely used as the high voltage transmission line due to its excellent tensile strength. In the aerospace and automobile industries, Mg-RE alloys are the first choice due to their excellent properties in all aspects and their low density.

1.3.2 Petroleum and Chemical Industry

REEs are used in these areas in the forms of inorganic RE compounds (oxides and salts) and organic RE complexes.

In petroleum cracking, REE containing molecular sieve catalyst can increase the capacity of the cracking unit by 20–30 % and increase the gasoline conversion by 10 %.

The organic REE complexes in gasoline and diesel improve their combustion performance, thus reducing fuel consumption and emission. The REE oxides and other REE compounds in lubricants increase their abrasion resistance and oxidation resistance.

In automobile exhaust purification units, cerium dioxide is widely used to help catalysis and stabilize the catalyst structure. Composite cerium and zirconium oxides have the high capacity for oxygen storage as an oxygen reservoir in the automobile exhaust purifier, thus can effectively control the carbon monoxide emission.

REE catalysts such as $[(C_5Me_3)_2LaH]_2$ and $[(C_5Me_5)_2NdH]_2$, are used in synthetic rubber industry due to their unique properties. The CPBR (*cis*-1,4-polybutadiene rubber) produced with REE catalysts has advanced anti-fatigue life, dynamic wear, and heat of formation properties over the traditional products.

REEs are also used as siccatives for paints and thermal stabilizers in the processing of plastic products.

1.3.3 Glass and Ceramics

Due to their intrinsic crystal structures, chemical reactivity and the colors of some REE cations, REEs are widely applied in glass clarification, decoloring/coloring, polishing as well as ceramic pigments. The REEs used in these areas include cerium, lanthanum, praseodymium, neodymium, samarium, erbium, and yttrium.

Cerium dioxide has good clarification and decoloring effect in the glass processing, thus increasing the transparency of the glass. These kinds of glass are used as the display screens of TV and computers, and optical glass. The toxic white arsenic (arsenic trioxide) for this use has been completely replaced by cerium dioxide.

REEs make the glass with different colors or with different specialties. Cerium/titanium oxides make the glass yellow, while praseodymium oxide makes the glass green and neodymium oxide makes the glass Turkey red. Lanthanum in low or no silicon glass makes the glass with a very high refractive index, a low dispersive index and a good chemical stability. Lanthanum glass is used for making lens. Addition of cerium oxide in the glass used for food containers can effectively prevent the food from ultraviolet irradiation. Cerium glass is also widely used in military and TV industry due to its anti-radiation property. It can keep its transparency under nuclear radiation. Samarium in the aluminosilicate glass can significantly increase the density, refractive index, micro-penetration hardness, thermal expansion coefficient, and the chemical durability of the glass. Neodymium glass is a very good laser material and is used in large-scale and high power laser devices.

Cerium-based REEs oxides are widely used as glass polishing materials with excellent grinding and polishing effect. More than 30 kinds of cerium-based polishing materials have been developed for the different glass with various compositions. The major applications are for the polishing of the display screens of TV and computers, optical instrument lens, jewels, and platy-monocrystals.

REEs in the ceramics and enamels enable the ceramics with rich colors and increase the cracking resistance of the enamels. As an example, addition of 3–5 % praseodymium-zircon yellow in the zirconium silicate-based enamel makes the enamel surface with a vivid lemon yellow. In addition, it increases the brightness, abrasion resistance, thermal stability of the enamel, as well as the yield.

After being partially stabilized with yttrium oxide, zirconium oxide is an excellent structural engineering ceramics. It is used in manufacturing special cutting tools and mechanical parts that require high hardness, good abrasion resistance and thermal stability. The RE oxides of erbium, lanthanum, and neodymium find applications in photoconductive fiber, PLZT (Lead–Lanthanum–Zirconate–Titanate) optical switches and ceramic capacitors, respectively.

1.3.4 Phosphors

Phosphors is the major application of the high purity single rare earth oxides. High purity oxides of yttrium, europium, and terbium (greater than 99.99 %) are used to make the red phosphor that is used in color TV and other display systems. The phosphor for tricolor fluorescent lamp application uses high purity oxides of yttrium, europium, terbium, and cerium (greater than 99.99 %). It has the high luminance, rich coloration, and long life. The oxides of lanthanum, cerium, and gadolinium are used to make the phosphors for X-ray intensifying screens.

1.3.5 Permanent Magnets

Rare earth permanent magnets include Sm–Co alloys and Nd–Fe–B alloys. The magnetic properties of the rare earth permanent magnets are outstandingly superior to those of the traditional magnets. For example, the magnetic energy product (BH product) of the rare earth permanent magnets is 4–10 times that of the traditional magnets. The rare earth permanent magnets are used in voice coil motors for computer disk drive, speakers, MRI (magnetic resonance imaging), and wind turbines. A 3 MW wind turbine can use up to 2700 kg of NdFeB magnets. The REEs used in permanent magnets include praseodymium, neodymium, samarium, terbium, and dysprosium.

REEs also have important defense applications, such as jet fighter engines, missile guidance systems, anti-missile defense systems, space-based satellites, and communication systems.

1.4 Rare Earth Minerals

Due to their electronic configuration, REEs are chemically active and do not occur in pure form naturally. Naturally, REEs occur as salts and are associated with other metals. They often occur together in a deposit. More than 250 rare earth minerals have been identified. However, only about a dozen of them have important industrial value (Table 1.5): bastnaesite, monazite, xenotime, fluocerite, parasite, fergusonite, gadolinite, aeschynite, synchysite, samarskite, polycrase, and loparite. Among these, the first four are the major industrial rare earth minerals (Wang and Chi 1996; Cheng et al. 2007).

In addition to the rare earth minerals in Table 1.5, ion-adsorption type rare earth clay is the major source of current heavy rare element production. It was first discovered in China's Jiangxi Province in 1969. The REEs are adsorbed on the surface of clays in the form of ions. The REEs are not soluble or hydrolyzed in water but follow ion-exchange laws. The REO grade in the ion-adsorbing type clay is 0.05–0.3 %. Of the REO up to 60 % are heavy rare earth elements such as Y_2O_3. The ore composition is relatively simple in comparison to the coastal placers. Sand

Table 1.5 Major rare earth minerals

Minerals	Chemical formula	REO%	Average density (g/cm^3)	Color
Bastnaesite	$Re[(CO_3)F]$	74.8	4.97	Yellow, reddish brown
Monazite	$(Re,Th)PO_4$	65.1	5.15	Brown, colorless, greenish, gray white, yellow
Xenotime	$Y(PO_4)$	62.0	4.75	Yellowish brown, greenish brown, gray, reddish brown, brown
Fluocerite	$(Ce, La)F_3$	83.4	6.13	Pale wax yellow, yellowish brown, reddish brown
Parisite	$Re_2Ca(CO_3)_3F_2$	60.3	4.36	Brown, brownish, grayish yellow, yellow
Fergusonite	$YNbO_4$	39.9	5.05	Black, brown, gray, yellow
Gadolinite	$Y_2FeBe_2Si_2O_{10}$	48.3	4.25	Brown, green, green black, light green, black
Aeschynite	$(Y,Ca,Fe)(Ti,Nb)_2(O,OH)_6$	24.6	4.99	Black, brownish black, brown, brownish yellow, yellowish
Euxenite	$(Y,Ca,Ce)(Nb,Ta,Ti)_2O_6$	24.3	4.84	Brownish black, brown, yellow, olive green
Synchysite	$Ca(Y,Ce)(CO_3)_2F$	49.6	5.27	White, reddish brown
Samarskite	$(Y,Fe,U)(Nb,Ta)_5O_4$	24.3	5.69	Black, brownish, yellowish brown
Polycrase	$(Y,Ca,Ce,U,Th)(Ti,Nb,Ta)_2O_6$	19.5	5.00	Black, brown
Loparite	$(Ce,Na,Ca)_2(Ti,Nb)_2O_6$	29.8	4.77	Black

content is low. The major components are clay minerals, quartz, and other rock-forming minerals. The clay minerals include halloysite, illite, kaolinite, and small amounts of montmorillonite (Chi and Tian 2011).

1.5 Resource and Production

Rare earth elements are not as rare as their name indicates; however, not many REE deposits are mineable. Most of the mineable deposits have been found in China, America, India, Middle Asian nations, South Africa, Australia, and Canada. Table 1.6 lists the world distribution of rare earth oxides (REOs). China, the Commonwealth of Independent States (CIS) region, and the USA have close to 80 % of the worldwide rare earth resources; however, China currently produces over 90 % of REOs worldwide.

1.6 Chemical Technologies for REE Separation

Separation of the REEs from impurities and from one another has been accomplished by a variety of technologies. Historically, fractional precipitation and fractional crystallization were used to produce small quantities of individual rare earths with purity up to 99.9 % (3 N). These technologies are no longer of interest due to their inefficiency and tedious process.

As one of the solid–liquid extraction separation technologies, ion exchange (IX) is used to separate and produce very high-purity rare earth products with purity up to 7 N but with limited quantities due to its low capacity and low efficiency.

Table 1.6 World REO resources and production

		Production (tREO)				
Country	Reserve (tREO)	2006	2007	2009	2010	2011
China	55,000,000	119,000	120,000	120,000	130,000	130,000
CIS region	19,000,000	N/A	N/A	N/A	N/A	N/A
USA	13,000,000	0	0	0	0	0
Australia	1,600,000	0	0	0	0	0
India	3,100,000	2,700	2,700	2,700	2,800	3,000
Brazil	48,000	730	730	650	550	550
Malaysia	30,000	200	200	380	30	30
Others	22,000,000	N/A	N/A	N/A	N/A	N/A
Total (rounded)	110,000,000	123,000	124,000	124,000	133,000	133,000

Source: Mineral commodities summary 2010 and 2012, US Geological Survey (*tREO* metric tonnes of rare earth oxide)

Solvent extraction (SX) or liquid–liquid extraction is a practical and effective technology for massive REE separation. It is the dominating process in the current industry practice. The purity of the REE products is up to 5 N.

Besides SX and IX technologies, there are chemical technologies that are used to separate the single REE from mixed REEs without separation of the whole REEs. Besides RE^{3+}, Ce, Sm, Eu, Yb, Pr, and Tb exist as tetravalent or divalent cations, Ce^{4+}, Sm^{2+}, Eu^{2+}, Yb^{2+}, Pr^{4+}, and Tb^{4+}, respectively at suitable oxidation-reduction conditions. The properties of these tetravalent or divalent RE cations are significantly different from those of the trivalent RE cations. With these property differences, these trivalent and divalent REEs can be easily and effectively separated from the trivalent REEs. The process has been applied in industrial practice to separate Ce^{4+}, Sm^{2+}, Eu^{2+}, Yb^{2+} from the trivalent REEs. While the technologies for separation of Pr^{4+} or Tb^{4+} are under development as neither Pr^{4+} nor Tb^{4+} is stable in aqueous solution due to their high oxidation-reduction potentials.

The following two sections are brief introductions to the separation of Ce and Eu from mixed REEs (Xu 2005).

1.6.1 *Ce Separation with Oxidation Process*

In the LREE resources CeO_2 accounts for 45–47 % in monazite and about 50 % in bastnaesite. However, the market demand for high purity Ce product is limited. Removal of Ce by chemical processes before SX separation can simplify the other REE separation and significantly reduce the volume of the feed to the SX plant by almost 50 %, thus reducing both the capital cost and the operation costs of the SX plant. In addition, the separated Ce oxide product is a low quality polishing material that is applied in the glass industry.

Ce is readily oxidized to Ce^{4+} and Ce^{4+} is stable in aqueous solution. Ce^{4+} begins hydrolysis and precipitates as $Ce(OH)_4$ at $pH = 0.8$, while other trivalent REEs are kept in solution due to their high hydrolysis pH (6–8), therefore, Ce can be separated from other REEs by this process. In addition, Ce^{4+} in aqueous solution can form coordination compounds that are easy to be extracted with SX. SX process is used in industrial practice to separate and purify Ce.

Oxygen, chlorine, ozone, potassium permanganate, and hydrogen peroxide, as well as electrolytic oxidation, are the commonly used oxidants for Ce oxidization.

Hydrogen peroxide has advantages over other oxidants. The non-Ce mixed REEs can be produced with hydrogen peroxide. In addition, there is no contamination to the mixed REEs after Ce removal.

The prepared mixed REE chloride solution contains 50 g REO/L and 1–50 % CeO_2/REO. The pH of the solution is adjusted to a $pH = 4$. H_2O_2 is added with agitation and the pH is kept between 5 and 6 during the reaction period at room temperature. NH_4OH or NaOH is added in order to maintain the required pH range. 1 mol H_2O_2 is required in order to oxidize 1 mol Ce^{3+} to Ce^{4+}.

$$2CeCl_3 + 3H_2O_2 + 6NH_4OH = 2Ce(OH)_3OOH \downarrow + 6NH_4Cl + 2H_2O$$

The dark red Ce peroxide precipitate is first formed. It is converted to yellow Ce dioxide after boiling.

$$Ce(OH)_3O \cdot OH = Ce(OH)_4 \downarrow + \frac{1}{2}O_2$$

After filtration and washing, $Ce(OH)_4$ containing about 90 % CeO_2 is produced. Almost 100 % of Ce is recovered into the product and non-Ce mixed REEs are left in the solution. The CeO_2 from the further treatment of the $Ce(OH)_4$ product can be sold as low quality polishing materials. The other REE loss into the Ce product is from 5–10 %.

1.6.2 *Phosphor Grade Eu_2O_3 Production Using Zinc Reduction-Alkalinity Process*

The Eu distribution in REE minerals is very low, Eu_2O_3/REO is from 0.07 to 0.2 % in monazite and bastnaesite, while Eu_2O_3/REO is about 0.5– 0.8 % in the ionic clay minerals. Historically, it was separated and purified using precipitation and IX technologies.

Now the industrial practice includes two steps, the first step is the production of Eu-rich REE chloride solution from the SX plant. Eu_2O_3/REO is up to 70 % in the Eu-rich REE chloride. The second step is to produce the high purity (phosphor grade) Eu_2O_3 by processing the Eu-rich REE chloride using the Zinc Reduction-Alkalinity Process. The alkalinity of Eu^{2+} is higher than that of the other trivalent REEs. Therefore, Eu^{2+} is left in solution when the other trivalent REEs are completely precipitated. The separation is very effective as the K_{sp} of $RE(OH)_3$ is from 10^{-19} to 10^{-24} for La to Lu. 4N Eu_2O_3 can be produced from the solution. The reactions in this process are shown below.

$$2EuCl_3 + Zn = 2EuCl_2 + ZnCl_2$$

$$RECl_3 + 3NH_4OH = RE(OH)_3 \downarrow + 3NH_4Cl$$

Addition of NH_4Cl before precipitation makes the course $RE(OH)_3$ precipitate, thus promoting the filtration. In addition, NH_4Cl can prevent the co-precipitate of small amounts of $Eu(OH)_2$ with $RE(OH)_3$ and increase the Eu recovery.

Eu^{2+} can be oxidized by oxygen in air and H^+ in solution. The filtration is operated under the protection environment with the inert gas.

References

Chen, J. Y. (2005). *Handbook of hydrometallurgy*. Metallurgical Industry Press, Beijing, China.

Cheng, J., Hou, Y., & Che, L. (2007). Making rational multipurpose use of resources of RE in baiyuan ebo deposit. *Chinese Rare Earths, 28*, 70–74. 1004-0277(2007)01-0070-05.

Chi, R., & Tian, J. (2011). *Weathered crust elution-deposited rare earth ores*. Nova Science Publishers.

Gupta, C. K., & Krishnamurthy, N. (2005). *Extractive metallurgy of rare earths*. CRC PRESS.

Li, H. G. (1990). *Rare Metal Metallurgy*. Metallurgical Industry Press, Beijing, China.

Li, L. C. (2011). *Rare earth extraction and separation*. Inner Mongolia Science and Technology Press, Chifeng, Inner Mongolia, China.

Taylor, S. R., & McLennan, S. M. (1985). *The continental crust: Its composition and evolution*. Blackwell Scientific Press, Oxford, UK.

Wang, D., & Chi, R. (1996). *Rare earth processing and extraction technology* (pp. p63–11). Scientific Press, Beijing, China.

Wu, W. Y. (2005). *Rare earth metallurgy*. Chemical Industry Press, Beijing, China.

Xu, G. X. (2005). *Rare earths (I)* (2nd ed.). Metallurgical Industry Press, Beijing, China.

Chapter 2
Rare Earth Beneficiation and Hydrometallurgical Processing

Prior to individual REE separation, the rare earth ore will go through a series of physical separation and hydrometallurgical processing. In this chapter, the rare earth ore beneficiation, mineral concentrate decomposition, and rare earth leaching are introduced briefly.

2.1 RE Mineral Processing Technology

Each rare earth deposit is unique and is always composed of a variety of minerals. For example, the Baiyun Obo deposit, the largest rare earth mine in production in China, contains bastnaesite, monazite, fluorite, magnetite, barite, calcite, quartz, feldspar, etc. (Cheng et al. 2007a, b). In order to make a rare earth project economically feasible, a series of ore beneficiation techniques are always employed to concentrate the rare earth minerals. The commonly used ore beneficiation technologies include gravity separation, flotation, and magnetic separation.

2.1.1 Gravity Separation

The successful application of gravity separation depends on the difference of specific gravity between the rare earth minerals and the major gangue materials. The particle size is also very important in gravity separation. The hinder-settling ratio as defined in Eq. (2.1) can be used to determine the suitability of gravity separation. In Eq. (2.1), SG_{RE} is the specific gravity of rare earth minerals, SG_M is the specific gravity of media, for example, water or heavy liquid, and SG_G is the specific gravity of associated gangue materials.

J. Zhang et al., *Separation Hydrometallurgy of Rare Earth Elements*,
DOI 10.1007/978-3-319-28235-0_2

$$\text{Hinder settling ratio} = (SG_{RE} - SG_M)/(SG_G - SG_M) \quad (2.1)$$

When the hinder-settling ratio is no less than 2.5, gravity separation will be effective; when the hinder-settling ratio is between 1.75 and 2.5, gravity separation can be used to separate those particles coarser than 0.15 mm; when the hinder-settling ratio is between 1.5 and 1.75, gravity separation can be used to separate those particles coarser than 1.6 mm; when the hinder-settling ratio is below 1.25, gravity separation is not recommended.

2.1.2 *Flotation*

Flotation is one of the most common beneficiation technologies in rare earth processing due to the fine liberation size of rare earth minerals. The development of rare earth flotation technology is always associated with the development of rare earth flotation reagents. Since the 1960s, extensive efforts have been devoted to the development of high efficiency rare earth flotation reagents. The rare earth flotation reagents include depressant, collector, and frother. Sodium silicate is the most common used gangue depressant in rare earth flotation. The major rare earth flotation collectors include hydroxamates, phosphonic acid, and carboxylates. The major rare earth collectors are shown in Table 2.1.

2.1.3 *Magnetic Separation*

Magnetic separation utilizes the difference of magnetic susceptibility between rare earth minerals and associated gangue materials. Materials can be classified into paramagnetic material and diamagnetic material based on the form of magnetism by an externally applied magnetic field.

Table 2.1 Major rare earth flotation collectors

Classification	Collector	General formula
Hydroxamate	C5-9 alkyl hydroxamate	RCONHOH
	C7-9 alkyl hydroxamate	RCONHOH
	Aromatic hydroxamate	$C_6H_5CONHOH$
	Salicylic hydroxamic acid	$C_6H_4OHCONHOH$
	H_2O_5	$C_8H_5NO_3$
Phosphonic Acid	Styrene phosphonic acid	$C_8H_9O_3P$
	Alky phosphate	RH_2PO_4
Carboxylate	Oleic acid	$C_{18}H_{34}O_2$
	Phthalate	$C_6H_4(COOCH_3)_2$
	Paraffin oxide soap	

In magnetic separation, the magnetic force needs to be larger than the mechanical force such as gravity force or centrifugal forces to achieve separation. This requires an uneven magnetic field with high intensity and large gradient. Also, the magnetic susceptibility difference between rare earth minerals and gangue materials must be sufficient. The ratio of magnetic susceptibility (K) as defined by Eq. (2.2) is often used to determine the suitability of magnetic separation, where x and x' are the magnetic susceptibility of magnetic material and non-magnetic material, respectively. When $K > 1$, magnetic separation can be an effective separation method.

$$K = x/x' \tag{2.2}$$

Due to the existence of other mechanical forces, for magnetic material separation, the magnetic force must be higher than the overall mechanical forces to achieve separation. For non-magnetic material separation, the magnetic force must be less than the overall mechanical forces to achieve separation.

Based on the magnetic susceptibility of minerals, they can be classified into: (1) high magnetic materials ($x \geq 3000 \times 10^{-9}$ m^3/kg), (2) low magnetic materials (15×10^{-9} m^3/kg $< x < 600 \times 10^{-9}$ m^3/kg), and (3) non-magnetic materials ($x < 15 \times 10^{-9}$ m^3/kg).

2.2 Rare Earth Ore Beneficiation

Besides the rare earth minerals, the ore always contains other minerals such as fluorite, magnetite, barite, calcite, quartz, and feldspar. These minerals have similar floatability, magnetic susceptibility, specific gravity, and electrical conductivity. The similarity of these physical properties poses significant challenges in separating the rare earth minerals. Often, more than one valuable mineral is present in a rare earth deposit. In addition to monazite and xenotime, other minerals such as ilmenite, rutile, zircon, and wolframite are also considered for recovery.

Even though more than 250 rare earth minerals have been identified, only a few of them have industrial value. Currently bastnaesite, monazite, xenotime minerals, and ion-adsorbing type rare earth clays are the major sources of rare earth production.

2.2.1 Bastnaesite

Bastnaesite is the most industrially important rare earth mineral, containing 67–73 % REO. It is the major source of light rare earth elements. Well known bastnaesite deposits are Mountain Pass in California, Baiyun Obo in Inner Mongolia, Eastern China Weishan (WS), and Western China Mianning (MN). Baiyun Obo is a complex rare earth deposit of bastnaesite and monazite. The others are bastnaesite deposits.

The chemical composition of bastnaesite is relatively simple in comparison with other rare earth minerals. A bastnaesite deposit is relatively easy to concentrate. The key is to manage the separation between bastnaesite and calcium and barium minerals. Flotation is the most common concentration process. A collector with good selectivity for bastnaesite and an effective depressant for other ore minerals are required. Bastnaesite concentrate can also be obtained through a combination of gravity-flotation or magnetic-flotation processes.

2.2.1.1 Eastern China WS Rare Earth Deposit

The WS rare earth deposit mineral veins contain bastnaesite, parisite, Ce-apatite, monazite, arfvedsonite, chalcopyrite, pyrite, molybdenite, galena, sphalerite, magnetite, and perovskite. The intergrown gangue includes carbonatite, barite, limonite, quartz, fluorite, and some muscovite. The major rare earth mineral is bastnaesite. It is easier to separate bastnaesite from silicate minerals than from barite and limonite because the latter have similar specific gravity and floatability to bastnaesite. In the WS rare earth deposit, bastnaesite is intimately associated with barite and limonite. Limonite is a weathered clay which makes the separation even harder. A gravity-flotation process can produce a rare earth concentrate, while avoiding the interference of clays in flotation. However, the gravity separation process is complex. A simple flotation process for the WS rare earth deposit is shown in Fig. 2.1.

The WS rare earth flotation process uses oleic acid and kerosene as collectors. Sulfuric acid is used to adjust the pH to 5.5–6. Oxidized paraffin soap (RCO_2Na) is used as the barite collector. Barite flotation is performed at pH 11.

Zeng (1993) has done extensive investigation of flotation of the WS rare earth ore. A modified reverse flotation process, shown in Fig. 2.2, was proposed to upgrade the rare earth from 3 to 7 % to 60 % rare earth oxide (REO) and barite from 25 to 92–95 %. Overall, rare earth recovery is 77–84 % and barite recovery is 61–68 %.

2.2.1.2 South West China MN Rare Earth Deposit

The South West China MN rare earth deposit is a bastnaesite deposit containing an average of 2.8 % REO. The major minerals include aegirineaugite, barite, biotite, and bastnaesite. Aegirineaugite has been weathered to a black–brown clay-type powder. Bastnaesite exists as chips with large grain sizes. There are more than 20 associated minerals including galena, molybdenite, pyrite, wulfenite, limonite, goethite, magnetite, arfvedsonite, quartz, and feldspar. De-sliming is required to remove clays to improve concentrate grade. The feed can be pre-concentrated

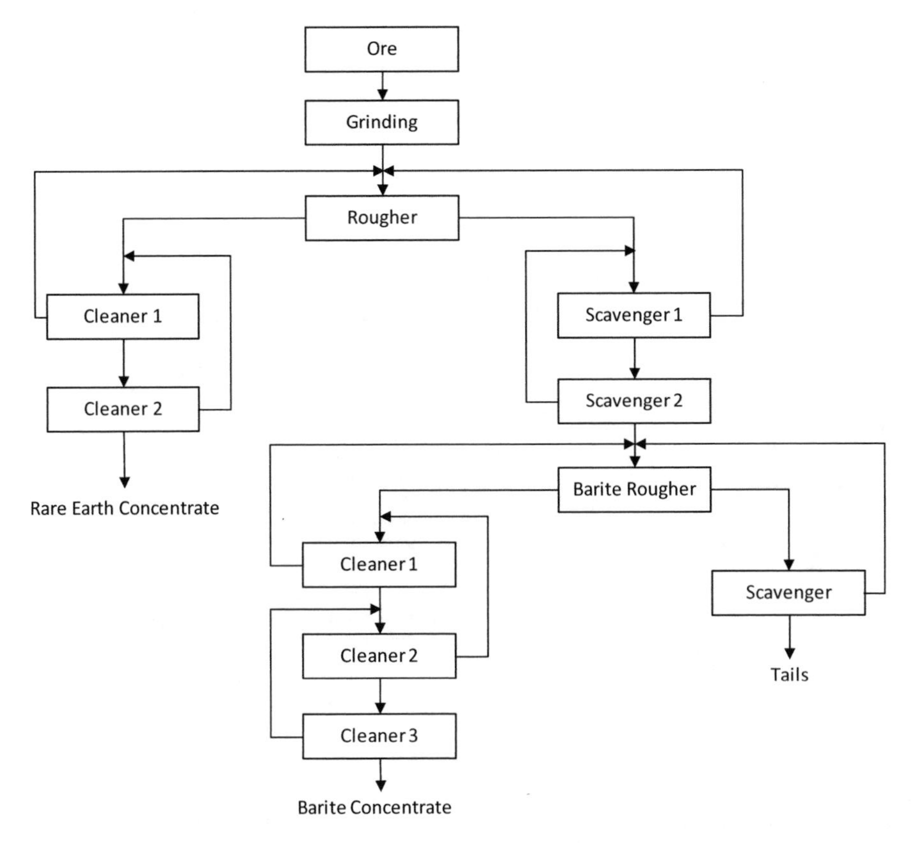

Fig. 2.1 Simple Eastern China WS rare earth flotation process (Huang 2006)

during de-sliming. Two methods of desliming have been investigated extensively (Wang and Chi 1996):

1. Attrition-scrubbing using water to wash clays from the feed before flotation.
2. Shaking table to remove clays as well as other low density gangue minerals.

Two processes have been developed for the MN rare earth deposit. One is a gravity-flotation separation process shown in Fig. 2.3. The flotation uses sodium silicate (Na_2SiO_3) as the depressant, soda ash (Na_2CO_3) as a modifier, and sodium alkyl hydroxamic acid as the rare earth collector. RE recovery is 53–67 % at a concentrate grade of 68–69 %.

The other is a desliming-flotation process. As shown in Fig. 2.4, the feed is washed using water to remove clays and then fed to flotation where Na_2SiO_3 is added as the depressant, Na_2CO_3 as the modifier, and sodium alkyl hydroxamic acid as the rare earth collector. Rare earth recovery is 40–67 % with a concentrate grade of 66–67 %.

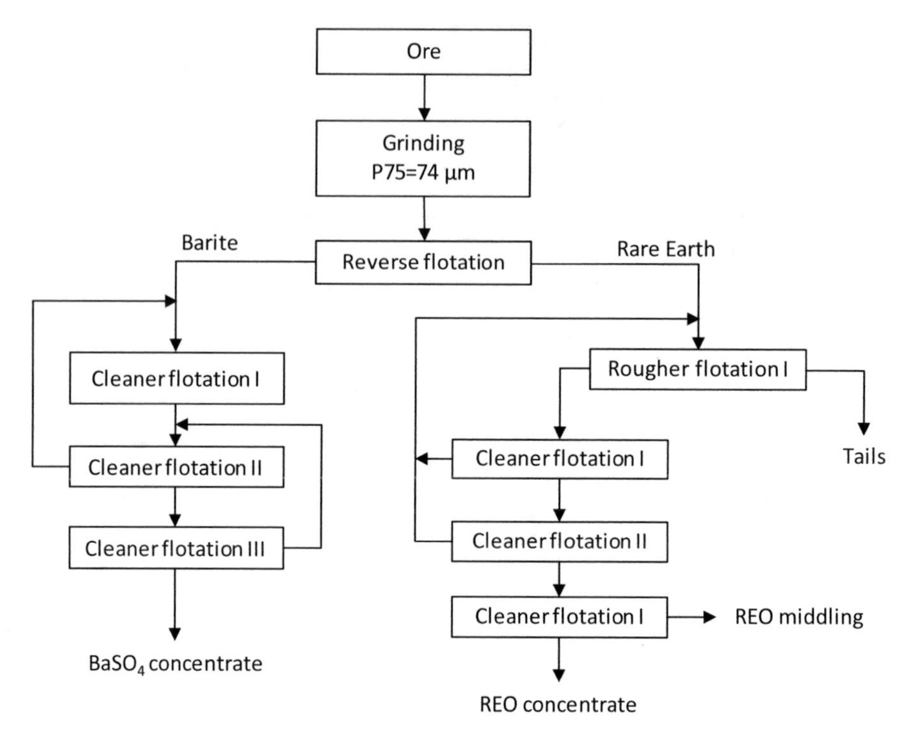

Fig. 2.2 Easter China WS rare earth reverse flotation process

2.2.2 *Monazite*

Monazite is one of the most important rare earth minerals and it was the first rare earth mineral recovered for industrial application. It is widely distributed across the world. Australia, Brazil, India, and China are all important monazite producers. There are a variety of endogenic and exogenic monazite mineral deposits. Monazite placers are exogenic mineral deposits. The China Baiyun Obo deposit and South African monazite deposits are endogenic mineral deposits.

2.2.2.1 Monazite Placers

Monazite exists mainly in placers, especially coastal placers. Many other minerals are commonly found in a monazite placer deposit. They are listed in Table 2.2. The separation of monazite from coastal placers is a complex process involving gravity separation, magnetic separation, electrostatic separation, flotation, and chemical separation. Magnetic and gravity separations are the major concentrating technologies for monazite. The usual equipment includes magnetic separators, shaking tables, spirals, and sluices. In a monazite separation process, there are typically several by-products, such as ilmenite, zircon, rutile, garnet, tinstone, and tungsten.

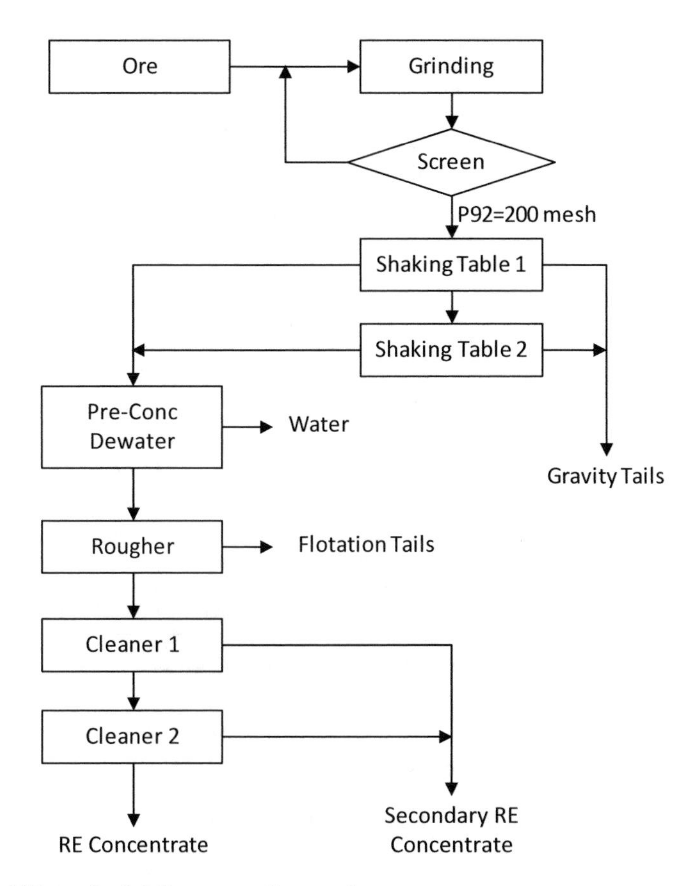

Fig. 2.3 MN gravity-flotation rare earth separation process

Figure 2.5 shows a general separation process for monazite placers. It removes the +1.00 mm gravels and −0.025 mm clays in the first step using attrition-scrubbing and settling to obtain the preliminary concentrate. Then, a series of magnetic separators with various magnetic intensities are utilized to separate the rare earth minerals and by-products.

2.2.2.2 Weathered Crust Monazite Deposit

Monazite is also found in weathered crust deposits. As listed in Table 2.3, many intergrown minerals can be found in a weathered crust deposit. The monazite separation process from a weathered crust deposit is more complicated in comparison to a monazite placer deposit. Figure 2.6 shows a general separation process for monazite from a weathered crust deposit. The first step is to remove coarse grain minerals and clay minerals, and then to remove light minerals such as silica, feldspar, and mica. After acid treatment, magnetic and gravity separation are used to separate the other minerals.

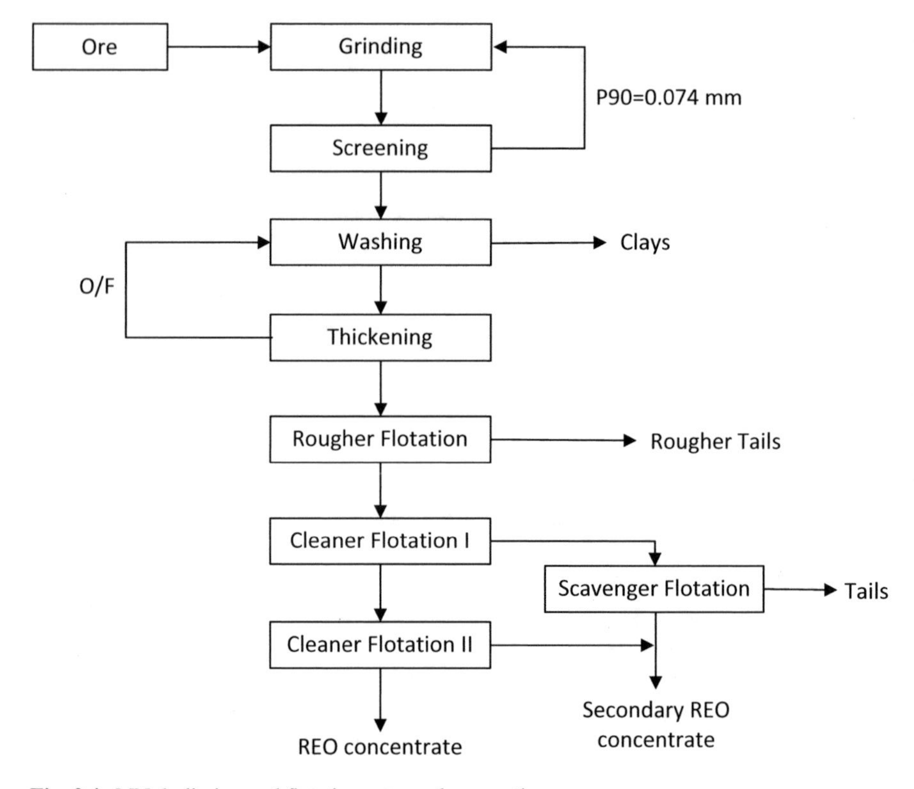

Fig. 2.4 MN desliming and flotation rare earth separation process

2.2.3 *Xenotime*

Xenotime is a rare earth phosphate mineral with its major component yttrium orthophosphate (YPO_4). It is one of the important sources of heavy rare earth elements. Occurring as a minor accessory mineral, xenotime is found in coastal placers, pegmatites, weathered crust, and igneous rocks. Associated minerals include zircon, magnetite, ilmenite, leucoxene, limonite, epidote, tourmaline, topaz, and scheelite. Major intergrown gangue materials are feldspar, silica, kaolin, and mica.

Yttrium is the major rare earth element in the xenotime. The total rare earth oxide content is higher than 42 % and yttrium accounts for 50–70 % of this.

The separation of xenotime uses a combination of gravity separation, electrostatic separation, and strong magnetic separation. However, it is difficult to separate fine grain ores through these separation processes, so flotation is always used for fine grain xenotime.

Table 2.2 Common monazite placer deposit minerals and their physical properties

Mineral	Color	Magnetic susceptibility (m^3/kg)	Density (g/cm^3)	Abrasive hardness	Conductivity (s/m)	Sorting voltage (v)
Magnetite	Brown	8,000	4.9–5.2	5.5–6.5	2.78	7,800
Ilmenite	Iron black	1,800–3,997	5.5–6.0	5.5–6.0	2.51	7,050
Garnet	Glass red	63	3.5	8.0	6.48	18,000
Monazite	Yellow to brown	14	4.9–5.5	5.0–5.5	2.34	6,552
Zircon	Purple or colorless	0.19	4.7	7.5	4.18	11,700
Rutile	Blood red, bluish, brownish yellow, brown red, violet	Nonmagnetic	4.2–4.3	6–6.5	$<10^{-8}$	8,000–25,000
Quartz	Colorless	0–10	2.7–2.8	7.0	3.57–5.30	8,890
Mica	Black	40	1.0–2.0	1.0–1.5	1.73	14,820

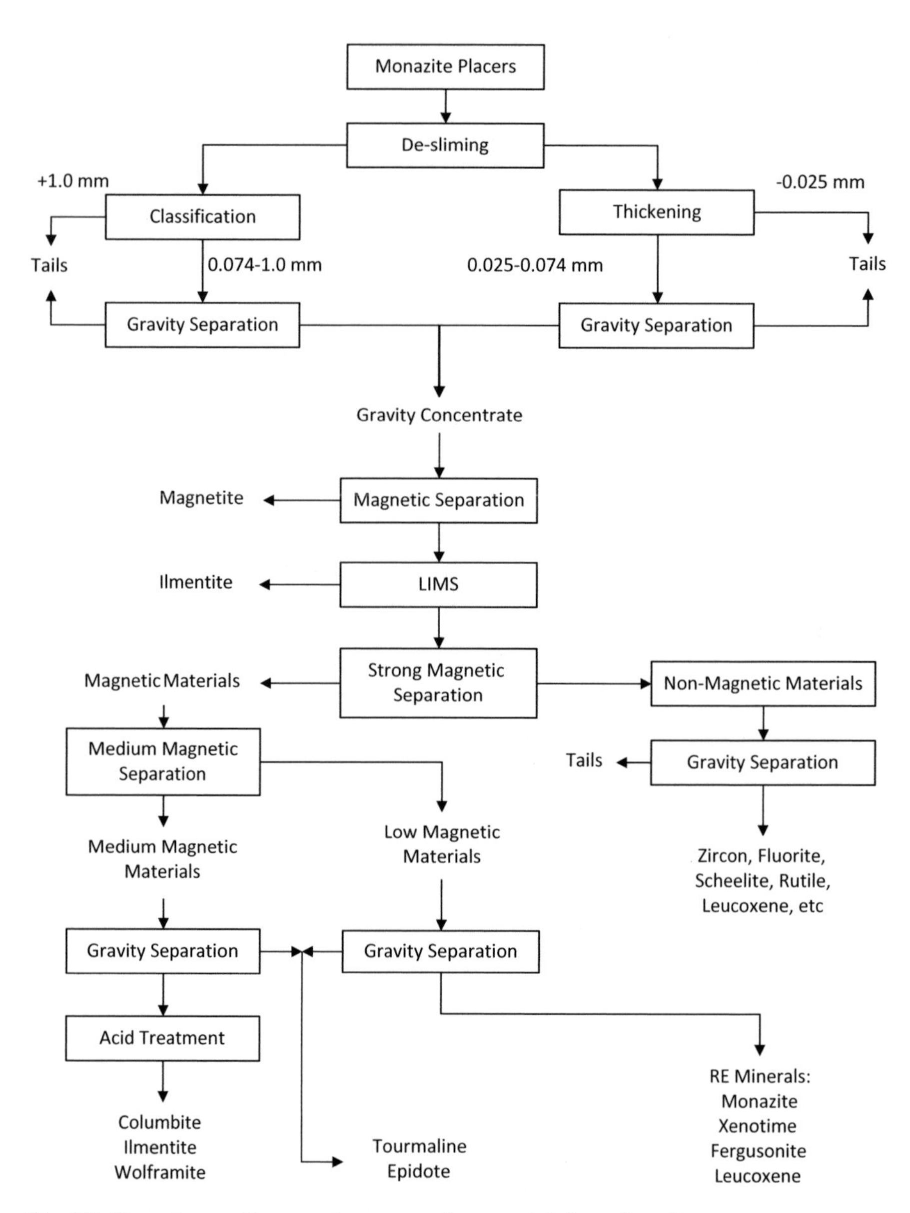

Fig. 2.5 General monazite separation process for a coastal placer deposit

2.2.3.1 Xenotime Separation from Coastal Placers

Coastal placers are exogenic mineral deposits containing many intergrown minerals. As shown in Table 2.4, a coastal placer deposit in southeastern China contains 0.0107 % xenotime, 0.0516 % monazite, and many other associated minerals. SiO_2 is the major component of the deposit and total rare oxide content is 0.086 %. The applicable separation process is shown in Fig. 2.7.

Table 2.3 Common minerals in a weathered crust monazite deposit

	Associated minerals		
Industrial minerals	Large amount	Minor amount	Trace amount
Monazite	Feldspar	Hematite	Columbite
Xenotime	Silica	Limonite	Fluorite
Fergusonite	Mica	Tourmaline	Apatite
Columbite-tantalite	Clay minerals	Pyrolusite	Aechynite
Microlite		Psilomelane	Cassiterite
Zircon		Topaz	Beryl, rutile
		Arsenopyrite	Ferrite
		Pyrite	Polycrase
		Wolframite	Thorium
		Epidote	Gadolinite

2.2.3.2 Xenotime Separation from Weathered Crust

Xenotime is also found in weathered crust deposits. The content of clay minerals is relatively high in a weathered crust type of xenotime deposit. If soluble rare earths are adsorbed on the clay, the recovery of the adsorbed rare earth must be considered as well in the separation process. This type of rare earth deposit is normally found in south China. REO content is 0.05–0.3 %. Figure 2.8 shows a recommended xenotime concentrating process from a weathered crust type of rare earth deposit in south China.

2.2.4 Ion-Adsorbed Type Rare Earth Deposits

An ion-adsorbing type rare earth deposit was first found in China's Jiangxi Province in 1969. The REEs are adsorbed on the surface of clays in the form of ions. The REEs are not soluble or hydrolyzed in water but follow ion-exchange laws. The grade of the ion-adsorbing type clay is 0.05–0.3 % rare earth oxide. Of the total rare earth up to 60 % are heavy rare earth elements including yttrium. The ore composition is relatively simple in comparison to the coastal placers. Sand content is low. The major components are clay minerals, quartz, and other rock-forming minerals. The clay minerals include halloysite, illite, kaolinite, and small amount of montmorillonite.

The separation process is relatively simple, including leaching, precipitation, and calcination. An alumina removal circuit can be installed to reduce the reagent consumption. The clarified solution can be precipitated by either oxalic acid or ammonium carbonate. Figure 2.9 shows an rare earth oxide (REO) production process from an ion-adsorbed type rare earth ore (Zhao et al. 2001).

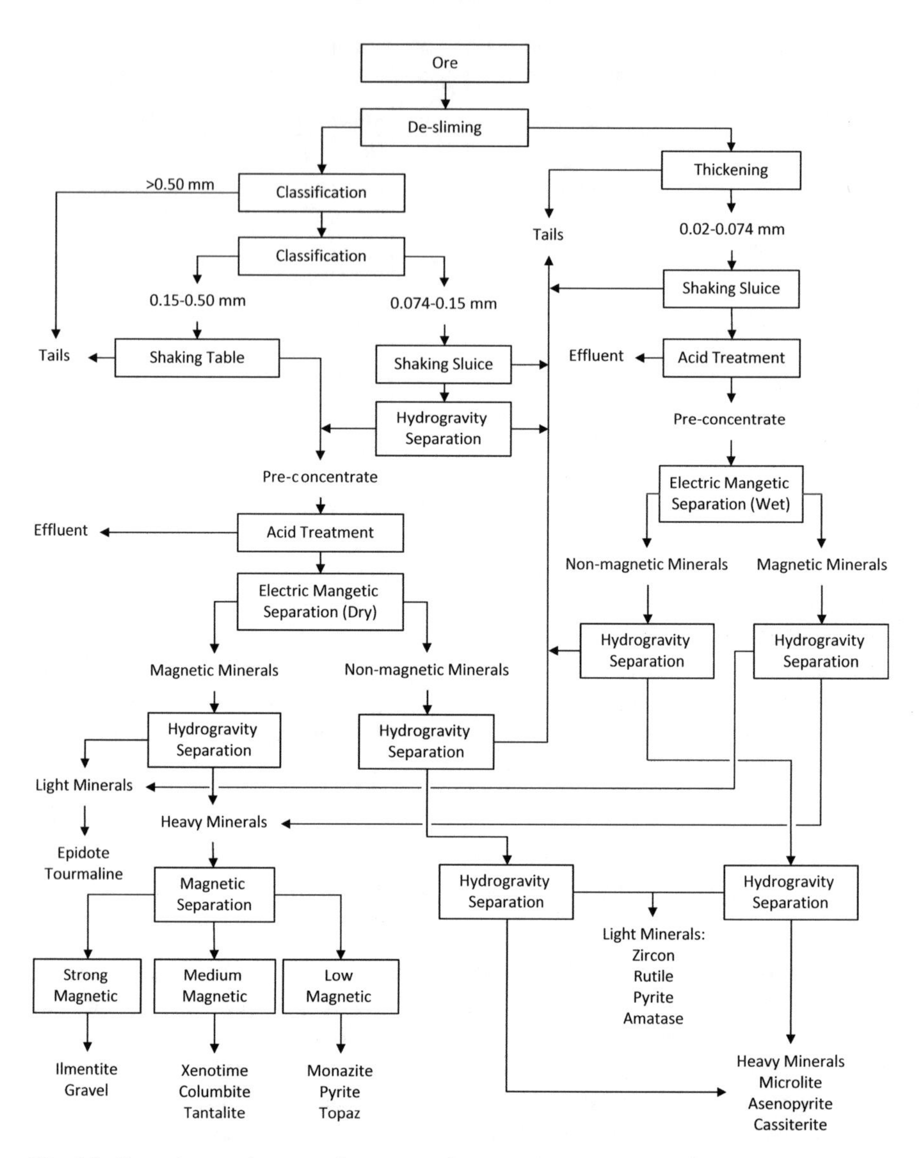

Fig. 2.6 General monazite separation process for a weathered crust deposit

Table 2.4 Common minerals in Chinese coastal placer deposits

Minerals	Content (%)	Minerals	Content (%)
Xenotime	0.0107	Ilmenite	0.1722
Monazite	0.0516	Marcasite	0.0730
Zircon	0.1355	Tourmaline	0.1663
Rutile and anatase	0.0280	Maghemite	0.0044
Topaz and staurolite	0.6758	Quartz, Feldspar, Mica	98.68

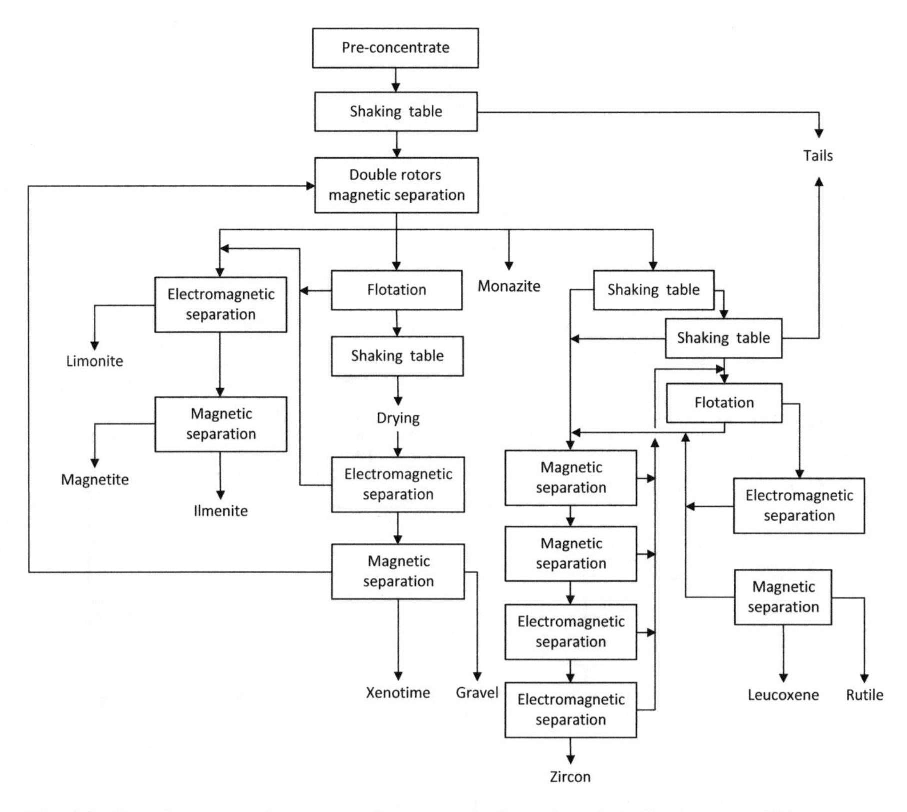

Fig. 2.7 Xenotime separation process for a coastal placer deposit in Southeastern China

2.2.5 Baiyun Obo Rare Earth Deposit

Baiyun Obo rare earth is a multi-metal mineral intergrown deposit. It is now the largest rare earth production mine in the world. It is located in Inner Mongolia, China. There are 71 elements and over 170 minerals in the Baiyun Obo ore. Fifteen rare earth minerals have been identified (Yu and Deng 1992; Xiang et al. 1985). However, bastnaesite and monazite are the dominant rare earth minerals. The valuable minerals are closely associated with each other with small grain sizes between 0.010 and 0.074 mm.

The processing of Baiyun Obo rare earth has gone through many stages of development (Zhang et al. 2002b; Yang 2005; Fang and Zhao 2003; Cheng et al. 2007a, b; Zhao et al. 2008; Cai et al. 2009). Milestones include: (1) Calcination-magnetic separation-flotation process; (2) Flotation-low magnetic separation process; (3) Low magnetic separation-flotation-high magnetic separation process; (4) Low magnetic separation-high magnetic separation-flotation process.

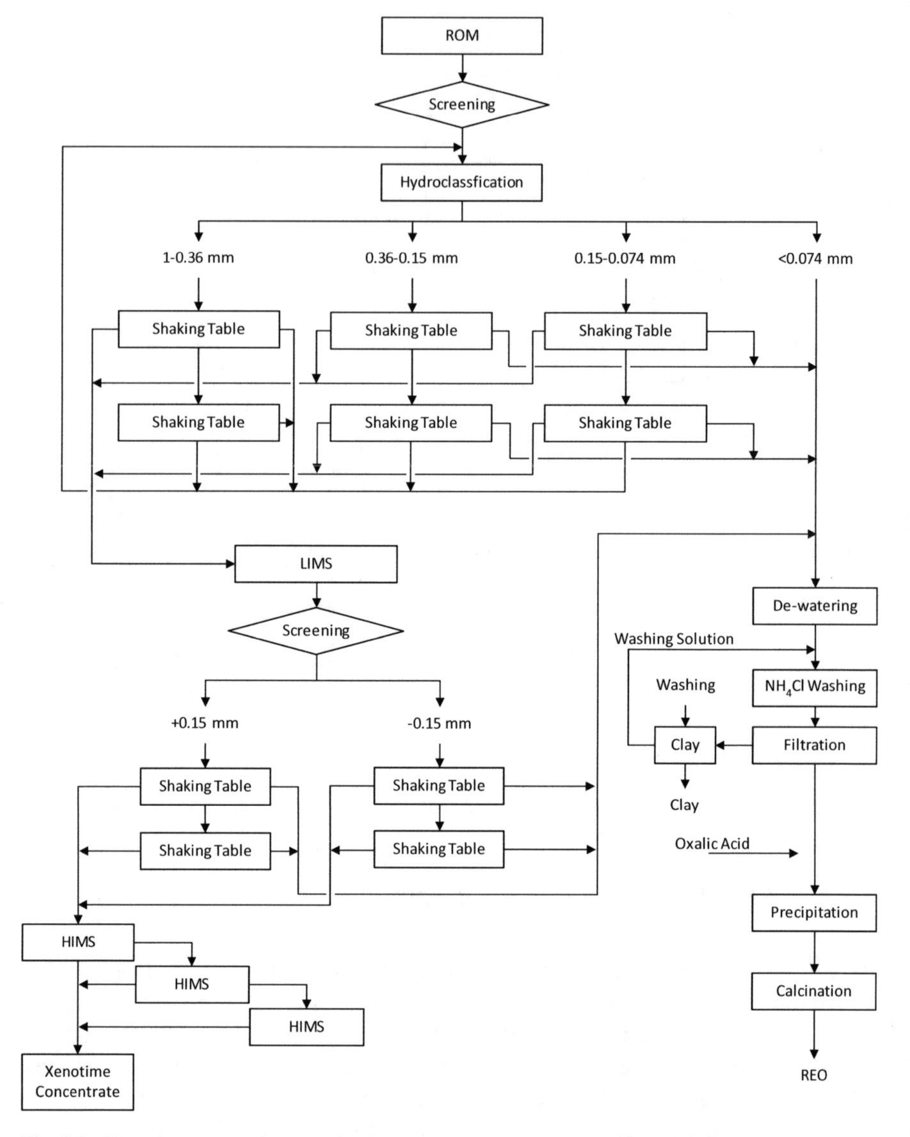

Fig. 2.8 Xenotime separation process for a weathered crust deposit in Southern China

2.2.5.1 Calcination-Magnetic Separation-Flotation Process

Ore is calcined at 650 °C with reducing reagents. The product is sent to magnetic separation. Iron minerals are recovered as magnetic materials. The rare earth and niobium minerals are left in the magnetic tails. The magnetic tails are sent to flotation to separate the rare earth minerals and niobium minerals. The process diagram is shown in Fig. 2.10 (Yu and Deng 1992). Of the total rare earth, 49.5 % can be recovered in the concentrate at a grade of 36 % rare earth oxide.

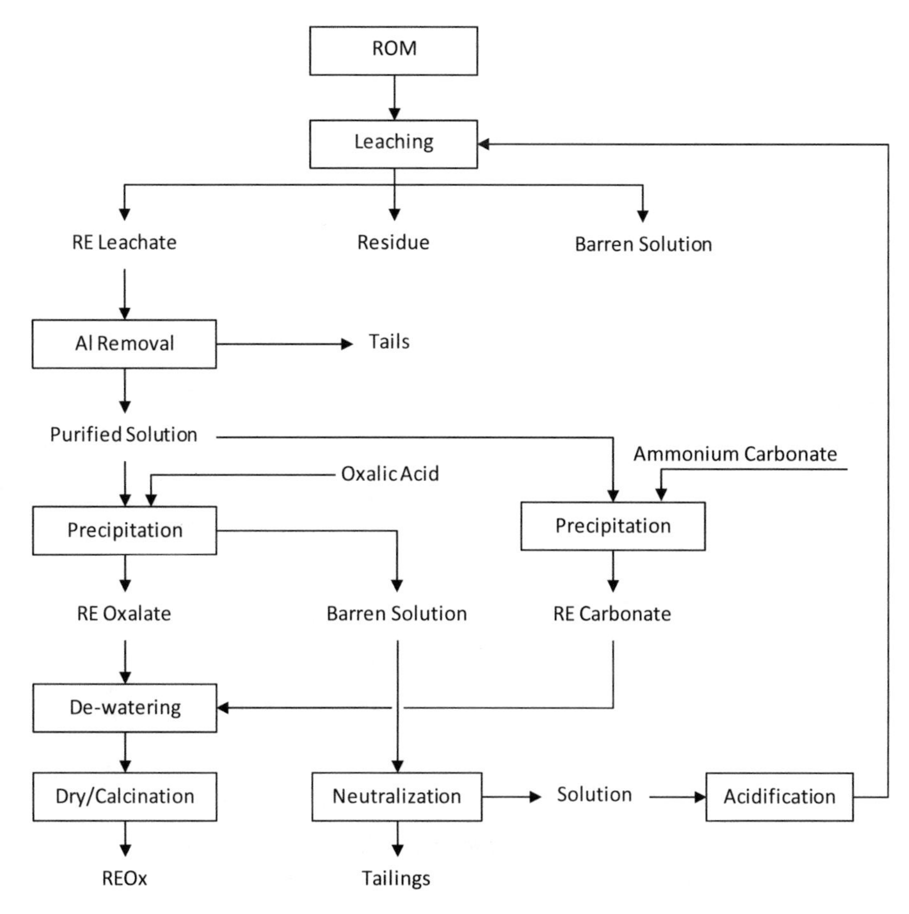

Fig. 2.9 Rare earth recovery from an ion-adsorbed type of clay deposit

2.2.5.2 Flotation-Low Magnetic Separation Process

As shown in Fig. 2.11, ore is ground to 95 % passing 74 μm and sent to flotation. At pH 11, polymerized sodium silicate is added to suppress rare earth minerals and iron minerals. Fluorite is floated first. Addition of an activator and pH adjustment to 8.5–9.0 floats rare earth minerals. Iron minerals are floated in an acidic environment using H_2SO_4 as activator and pH modifier. Low intensity magnetic separation is applied to recover magnetite from the iron flotation tails. Of the rare earth, 37.3 % can be recovered at a grade of 24.8 % rare earth oxide. Of the iron 80.8 % can be recovered at an iron grade of 56.0 %.

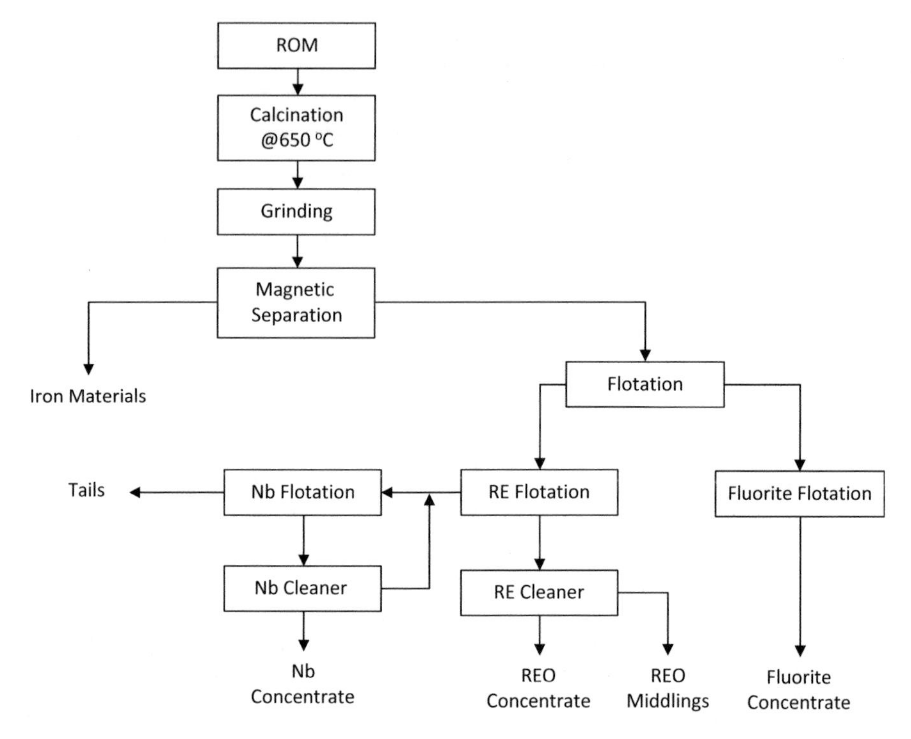

Fig. 2.10 Calcination, magnetic separation, and flotation process for the Baiyun Obo ore

2.2.5.3 Low Magnetic Separation-Flotation-High Magnetic Separation Process

Low intensity magnetic separation is used to separate iron-rich minerals. The magnetic separation tails are sent to flotation to separate fluorite and rare earth products. The flotation tails are fed to high intensity magnetic separation to recover hematite. The flow sheet is shown in Fig. 2.12.

A permanent magnet magnetic separator is used in the high intensity magnetic separation circuit with the magnetic field intensity 5500–6000 G. The rare earth recovery is between 20 and 30 % at a grade of 14–18 % REO. Of the iron, 65.4 % can be recovered at a grade of 55 %.

2.2.5.4 Low Magnetic Separation-High Magnetic Separation-Flotation Process

Yuand his colleagues developed the low magnetic-high magnetic-flotation process (Yu and Deng 1992; Yu and Che 2006). As shown in Fig. 2.13, the ore is ground to 95 % passing 0.074 mm and then separated into three fractions based on the

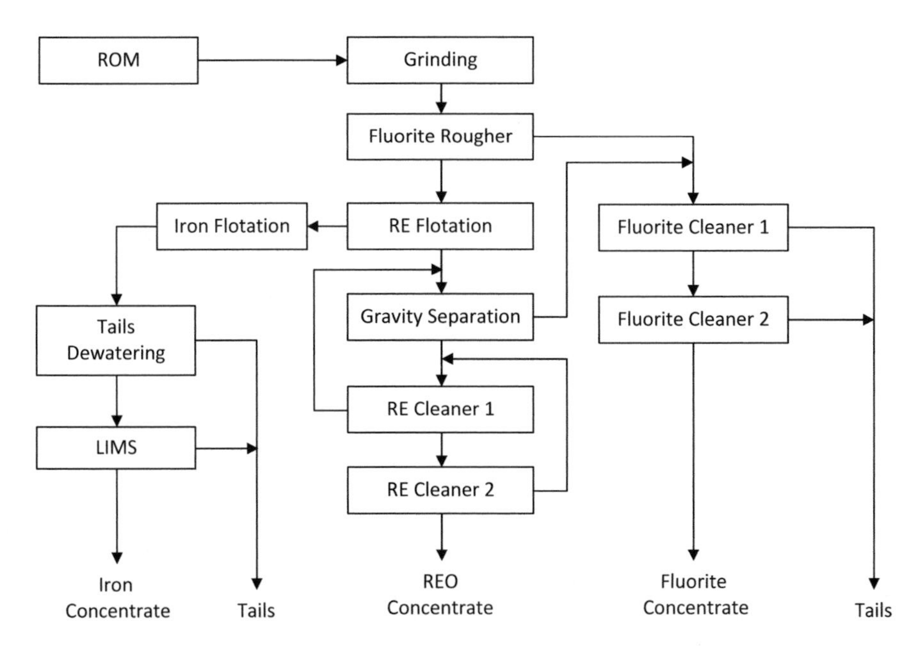

Fig. 2.11 Flotation and low magnetic separation process

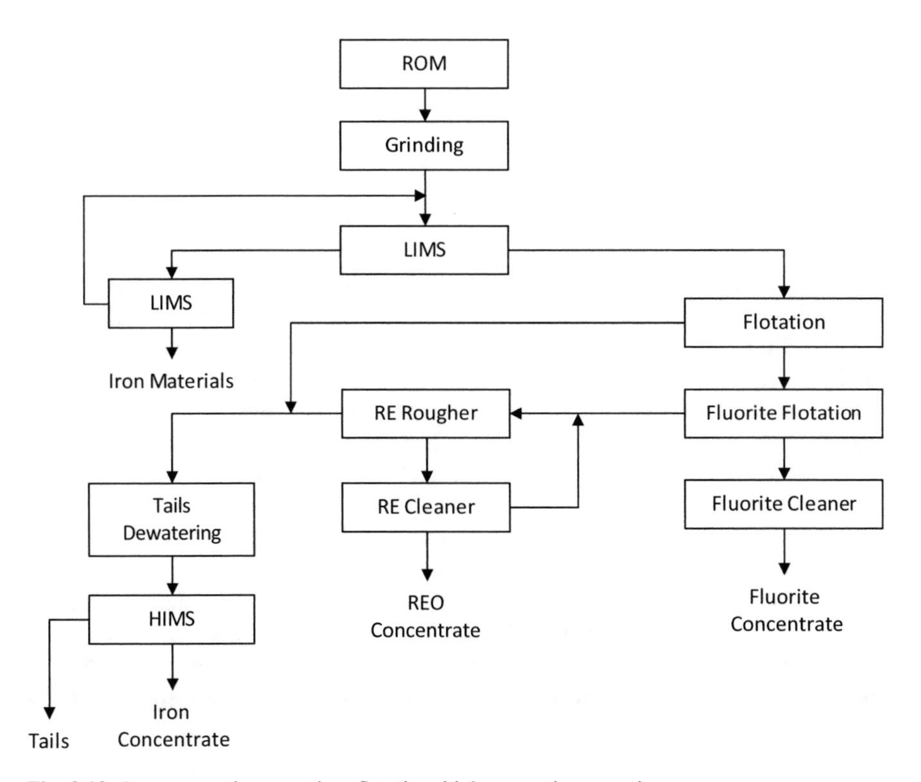

Fig. 2.12 Low magnetic separation, flotation, high magnetic separation process

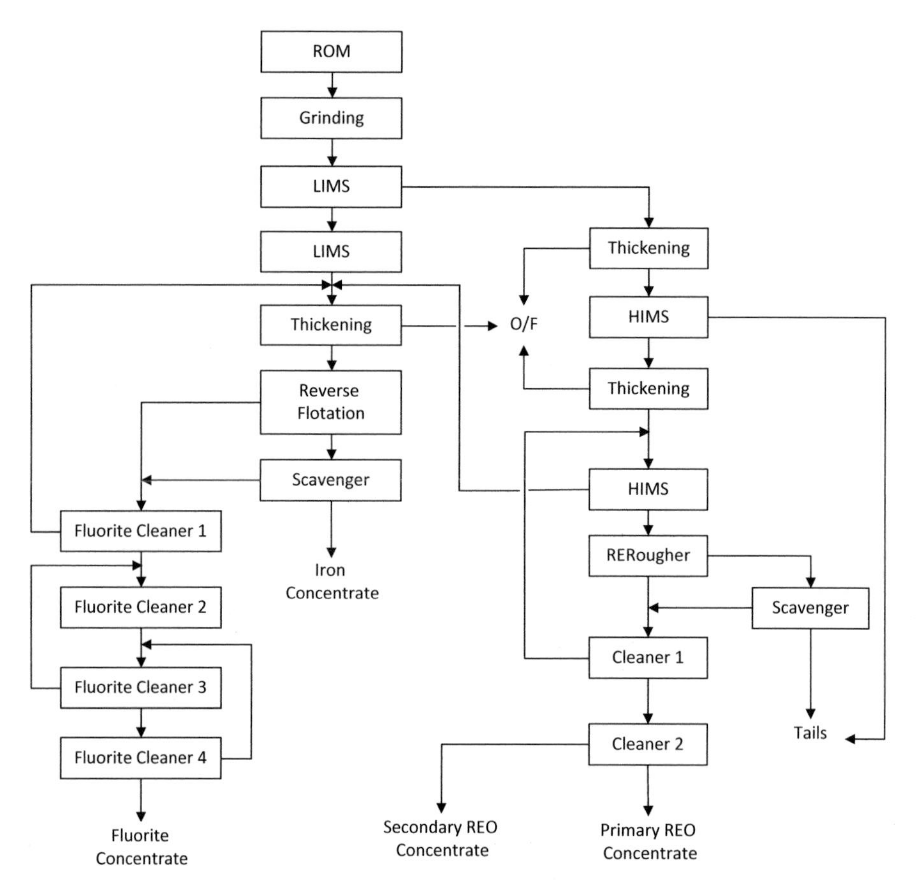

Fig. 2.13 Low magnetic separation, high magnetic separation, flotation process

difference on magnetic susceptibility. The ferromagnetic fraction is recovered using low intensity magnetic separation and the martite as well as primary hematite are recovered through high intensity magnetic separation. The rare earth minerals are preliminarily concentrated in the middlings of high intensity magnetic separation. The amounts of minerals such as fluorite, barite, and apatite, which have similar floatability to the rare earth minerals, are reduced significantly through high intensity magnetic separation. This enables more efficient flotation of the rare earth minerals.

Sodium silicate is used as a depressant and H_2O_5 as collector. H205 is a kind of hydroxamic acid with amide as the polar end and aromatic hydrocarbon as the non-polar end. The flotation pH is controlled at 9.5, temperature at 35–45 °C, and feed solids at 35–45 wt%. The process with one stage of rougher flotation, plus one scavenger flotation as well as two cleaner flotation stages produces a primary rare earth concentrate and a secondary rare earth concentrate. In this process, 12.6 % of rare earth is recovered at a REO grade of 55.6 and 6.0 % is recovered at a REO grade of 34.1 %, respectively.

2.2.6 WS Rare Earth Deposit

The WS rare earth deposit is one of three big light rare earth sources in China. It is located in Shandong Province in eastern China. The deposit is composed of rare earth minerals, iron minerals, barite, carbonatite, quartz, feldspar, mica, amphibole, and other minor minerals. Its main feature is NW-trending bastnaesite-baryte-carbonate veins associated with a quartz-syenite complex emplaced in Archaean gneisses (Jones et al. 1996). Bastnaesite is the only mineral of interest, although monazite, allanite, pyrochlore, aeschynite, chevkinite, columbite, and thorite are also present.

Flotation has been used dominantly in the processing of the WS rare earth deposit (Zeng et al. 1992). As shown in Fig. 2.14, the ground ore is first fed to the rougher flotation. The first cleaner tails and scavenger concentrate are recycled. The third cleaner tails are recovered as secondary rare earth concentrate to ensure the quality of primary rare earth concentrate and maintain overall rare earth

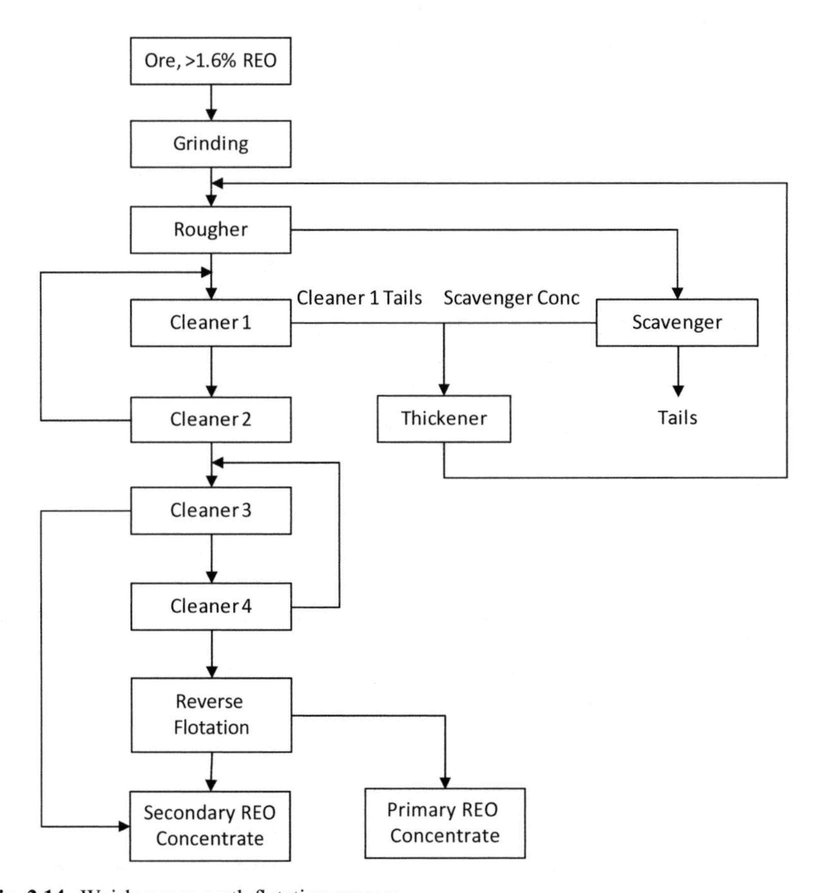

Fig. 2.14 Weishan rare earth flotation process

recovery. Reverse flotation is performed on the fourth cleaner concentrate to remove gangue minerals such as barite to ensure the quality of primary concentrate (Pan and Feng 1992).

The flotation is performed in an alkaline environment with sodium silicate and alum as depressants and H_2O_5 as collector. Of the total rare earth, 59.6 % can be recovered at a REO grade of 61.2 % in the primary concentrate and 23.8 % at a REO grade of 33.5 % in the secondary concentrate.

2.2.7 MN Maoniuping Rare Earth Deposit

The MN Maoniuping rare earth deposit is located in Sichuang Province, south western China. It is NNE-trending bastnaesite-baryte-carbonate vein associated with nordmarkite intruding a granite batholith (Pu 1988). Major minerals include bastnaesite, barite, celestine, fluorite, aegirine, quartz, feldspar, arfvedsonite, biotite, pyrite, galena, magnetite, hematite, limonite, and calcite. In addition to bastnaesite, barite, fluorite, and galena can be recovered as by-products. The disseminated grain sizes of the bastnaesite vary between 0.03 and 2.0 mm (Li and Zeng 2003a, b).

A combination of gravity separation, magnetic separation and flotation processes are used for the separation of the Maoniuping rare earths (Xiong 2002; Li and Miu 2002; Li and Zeng 2003a, b; Lin 2005, 2007; Yu and Che 2006; Cheng et al. 2007a, b). As shown in Fig. 2.15, the run of mine ore is ground to 65 % passing 0.150 mm and sent to a shaking table where the coarse grain bastneasite concentrate is recovered. The shaking table middlings are fed to the magnetic separation circuit to remove iron minerals and recover the magnetic rare earth concentrate. The shaking table tails and the magnetic tails are combined and classified to remove coarse grain tails. The classified products are ground to 75 % passing 0.074 mm and mixed with water. The clay is removed by de-sliming. Flotation is performed at the end of the rare earth separation process to obtain a flotation concentrate.

Starting with REO ore grading 5.46 %, the shaking table produces concentrate grading 57.7 % REO. Magnetic and flotation concentrates are 73.2 % REO and 64.3 % REO, respectively. The overall REO recovery is 83.1 %.

A relatively simple flotation process for the MN Maoniuping rare earth ore was reported by Xiong and Chen (2009). As shown in Fig. 2.16, a closed flotation circuit was tested. REO recovery was 87.0 % at a grade of 62.1 %.

2.2.8 Mountain Pass Rare Earth Deposit

The Mountain Pass mine in California used to be the largest rare earth concentrating facility in the world. However it was closed in 2003 due to environmental restrictions and international competition. The ore body contains bastnaesite rich in

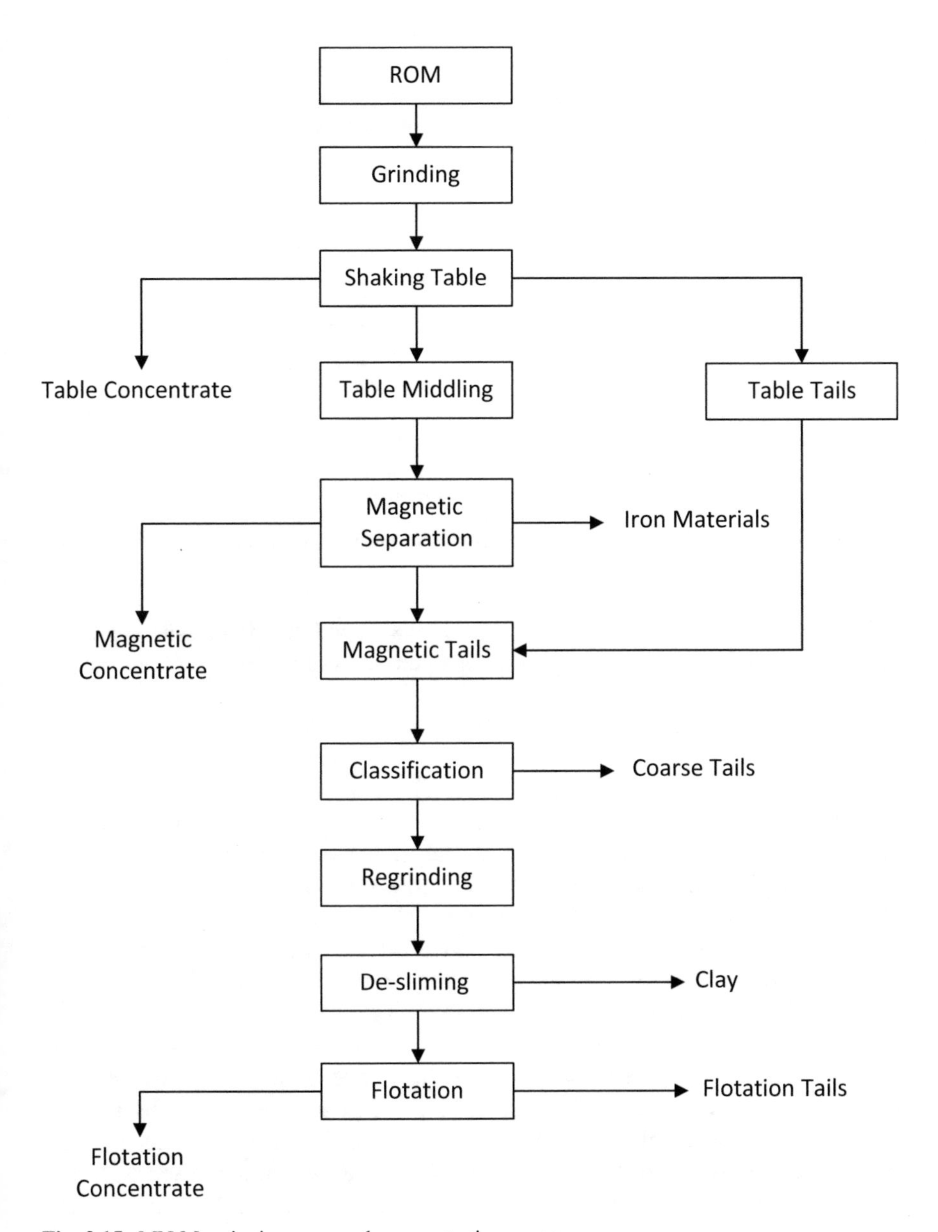

Fig. 2.15 MN Maoniuping rare earth concentrating process

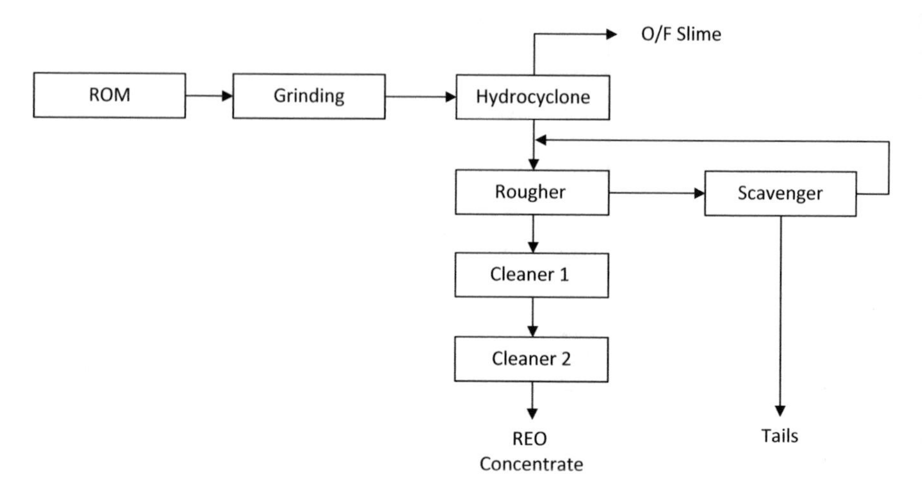

Fig. 2.16 Flowsheet of a simplified flotation circuit for the MN Maoniuping rare earth deposit

cerium group REEs as well as monazite. Associated with the bastnaesite are barite, calcite, strontanite, and silica as well as small amounts of apatite, hematite, galena, talcum, and phlogopite.

Flotation was used to separate the Mountain Pass rare earth ore. As shown in Fig. 2.17 (Johnson 1966), the ore was crushed and ground to minus 100 mesh and sent to conditioning at 55 % solids slurry. The staged conditioning was done in four stages of agitated tanks. Steam was injected to the first stage tank to raise the slurry temperature to 60 °C. Soda ash was added to control the pH at 8.95. Orzan, a kind of lignin sulfonate, was added as depressant in the second stage where temperature was raised to 80 °C using steam. The third stage was used to raise the slurry to boiling and to add N-80 oleic acid as collector and promotor. The fourth stage conditioner was used to cool the slurry to 60 °C for pumping to rougher flotation. One stage of rougher flotation, four stages of cleaner flotation, and one stage of scavenger flotation were used to concentrate the REO to 60 %. The overall REO recovery was between 65 and 70 %. The flotation concentrate could be sent to leaching using 10 % HCl. Lime was leached out and the bastnaesite stayed in the solids. Thickening and filtration were then used to remove excess water. REO at 70 % grade was produced after drying the leached solids. Calcination was also used to decompose the carbonate and obtain 90 % grade REO product.

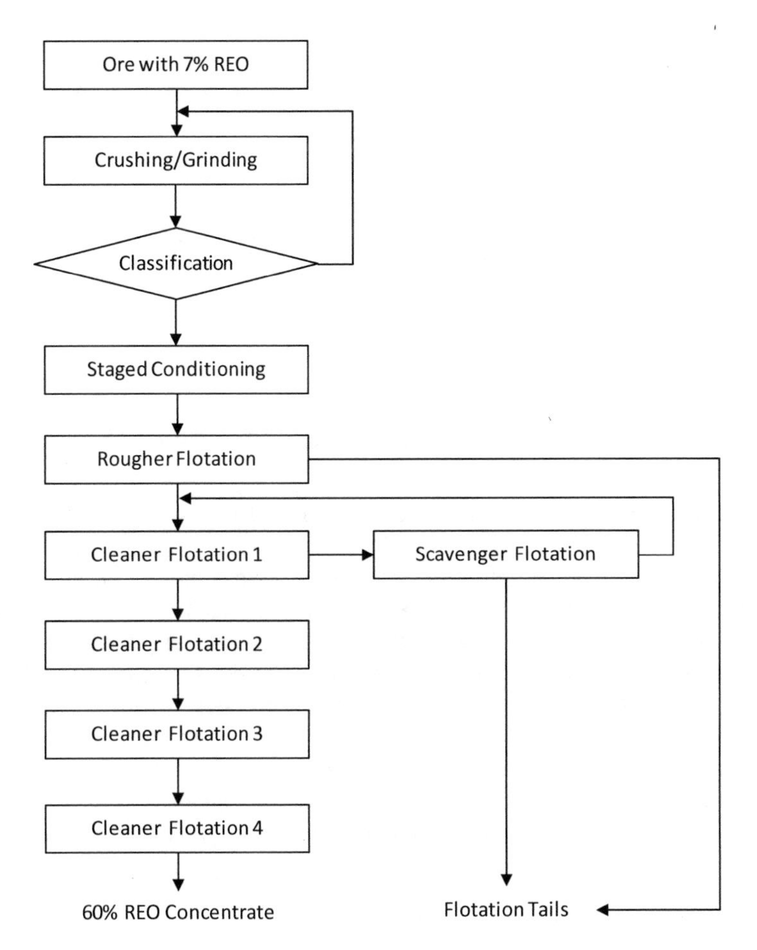

Fig. 2.17 Mountin pass rare earth concentrating process

2.2.9 *Mount Weld*

The Mount Weld deposit is located in Western Australia and is a deeply weathered volcanic carbonatite structure that contains approximately two million tons of rare earth oxides at an average grade of 20 % (Guy et al. 2000). Two main types of ore are identified as limonitic siltstone (CZ) and nodular limonitic ironstone (LI). The lower grade LI-type ore overlies the higher grade and more extensive CZ-type ore. The deposit contains a complex association of carbonates, phosphates, and iron oxides. The rare earth minerals include monazite, cheralite, cerianite, rhabdophane, florencite, and bastnaesite.

Two processing options for treatment of CZ-type ore were explored by Lynas Corporation Ltd. for pilot plant consideration. Option 1 includes gravity concentration and flotation. Option 2 includes finer grinding and four stages of open circuit cleaner flotation. The Option 1 flowsheet is shown in Fig. 2.18. Typical metallurgical results are a final concentrate grade of 49 % with recovery of about 35 %.

The Option 2 flowsheet is shown in Figure 2.19. The Option 2 process has similar metallurgical results. The Mount Weld mine officially opened in 2011. The process used is not public information. It is probably a hybrid of the two process schemes, incorporating the best features of each. Mount Weld intends to supply 11,000 ton REO in phase I and 22,000 ton REO in phase II. Full capacity production from phase I is expected in 2012 (Latimer 2011).

2.2.10 Summary of the Rare Earth Beneficiation Processes

The unit operations used in rare earth mineral processing are well-known and conventional. However, the above flowsheets demonstrate that for rare earth mineral processing these are employed in unusual and often complex arrangements. In addition, the rare earth mineral processing reagents tend to be specific to this industry and, as always, the optimal unit operations and reagents are specific to each mineral deposit, so their selection requires skilled testwork. Increasing demand for rare earth minerals, coupled with uncertainties over the reliability of supply from China, are expected to drive an increased effort to discover and process rare earth minerals around the world. However, a significant portion of the new discoveries belongs to challenging low grade deposit which will need the development of innovative processing technologies including specialized equipment and selective reagents.

2.3 Hydrometallurgical Processing of Rare Earth Mineral Concentrate

The processes for rare earth extraction include the decomposition of rare earth minerals, and the concurrent or subsequent leaching of the rare earth elements from the decomposed minerals. Ore beneficiation concentrate can be decomposed by, for example, acid roasting, caustic cracking, and chlorination. The rare earth elements can be selectively extracted. The nature of the rare earth extraction process depends on the type of minerals in the concentrate, the grade of the concentrate and the targeted products. This section reviews rare earth element extraction from the major rare earth minerals, including bastnaesite, monazite, xenotime, ion-adsorption clay, allanite, cerite, and eudialyte.

Different with the physical beneficiation introduced in Sect. 2.2, rare earth extraction is chemical processing, which converts the rare earth mineral concentrate to a rare earth compound which is either an end product or an intermediate product

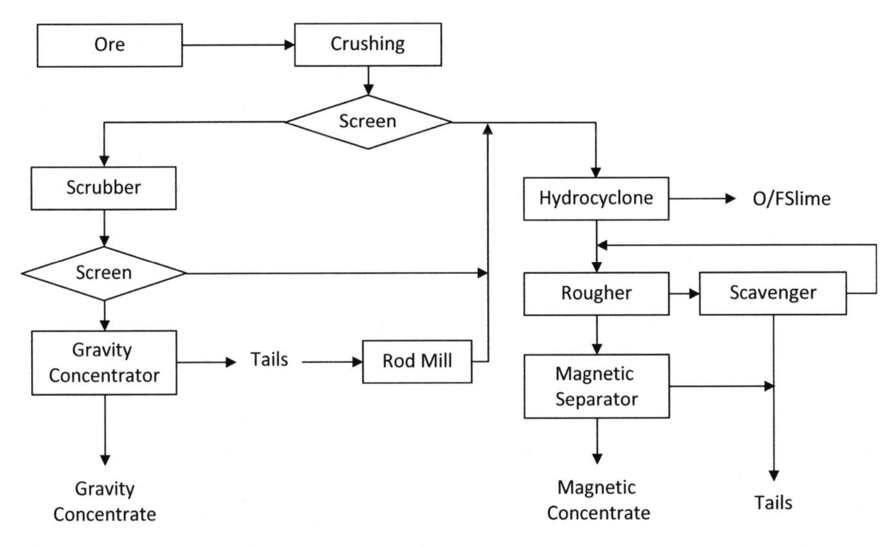

Fig. 2.18 Option 1 flowsheet development for CZ-type of ore

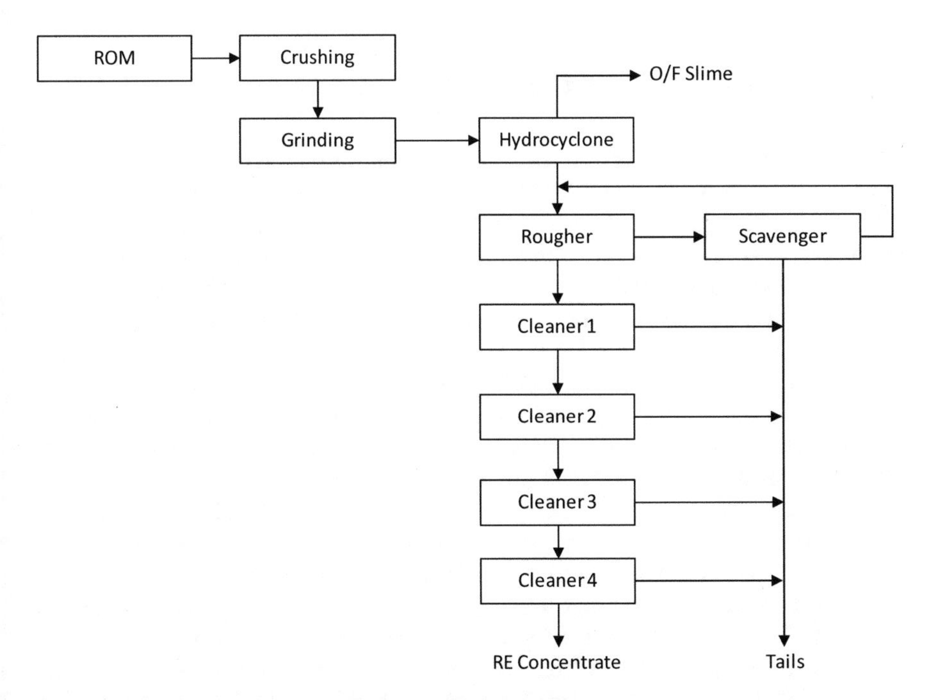

Fig. 2.19 Option 2 flowsheet development for CZ-type of ore

for the subsequent production of individual rare earth elements or other compounds. The rare earth extraction process uses one or more reagents to decompose the minerals and to leach the rare earth elements into solution. The rare earth element separation process uses solvent extraction, ion-exchange, or chemical precipitation to produce either mixed rare earth oxides or individual rare earth oxides. The varieties of rare earth extraction methods use many different reagents. These reagents are mainly inorganic acids, alkalis, electrolytes, and chlorine gas. The commonly used acids include sulfuric acid (H_2SO_4), hydrochloric acid (HCl), and nitric acid (HNO_3). Sodium hydroxide (NaOH) and sodium carbonate (Na_2CO_3) are the most common alkalis. Electrolytes include ammonium sulfate (($NH_4)_2SO_4$, ammonium chloride (NH_4Cl), and sodium chloride (NaCl). HCl, HNO_3, and H_2SO_4 are commonly used to extract RE from silicate ore minerals such as gadolinite, eudialite and allanite. The alkalis and H_2SO_4 are mainly used to leach rare earth elements from phosphate ore minerals like monazite and xenotime. The carbonatite rare earth minerals such as bastnaesite are treated using either H_2SO_4 or alkalis. HNO_3 is mainly used to leach eudialyte and apatite. The extraction of rare earth elements from ion-adsorption clay deposits uses electrolyte solutions. Chlorine gas (Cl_2) is exclusively used in the chlorination process which can be used to treat most of the rare earth minerals. In the following sections, the decomposition and leaching of rare earth mineral concentrates are described.

2.3.1 H_2SO_4 Acid-Roasting and Water Leaching

Acid-roasting is a major RE mineral decomposition process, classified into low-temperature roasting (<300 °C) or high-temperature roasting (>300 °C). The low-temperature process, formerly used in the 1970s to treat low-grade RE concentrates, is normally followed by long and complex processes to remove impurities from the leach solution. To inhibit the generation of soluble impurities, the high-temperature acid roasting process was developed in the 1980s to treat high-grade RE concentrates (Shi 2009). Currently, the majority of acid-roasting RE extraction operations use the high-temperature process. The high-temperature acid-roasting process is relatively simple but generates potentially hazardous exhaust gases such as hydrogen fluoride (HF), sulfur dioxide (SO_2), sulfur trioxide (SO_3) and silicon tetrafluoride (SiF_4). Normally, a water scrubber is used initially to capture most of the exhaust gases. A mixed acid containing HF, H_2SO_4, and H_2SiF_6 can be recovered in this initial scrub. A second scrubber using diluted sodium carbonate solution is used to purify the exhaust gas before it is released.

A general acid-roasting and leaching process for bastnaesite and monazite is shown in Fig. 2.20. It consists of grinding, mixing, roasting, leaching, and solid–liquid separation. The RE mineral concentrate is normally ground to less than 100 μm (150 mesh) before mixing with concentrated acid. A rotary kiln is employed for the acid roasting. Either a filter or thickener can be used to separate the leach solution from the residue. At higher acid/ore ratio more REEs plus

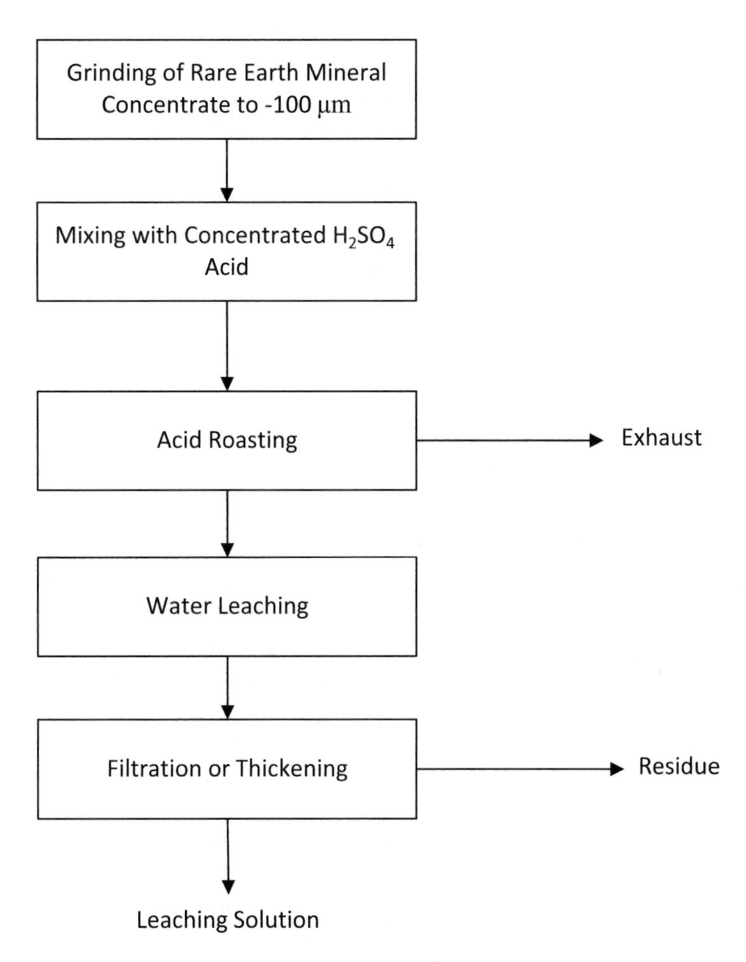

Fig. 2.20 General acid-roasting and leaching process for bastnaesite and monazite

thorium are solubilized. A lower ratio permits selective dissolution. With the increase in temperature higher than 300 °C, the decomposition rate is compromised but the leaching of thorium is also reduced due to the formation of insoluble ThP_2O_7. This is utilized to leave the thorium in the residue in some rare earth processing plants.

The REEs, thorium, and uranium are converted to soluble sulfates during H_2SO_4 acid-roasting. The major reactions include (Chi and Wang 1996):

$$2RECO_3F + 3H_2SO_4 = RE_2(SO_4)_3 + 2HF\uparrow + 2CO_2\uparrow + 2H_2O \quad (2.3)$$

$$2REPO_4 + 3H_2SO_4 = RE_2(SO_4)_3 + 2H_3PO_4 \quad (2.4)$$

$$ThO_2 + 2H_2SO_4 = Th(SO_4)_2 + 2H_2O \uparrow \tag{2.5}$$

$$2U_3O_8 + O_2 + 6H_2SO_4 = 6UO_2SO_4 + 6H_2O \uparrow \tag{2.6}$$

$$RE_2O_3 + 3H_2SO_4 = RE_2(SO_4)_3 + 3H_2O \uparrow \tag{2.7}$$

The major side reactions include:

$$CaF_2 + H_2SO_4 = CaSO_4 + 2HF \uparrow \tag{2.8}$$

$$Fe_2O_3 + 3H_2SO_4 = Fe_2(SO_4)_3 + 3H_2O \uparrow \tag{2.9}$$

$$SiO_2 + 2H_2SO_4 = H_2SiO_3 + H_2O \uparrow + 2SO_3 \uparrow \tag{2.10}$$

$$4HF + SiO_2 = SiF_4 \uparrow + 2H_2O \uparrow \tag{2.11}$$

$$H_2SiO_3 = SiO_2 + H_2O \uparrow \tag{2.12}$$

$$2H_3PO_4 = H_4P_2O_7 + H_2O \uparrow \tag{2.13}$$

$$Th(SO_4)_2 + H_4P_2O_7 = ThP_2O_7 + 2H_2SO_4 \tag{2.14}$$

At the Saskatchewan Research Council, numerous H_2SO_4 roasting and leaching tests have been conducted. Normally the acid/concentrate mass ratio is between 1/1 and 2/1 depending on the grade of the rare earth mineral concentrate and the gangue minerals occurring in the concentrate. When the grade is high, the acid consumption is relatively low. Acid consumption will increase if the content of carbonate, fluoride and/or iron minerals is high. As well, the acid consumption is affected by roasting time and temperature. A higher roasting temperature and shorter roasting time can be used to reduce acid consumption while maintaining the same decomposition efficiency. However, higher roasting temperature may reduce the conversion of rare earth elements to soluble sulfates and reduce the rare earth recovery. A white or reddish color of roasted sample is one of the indicators of over-roasting. However, a dark green color indicates under-roasting. Roasting between 180 and 300 °C for 2–4 h is often successful.

The solubility of rare earth sulfates will also drop with an increase in leaching temperature. Therefore, water leaching is normally performed at ambient temperatures. To reduce dissolved rare earth loss from adsorption in the leach residue, the leach is relatively dilute, at a water/concentrate mass ratio between 7 and 15.

In addition to bastnaesite and monazite, the acid-roasting and water leaching process can also be used to process xenotime, aeschynite, and RE silicates.

2.3.2 HCl Acid Leaching

Dilute HCl is used to dissolve calcium carbonate and concentrated HCl is used to decompose bastnaesite. The major reaction of the latter type is shown by Eq. (2.15).

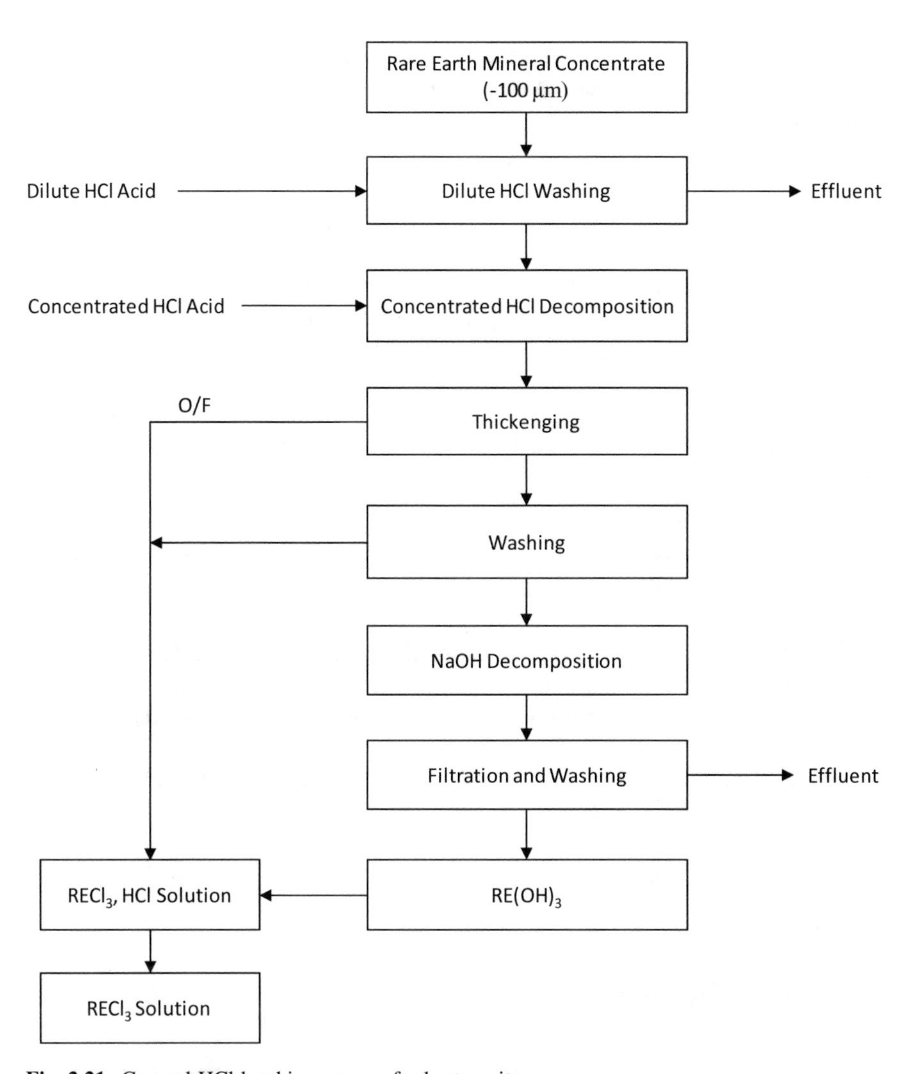

Fig. 2.21 General HCl leaching process for bastnaesite

$$3RECO_3F + 6HCl = 2RECl_3 + REF_3 \downarrow + 3H_2O + 3CO_2 \uparrow \tag{2.15}$$

RE fluorides remain in the residue after HCl decomposition. After an initial solid–liquid separation, NaOH is used to convert them to RE hydroxides as shown by Eq. (2.16).

$$REF_3 + 3NaOH = REOH \downarrow + 3NaF \tag{2.16}$$

After a second solid–liquid separation, the RE hydroxides are combined with the $RECl_3$ solution from the major reaction and are dissolved by the excess HCl. This process is diagrammed in Fig. 2.21.

Concentrated HCl is also often used to decompose allanite, cerite, and gadolinite. The reaction between concentrated HCl and gadolinite is expressed by Eq. (2.17).

$$RE_2FeBe_2(SiO_4)_2O_2 + 12HCl = 2RECl_3 + FeCl_2 + 2BeCl_2 + 2SiO_2 + 6H_2O \quad (2.17)$$

HCl can also be used to dissolve rare earth oxide, rare earth carbonate and other rare earth intermediate products in subsequent process stages.

2.3.3 *HNO_3 Acid Leaching*

Eudialyte is a rare earth silicate. A eudialyte mineral concentrate can be decomposed using 30–50 % nitric acid (HNO_3). In addition to rare earth elements, zirconium (Zr), tantalum (Ta), and niobium (Nb) can also be recovered from eudialyte.

REEs can enter the crystal matrix of apatite through isomorphous replacement. The rare earth content in apatite is normally very low. However, apatite is one of the important rare earth element sources because it is a relatively common and widely occurring mineral. Apatite can be decomposed by HNO_3, HCl, or H_2SO_4, releasing rare earth elements into the leach solution. Apatite readily dissolves in 50–60 % HNO_3 acid at 60–70 °C as shown in Eq. (2.18).

$$Ca_5(PO_4)_3F + 10HNO_3 = 3H_3PO_3 + 5Ca(NO_3)_2 + HF \quad (2.18)$$

2.3.4 *Na_2CO_3 Roasting*

Low grade (20–30 % REO) bastnaesite and monazite concentrate can be decomposed through 350–550 °C roasting with sodium carbonate at a mass ratio of concentrate to sodium carbonate from 6:1 to 3:1. The reactions during roasting include (Huang 2006):

$$RECO_3F = REOF + CO_2 \uparrow \quad (2.19)$$

$$2CeCO_3F + \frac{1}{2}O_2 = Ce_2O_3F_2 + 2CO_2 \uparrow \quad (2.20)$$

With an increase in roasting temperature to 600–700 °C, the following additional reactions will occur.

$$2REOF + Na_2CO_3 = RE_2O_3 + CO_2 \uparrow + 2NaF \tag{2.21}$$

$$Ce_2O_3F_2 + Na_2CO_3 = 2CeO_2 + 2NaF + CO_2 \uparrow \tag{2.22}$$

$$2REPO_4 + 3Na_2CO_3 = RE_2(CO_3)_3 + 2Na_3PO_4 \tag{2.23}$$

$$2REPO_4 + 3Na_2CO_3 = 2RE_2O_3 + 3CO_2 \uparrow + 2Na_3PO_4 \tag{2.24}$$

$$CaF_2 + Na_2CO_3 = CaCO_3 + 2NaF \tag{2.25}$$

$$BaSO_4 + Na_2CO_3 = BaCO_3 + 2Na_2SO_4 \tag{2.26}$$

$$Ca_3(PO_4)_2 + 3Na_2CO_3 = 3CaCO_3 + 2Na_3PO_4 \tag{2.27}$$

After roasting, the soluble salts can be washed out using water or dilute acid. The rare earth oxide will stay in the solids and the grade can be increased from 20–30 % to 50–60 %.

2.3.5 *NaOH Decomposing*

Concentrated NaOH solution (>50 % by weight) can be used to decompose rare earth mineral concentrates. When it is used to decompose bastnaesite and monazite, the following reactions occur (Chi and Wang 1996):

$$RECO_3F + 3NaOH = RE(OH)_3 + NaF + Na_2CO_3 \tag{2.28}$$

$$REPO_4 + 3NaOH = RE(OH)_3 + Na_3PO_4 \tag{2.29}$$

At 140 °C and a mass ratio of NaOH/rare earth concentrate of 1.2–1.4, over 90 % rare earth elements can be extracted within 5 h. The concentration of NaOH solution has significant effects on the extraction kinetics and alkali consumption. For example, the reaction time can be reduced from 5 h to less than 1 h with NaOH concentration increased from 50 to 60 %. Concurrently, the NaOH consumption is reduced from 1,200–1,400 to 800–900 kg/t concentrate. Particle size and agitation can also play important roles. For example, in monazite decomposition 100 % passing 325 mesh or 43 μm and sufficient agitation are necessary due to the fact that the produced RE hydroxide attaches to the surface of the concentrate particles and forms a layer around the reacting core thus increasing the mass transfer resistance and the reaction time. Furthermore, the increase in reaction temperature improves the subsequent $RE(OH)_3$ dissolution in hydrochloric acid.

Rare earth mineral concentrate can also be decomposed with fused NaOH at higher temperatures of approximately 350 °C or higher. However, the rare earth products of fused NaOH decomposition are difficult to dissolve. For this reason, this process has been abandoned by the rare earth industry.

To reduce cost and improve operational safety, an alternating electric field NaOH decomposing process was developed in 1980s by the Beijing General Research Institute of Nonferrous Metals. The major improvement with this process is using electrode heating to replace steam heating of NaOH. The decomposition reaction time is reduced from 6–12 h to 2 h. As well, the alkali consumption is low in comparison with the traditional NaOH decomposing process.

2.3.6 *Chlorination*

The chlorination process uses chlorine gas (Cl_2), at high temperatures (600–1200 °C) in the presence of carbon (C), to decompose RE mineral concentrates. The REEs and some other elements are converted to chlorides. $SiCl_4$ is used as required to remove fluorine through the formation of gaseous SiF_4. The major RE mineral decomposition and chlorination reactions during chlorination include (Zhang et al. 2002a; Zeng et al. 2007):

$$RECO_3F = REOF + CO_2\uparrow \tag{2.30}$$

$$REOF + C + Cl_2 + \frac{1}{4}SiCl_4 = RECl_3 + \frac{1}{4}SiF_4\uparrow + CO\uparrow \tag{2.31}$$

$$REOF + \frac{1}{2}C + Cl_2 + \frac{1}{4}SiCl_4 = RECl_3 + \frac{1}{4}SiF_4\uparrow + \frac{1}{2}CO_2\uparrow \tag{2.32}$$

$$\frac{1}{3}REPO_4 + C + Cl_2 = \frac{1}{3}RECl_3 + \frac{1}{3}POCl_3\uparrow + CO\uparrow \tag{2.33}$$

$$\frac{1}{3}REPO_4 + \frac{1}{2}C + Cl_2 = \frac{1}{3}RECl_3 + \frac{1}{3}POCl_3\uparrow + \frac{1}{2}CO_2\uparrow \tag{2.34}$$

As well, some gangue minerals also undergo chlorination reactions.

$$\frac{1}{3}Fe_2O_3 + C + Cl_2 = \frac{2}{3}FeCl_3 + CO\uparrow \tag{2.35}$$

$$\frac{1}{3}Fe_2O_3 + \frac{1}{2}C + Cl_2 = \frac{2}{3}FeCl_3 + \frac{1}{2}CO_2\uparrow \tag{2.36}$$

$$BaSO_4 + C + Cl_2 = BaCl_2 + CO_2\uparrow + SO_2\uparrow \tag{2.37}$$

$$CaF_2 + \frac{1}{2}SiCl_4 = CaCl_2 + \frac{1}{2}SiF_4\uparrow \tag{2.38}$$

The resulting chlorination products have a range of melting and boiling points. The lower boiling point products, such as $FeCl_3$, will exit the process in the exhaust gases, while the higher melting point products will remain in the solid residue. The chlorides of rare earth elements, Ca, Ba, and other alkali earth metals are collected as melts, which go through impurity removal, separation, and purification to produce rare earth products.

The major stages of the chlorination process are feed preparation, chlorination, and separation of chlorides. The reaction rate of chlorination increases with temperature.

Chlorination can also be performed using ammonium chloride as shown by the following equation (Shi and Zhou 2003).

$$RE_2O_3 + 6NH_4Cl_4 = 2RECl_3 + 6NH_3 \uparrow + 3H_2O \uparrow \tag{2.39}$$

Chlorination can be used to treat a variety of rare earth minerals like bastnaesite, monazite, xenotime, allanite, cerite, euxenite, fergusonite, and gadolinite. The REO grade of chlorination products is relatively high. However, rare earth mineral concentrate chlorination products still require subsequent extraction and separation processes. Currently, there is no industrial application of the rare earth chlorination process.

2.3.7 Leaching of Ion-adsorbing Type of Rare Earth Clay

An ion-adsorbing type rare earth deposit is normally the weathered crust of a magmatic-type primary deposit. The rare earth elements are adsorbed on the surface of clays in the form of ions. The rare earth elements are not soluble or hydrolyzed in water but follow ion-exchange laws (Lu et al. 1997; Tian and Yin 1996; Li 1993; Chi 1989).

$$[Al_2Si_2O_5(OH)_4]_m \times nRE + 3nMe^+ = [Al_2Si_2O_5(OH)_4]_m \times 3nMe + nRE^{3+} \tag{2.40}$$

where, Me could be Na^+ or NH_4^+.

The ion-adsorption type of rare earth deposit has unique features: (1) Mining is simple due to the shallowness of the deposit; (2) No grinding is required since the rare earth elements are adsorbed and concentrated on the surface of the silicate clays; (3) Rare earth element grade is low from 0.05 to 0.3 %; (4) Almost all of the rare earth elements can be found in the ion-adsorbing type of deposit and HRE accounts for approximately 40 % of the TRE; (5) There is a low concentration of radioactive impurities. These features of the ion-adsorbing type of deposits make them a very important and an economical REE source, especially as a HREE source.

The processes for REE extraction from the ion-adsorption type of deposits include heap leaching, agitated leaching and in situ leaching (or direct leaching). In heap leaching, the ore is stacked to about 1.5 m in height on a solution collection membrane. An electrolyte solution such as NaCl or $(NH_4)_2SO_4$ is sprayed onto the stack and trickles down through the stack to the collection membrane. In the stack Na^+ or NH^{4+} exchanges the RE^{3+} to the solution. Temperature and pressure have no significant impact on the ion exchange efficiency. The ion exchange rate is affected by the concentration of the electrolyte solution, pH, and solution wash rate. The ion exchange rate between Na^+/NH_4^+ and RE^{3+} increases with electrolyte concentration and wash rate. A pH4 electrolyte solution is normally employed. A higher pH causes RE hydrolyzation thus decreases the leaching rate while a lower pH will dissolve more impurities such as aluminum and iron.

If the deposit is mined by a wet process, the permeability of the ore required for heap leaching will be destroyed. In this case, agitated leaching will be used as the extraction process. Because ion exchange is a reversible process, the extraction efficiency with agitated leaching is normally lower than with heap leaching. However, both agitated leaching and heap leaching involve mining a large amount of material and producing large amount of tailings, which pose a serious environment protection challenge. In-situ leaching was developed and implemented as the second generation leaching process to reduce the environmental impacts associated with the development of ion-adsorption type of rare earth deposits. The overall resources utilization was also improved from no more than 50 % in agitated leaching and heap leaching to over 70 % in in situ leaching (Tang and Li 1997).

2.3.8 Summary

The decomposition and leaching of rare earth mineral concentrates involve the use of relatively large amounts of reagents and highly specific, relatively high cost reaction conditions. Except for the leaching of ion-adsorbing type of clay, the lack of selectivity of decomposition and leaching of RE minerals makes the subsequent impurity removal and rare earth element separation from the leaching solution of low grade rare earth mineral concentrate with higher reagent consumption than the processing of high grade concentrate. Therefore, the rare earth mineral concentrate grade plays a significant role in the economics of a rare earth extraction process.

References

Cai, Z., Cao, M., Che, L., Yu, Y., & Hu, H. (2009). Study on beneficiation process for recovering rare earth from the LIMS tails of HIMS rougher concentrate after magnetizing roasting in Baogang concentrator. *Metal Mine, 397*, 155–157.

Cheng, J., Hou, Y., & Che, L. (2007a). Flotation separation on rare earth minerals and gangues. *Journal of Rare Earths, 25*, 62–66. 10002-0721(2007)-0062-05.

Cheng, J., Hou, Y., & Che, L. (2007b). Making rational multipurpose use of resources of RE in baiyuan ebo deposit. *Chinese Rare Earths, 28*(1), 70–74. 1004-0277(2007)01-0070-05.

Chi, R. (1989). Investigation on processes of ion-absorption type of rare earth. *Hunan Mining and Metallurgy, 5*, 38–41.

Chi, R., & Wang, D. (1996). *Rare earth ore processing and extraction technology*. Science Publishing Company, Beijing, pp. 305–332.

Fang, J., & Zhao, D. (2003). Separation of rare earth from tails of magnetite separation in Bao Steel's concentrator. *Metal Mine, 321*, 47–49.
Guy, P., Bruckard, W., & Vaisey, M. (2000). Beneficiation of Mt Weld rare earth oxides by gravity concentration, flotation, and magnetic separation. *Senventh Mill Operator's Conference*, Kalgoorlie, WA. pp. 197–206.
Huang, L. (2006). Rare earth extraction technology. Metallurgy Industry Publishing Company. pp. 113–168.
Jones, A. P., Wall, F., & Williams, C. T. (1996). *Rare earth minerals-Chemistry, origin, and ore deposits* (pp. 181–188). London: Chapman & Hall.
Johnson, N. L. (1966). Rare earth concentration at Molybdenum Corporation of America. *Deco Trefoil, 3*, 9–16.
Latimer, C. (2011). *Mt Weld rare earths mine officially open*. Retrieved 15 Aug, 2011 from http://www.miningaustralia.com.au/news/mt-weld-rare-earths-mine-officially-open.
Li, F., & Miu, W. (2002). The investigation on Maoniuping bastnaesite ore processing. *Rare Earth Mining and Metallurgy, 2*, 7–12.
Li, F., & Zeng, X. (2003a). Bastnasite separation process based on size fractions. *Chinese Journal of Rare Metals, 27*, 482–485. 0258-7076 (2003)04-0482-04.
Li, F., & Zeng, X. (2003b). Maoniuping bastnasite separation process. *Journal of Shanhai Second Polytechnic University, 1*, 11–16.
Li, Q. (1993). *New extraction technology of the ion-absorption rare earth*. Guanxi Chemical Engineering. pp. 8–10.
Lin, H. (2005). Design and its practice for rare earth smelter. *World Nonferrous Metals, 12*, 15–17.
Lin, H. (2007). The design of high grade rare earth concentrator and operation practice. *Sichuan Rare Earths, 1*, 6–11.
Lu, S., Lu, C., Wu, N., & Xiao, Q. (1997). Washing leaching process of ion-absorption type of rare earth. *Hydrometallurgy*, 34–38.
Pan, M., & Feng, J. (1992). Study on Weishan rare earth concentrating processes. *Shandong Mining and Metallurgy, 14*, 31–34.
Pu, G. (1988). Discovery of an alkali pegmatite-carbonatite complex zone in Maoniuping. *Geology Review, 34*, 82–92.
Shi, F. (2009). *Rare earth metallurgy*. Metallurgy Industry Publishing Company. pp. 24–59.
Shi, W., & Zhou, G. (2003). Kinetics on chlorinating rare earth of Weishan mid-grade bastnasite in Shandong after fixed fluorine treatment. *Chinese Journal of Process Engineering*, 278–282.
Tang, X., & Li, M. (1997). In-situ leaching mining of ion-adsorbed rare-earth mineral. *Mining R & D*, 1–4.
Tian, J., & Yin, J. (1996). A kinetic analysis of leaching a South China heavy rare earth ore. *Jiangxi Science*, 81–86.
Wang, D., & Chi, R. (1996). *Rare earth processing and extraction technology* (pp. 63–117). Beijing: Ke Xue Chu Ban She.
Xiang, J., Jin, X., & Wang, Y. (1985). Study on bastnaesite separation from complex ore deposit. *The Complex Utilization of Mineral Resources, 3*, 71–74.
Xiong, S. (2002). An experimental research on combined gravity-flotation flowsheet for a rare earth ore in ShiChuan. *Multipurpose Utilization of Mineral Resources, 5*, 3–6. 1000-6532 (2002)05-0003-04.
Xiong, W., & Chen, B. (2009). Beneficiation study on rare earth ore in Mianning of Sichuan. *Chinese Rare Earths, 30*, 89–92. 1004-0277(2009)03-0089-04.
Yang, L. (2005). Investigation on improving flow sheet of rare earth flotation. *Chinese Rare Earths, 26*, 75–77. 1004-0277(2005)01-0075-03.
Yu, Y., & Che, L. (2006). The current rare earth production practice and the development of mineral processing technology in China. *Chinese Rare Earths, 27*, 96–102. 1004-0277(2006)01-0095-08.
Yu, Y., & Deng, Y. (1992). Baiyun Obo low and middle grade rare earth oxide ore processing and recovery of iron. *Metal Mine, 1*, 39–44.

Zeng, F., Yu, X., Liu, J., Zhang, J., Zhang, Q., & Wang, Z. (2007). Recovery of rare earth from gravity concentrated tailings by carbochlorination. *The Chinese Journal of Nonferrous Metals*, 1195–1200. 1004-0609 (2007)07-1195-06.

Zeng, X. (1993). WS rare earth resources and its development. *National Rare Earth Resources and Application Conference*, Xichang. pp. 21–25.

Zeng, X., Li, F., & Liu, J. (1992). The progress of RE beneficiation technology in China. *Journal of Shanhai Second Polytechnic University, 2*, 27–31.

Zhang, L., Wang, Z., You, J., Chi, M., Yang, D., & Lei, P. (2002a). Study on characteristics of carbochlorination of mixed bastnaesite-monazite concentrate. *Journal of the Chinese Rare Earth Society*, 193–196. 1000-4343(2002)-0193-04.

Zhang, W., Zheng, C., & Qin, Y. (2002b). Concentrating of rare earth in tails from concentrating mill of Baotou Iron and Steel Co. *Hydrometallurgy, 21*, 36–38. 1009-2617(2002)01-0036-03.

Zhao, J., Tang, X., & Wu, C. (2001). Status of mining and recovering technologies for ion-adsorbed rare earth deposits in China. *Yunnan Metallurgy, 30*, 9–13.

Zhao, R., Zhang, B., & Li, B. (2008). Experimental research on separation of iron from the flotation tailings of Bayun Obo rare earth minerals. *Multipurpose Utilization of Mineral Resources, 4*, 34–37. 1000-6532-(2008)04-0034-05.

Chapter 3
Solvent Extraction in Metal Hydrometallurgy

3.1 Solvent Extraction Basics

The rare earth produced from different ores or solutions is a mixed product containing various individual rare earth elements. To separate them to individual elements presents one of the most challenging separation tasks in metal hydrometallurgy. In comparison with the complex and lengthy processes such as fractional precipitation and crystallization, solvent extraction, also called liquid–liquid extraction, offers a more flexible approach that allows for the production of individual rare earth elements at required purities and quantities. Solvent extraction is currently the dominant technology to separate and purify the individual rare earth elements.

Like the solvent extraction of all other metals, the solvent extraction of rare earth utilizes the partition of rare earth elements between two immiscible phases to realize the separation and concentration of one or one group of rare earth elements from the others. A basic solvent extraction process includes extraction, scrubbing/washing, and stripping.

3.1.1 Extraction

The ability of metal ions to distribute themselves between an aqueous solution and an immiscible organic solution has long been utilized by hydrometallurgists to transfer the valuable metals from the aqueous solution to the organic solution and leave the unwanted metals and impurities in the aqueous solution.

The aqueous solution, normally the feed, contains the valuable metals, free acid, and impurities dissolved in water.

The organic solution usually contains an extractant and a modifier dissolved in diluent or carrier. Diluent or carrier is the liquid or homogeneous mixture of liquids in which extractant and modifier are dissolved to form the major component of the

J. Zhang et al., *Separation Hydrometallurgy of Rare Earth Elements*,
DOI 10.1007/978-3-319-28235-0_3

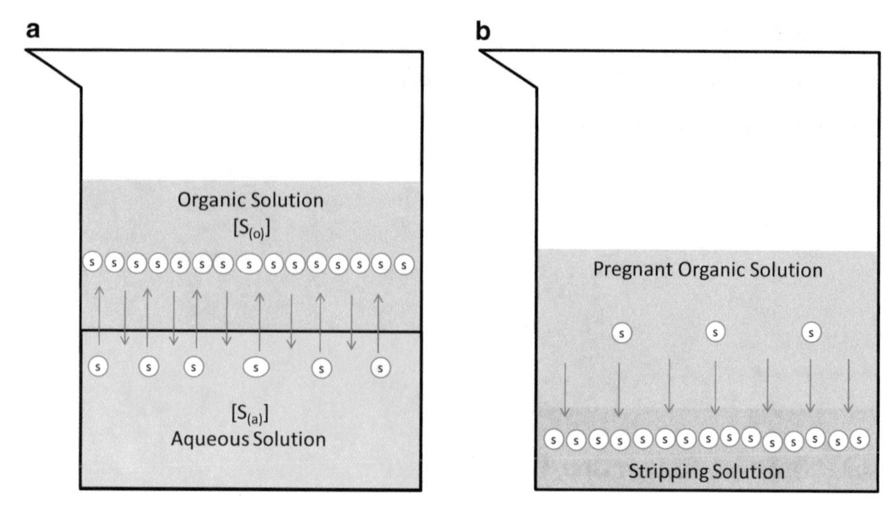

Fig. 3.1 Schematic representative of solvent extraction principle (**a**) Extraction and (**b**) Stripping

organic phase. The majority of diluents used in solvent extraction are crude oil fractions with various proportions of aliphatic, naphthenic, and aromatic components.

The extractant is the active component primarily responsible for the transfer of the metals from the aqueous solution to the organic solution. In order to transfer the metals from the aqueous solution to the organic solution, metal–extractant complex is often formed to increase the solubility of the metal in the organic solution.

The modifier is added to improve the solubility of the metal–extractant complex in the diluent to improve phase disengagement. A third phase between aqueous and organic phases is often formed in solvent extraction operation which is related to the solubility of the metal–extractant complex in the organic solution. The third phase can cause significant processing challenges and lead to organic loss. The addition of a modifier to the organic phase is an effective approach to minimize the formation of the third phase.

The principle of extraction is shown in Fig. 3.1a. The beaker contains two phases of liquids. The top (lighter) one is generally organic solution (o), the bottom (heavier) one is generally aqueous solution (a). The solute is initially dissolved in the aqueous solution but eventually distributed between the two solutions. When the solute distribution between the two solutions reaches equilibrium, the solute is at the concentration $[S_{(a)}]$ in the aqueous solution and at the concentration $[S_{(o)}]$ in the organic solution. The distribution ratio (D) of the solute is expressed by Eq. 3.1.

$$D = \frac{[S_{(o)}]}{[S_{(a)}]} \tag{3.1}$$

The distribution ratio is also referred to as the distribution coefficient or distribution factor. The difference of distribution ratios of solutes allows the separation of one from the other. The objective of extraction is to partially or entirely transfer the valuable solute from the aqueous solution to the organic solution and leave the undesirable solutes in the aqueous solution. A solute with a higher distribution ratio is easier to be extracted. A single stage extraction requires one time of contact (or mixing) of organic and aqueous solutions. A two stage extraction requires two times of contact of the two solutions. A multi stage extraction needs multi times of contact of the two solutions in order to complete the extraction of the valuable solute. The organic solution will be loaded with the valuable solute after extraction and usually is called pregnant organic solution. There will be only a trace amount of the valuable solute left in the aqueous solution after extraction. The aqueous solution containing only undesirable solutes and impurities is called raffinate.

3.1.2 Scrubbing/Washing

Following extraction, a scrubbing stage, sometimes also called washing stage, is always performed by contacting the pregnant organic solution to remove any undesirable solutes that are entrained in the organic solution. Scrubbing can improve the purity of the valuable metal. Normally scrubbing is undertaken with water, dilute acid or base solution. The scrubbing solution may contain a relatively high amount of valuable metal. For this reason it is normally recycled back to extraction and mixed with the feed aqueous solution.

3.1.3 Stripping

The general principles established for extraction apply to stripping. It is the reverse operation of extraction. As shown in Fig. 3.1b, it is used to transfer the metal from the pregnant organic solution to a stripping solution. A stripping solution normally contains concentrated acid, alkaline, or salt in order to attract the metal from the organic solution. The valuable metal is stripped from the organic solution and is concentrated in the stripping solution. The organic solution can be reused for extraction. The metal in the stripping solution can be recovered as the final product after a series of downstream treatments. The objectives of stripping are to transfer the valuable metal from the pregnant organic solution to the stripping solution as completely as possible and to concentrate the metal to a higher concentration. A metal with a lower distribution ratio is easier to be stripped. Similar to extraction, stripping can be performed by one stage, two stages, or multi stages.

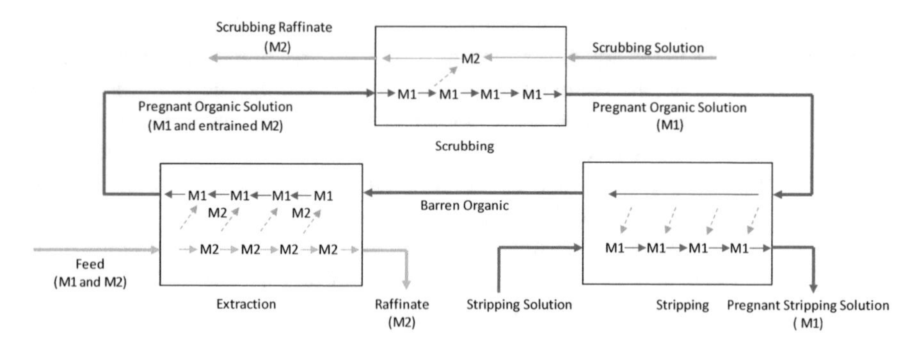

Fig. 3.2 Basic aqueous–organic solvent extraction separation process

3.1.4 Basic Solvent Extraction Separation Process

Solvent extraction separation is the process where the valuable metal(s) are partially or entirely transferred from the original aqueous solution to the organic solution and then stripped to an aqueous solution for recovery. To achieve the separation of one valuable metal from the unwanted others in a continuous manner, extraction, scrubbing, and stripping are performed in conjunction. An example of a solvent extraction separation process is shown in Fig. 3.2 for the separation of M1 and M2. It includes extraction, scrubbing, and stripping.

In extraction, the valuable solute M1 is extracted from the aqueous feed solution to the organic solution. The organic solution leaves extraction as pregnant organic loaded with M1 and some entrained M2. The aqueous solution leaves extraction as raffinate with no M1 left.

In scrubbing, the entrained M2 in the pregnant organic is washed. The pregnant organic leaves scrubbing with purified M1. The scrubbing solution leaves scrubbing as scrubbing raffinate with M2 and some M1.

In stripping, a stripping solution strips the M1 from the organic solution. The organic solution leaves stripping as barren organic which will be recycled to extraction to start the next cycle of solvent extraction separation. The stripping solution leaves stripping as pregnant stripping solution loaded with M1 which can be recovered as the final product after a series of downstream processes.

3.1.5 Cascade Solvent Extraction Process

To achieve the targeted separation results, multiple stages of extraction are often used. A solvent extraction process with multiple stages is called a cascade extraction process. According to the configurations of organic flow and aqueous flow, a cascade extraction process is commonly classified as countercurrent extraction, cross flow extraction, fractional extraction, and recirculating extraction.

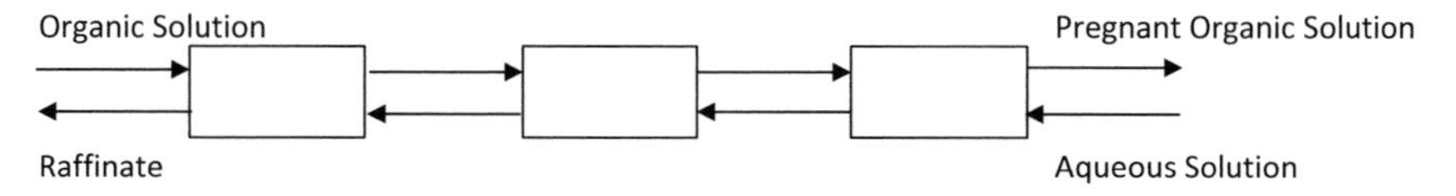

Fig. 3.3 Representation of countercurrent extraction process

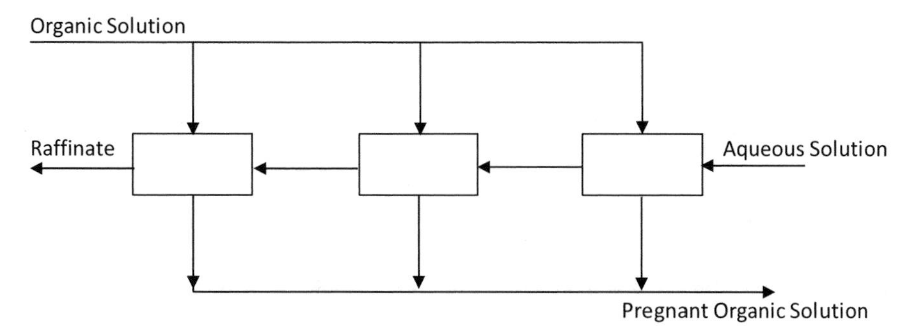

Fig. 3.4 Representation of cross flow extraction process

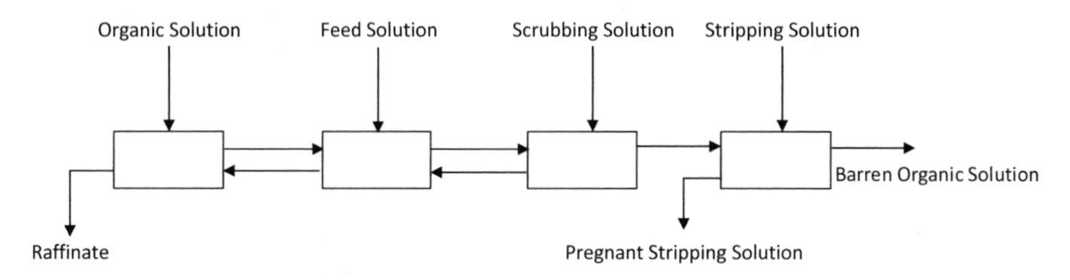

Fig. 3.5 Representation of fractional extraction process

As shown in Fig. 3.3, in countercurrent extraction, aqueous solution and organic solution flow in opposite directions to each other. Countercurrent enables the contacts of low concentration aqueous solution with low concentration organic solution and high concentration aqueous solution with high concentration organic solution. A greater amount of extraction can be obtained because countercurrent maintains a slowly declining concentration difference.

Figure 3.4 shows the concept of cross flow extraction. The feed is passed from the first stage to the last stage in series. Fresh organic solution is fed to each stage in parallel.

In the fractional extraction process as shown in Fig. 3.5, the feed is introduced in a middle stage while the organic solution and scrubbing solution are fed from each end stage.

As shown in Fig. 3.6, a recirculating extraction process is the fractional extraction process with a portion of pregnant stripping solution recycled back to the last stage of extraction along with scrubbing solution.

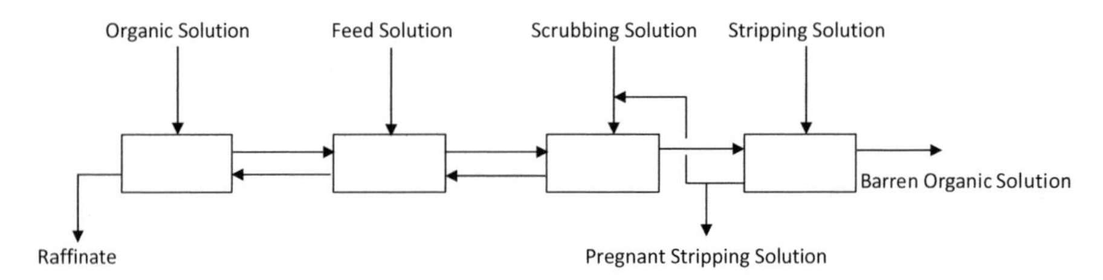

Fig. 3.6 Representation of recirculating extraction process

3.2 Liquid–Liquid Equilibrium

3.2.1 Equilibrium and Nernst's Distribution Law

Solvent Extraction utilizes the distribution of solutes in two immiscible phases to realize separation. The distribution of solutes between the aqueous phase and the organic phase is a dynamic competing process. At equilibrium, the system is in a condition that all competing influences are balanced. In a heterogeneous system comprising solutes and phases, the conditions for thermodynamic equilibrium are the temperature (T), pressure (P), and chemical potential (μ) of each component being uniform throughout the system (Rydberg et al. 2004; Kislik 2012).

Distribution law or Nernst's distribution law at equilibrium was given by Walther Nernst in the 1890s. It gives the distribution of a solute between two immiscible solvents at equilibrium. As expressed by Eq. 3.2, it states if a solute M distributes itself in two immiscible solvents A and B at constant temperature and the solute is in the same molecular condition in both solvents, then the ratio of the concentrations of the solute M in solvents A and B is a constant at equilibrium. The constant is called the distribution constant or the partition constant.

$$K_{\mathrm{d}} = \frac{[M_{(\mathrm{A})}]}{[M_{(\mathrm{B})}]} \tag{3.2}$$

where K_d is the distribution constant, $[M_{(\mathrm{A})}]$ is the concentration of the solute M in solvent A at equilibrium, and $[M_{(\mathrm{B})}]$ is the concentration of the solute M in solvent B at equilibrium.

Strictly, this is only valid with immiscible pure solvents. However, in reality, the solvents always contain molecules of each other. In a typical aqueous–organic extraction system, the aqueous solution is saturated with the organic solvent and the organic solution is saturated with water. Also, the solute M may exist differently in the two solutions. The distribution law is valid only when the mutual solubility of the solvents is negligible and the activity factors of the system are constant. Therefore, the distribution ratio is not the distribution constant.

3.2.1.1 Distribution Constant and Distribution Ratio

At constant temperature and pressure when the distribution of solute M in the two solvents A and B reaches equilibrium, the chemical potential (μ) of the solute M in the two solvents must be equivalent.

$$\mu_{\mathrm{A}} = \mu_{\mathrm{B}} \tag{3.3}$$

The chemical potential of the solute M is related to its activity (α) in each solvent.

$$\mu_{\mathrm{A}} = \mu_{\mathrm{A}}^{*} + RTln\alpha_{\mathrm{A}} \tag{3.4}$$

$$\mu_{\mathrm{B}} = \mu_{\mathrm{B}}^{*} + RTln\alpha_{\mathrm{B}} \tag{3.5}$$

where μ_{A}^{*} and μ_{B}^{*} are the standard chemical potentials of solute M in solvent A and B.

The activity of solute M in each solvent is the product of the mole fraction (x) and the activity coefficient (γ).

$$\alpha_{\mathrm{A}} = x_{\mathrm{A}}\gamma_{\mathrm{A}} \tag{3.6}$$

$$\alpha_{\mathrm{B}} = x_{\mathrm{B}}\gamma_{\mathrm{B}} \tag{3.7}$$

In a 1M (or mol/L) system,

$$\frac{[M_{(\mathrm{A})}]}{[M_{(\mathrm{B})}]} = \frac{x_{\mathrm{A}}}{x_{\mathrm{B}}} \tag{3.8}$$

Therefore,

$$K_{\mathrm{d}}^{*} = \frac{\alpha_{\mathrm{A}}}{\alpha_{\mathrm{B}}} = e^{-(\mu_{\mathrm{A}}^{*} - \mu_{\mathrm{B}}^{*})/RT} = \frac{[M_{(\mathrm{A})}]\gamma_{\mathrm{A}}}{[M_{(\mathrm{A})}]\gamma_{\mathrm{B}}} = K_{d}\frac{\gamma_{\mathrm{A}}}{\gamma_{\mathrm{B}}} \tag{3.9}$$

where K_{d}^{*} is the Nernst distribution constant at thermodynamic equilibrium.

Strictly, only when the ratio of the activity coefficient of the solute M in the two solvents is independent of its total concentration, K_d is a real constant from a thermodynamics perspective.

In industrial solvent extraction, the deviation from the Nernst's distribution law can lead to considerable errors for new solvent extraction process development. The distribution constant has to be distinguished from the distribution ratio. The distribution constant cannot be obtained directly. However, the distribution ratio can be measured at defined conditions. If a system contains various solute species, the distribution ratio is defined as:

$$D = \frac{\text{Total Solute Concentration in Organic Phase, } \sum [M_{(\mathrm{o})}]}{\text{Total Solute Concentration in Aqueous Phase, } \sum [M_{(\mathrm{a})}]} \tag{3.10}$$

The distribution ratio is not a constant. It is affected by various system variables like solution acidity, temperature, flow rate, free ligand, impurities, etc.

3.2.2 Distribution Data

Distribution data or liquid–liquid equilibrium data are the basis of solvent extraction process development and design. There are two approaches to obtain the distribution data: distribution isotherm method and mathematic model method.

3.2.2.1 Distribution Isotherm Method

Distribution isotherm is the equilibrium relation between the solute concentration in organic phase and the solute concentration in aqueous phase at constant temperature and other conditions. It is expressed as a function:

$$\left[M_{(\mathrm{o})}\right] = f\left[M_{(\mathrm{a})}\right] \tag{3.11}$$

The isotherm of a solvent extraction system can be measured directly and presented as an isotherm curve with two dimension coordinates. A representation of an isotherm curve is shown in Fig. 3.7. The development of an isotherm curve is normally done by varying the organic/aqueous phase (O/A) ratio from 1/10 to 10/1 in separation funnel shaking out tests (Ritcey and Ashbrook 1979). For example, an organic solution and an aqueous solution are contacted at a predetermined O/A ratio until equilibrium is reached. After phase disengagement, they are separated and analyzed for the concentration of the solute. The same procedures are repeated until all shake out tests are completed at the predetermined O/A ratios. The solute concentrations in the organic phase and the solution concentrations in the aqueous

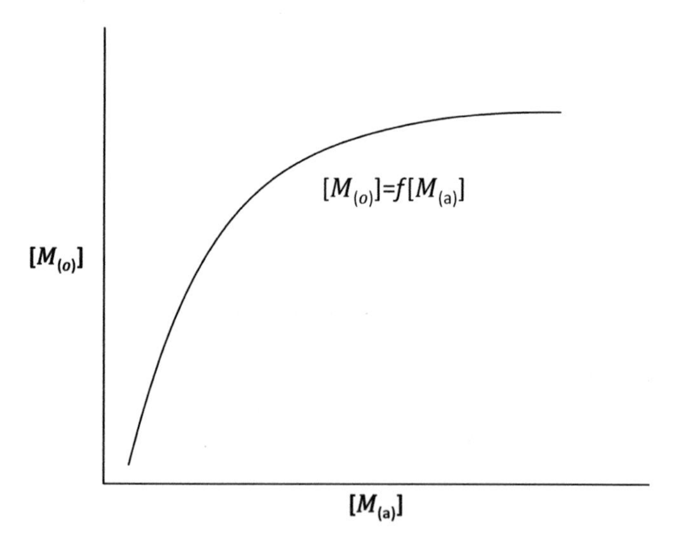

Fig. 3.7 Representation of distribution isotherm curve

phase can be used to construct the distribution isotherm curve as shown in Fig. 3.7. Based on the isotherm curve, the loading capacity of the organic solution at the testing conditions can be obtained. The solute concentration in one phase can be determined based on the concentration in another phase. The distribution ratios can be calculated using the isotherm data.

3.2.2.2 Mathematic Model Method

Extraction equilibrium data can also be obtained by computation based on mathematic models. The empirical model, chemistry model, and thermodynamic model are most often used in solvent extraction process simulation and optimization (Zhang and Zhang 1995).

Empirical models are developed purely on data by considering physical and chemical equilibrium. For example, an empirical model for the rare earth solvent extraction using D2EHPA was developed to calculate the individual rare earth element equilibrium data:

$$\left[M_{(\mathrm{o})}\right] = \alpha_1 \left[M_{(\mathrm{a})}\right]^{\alpha_2} \mathrm{e}^{\alpha_2 \left[M_{(\mathrm{a})}\right]} H^{\left(\alpha_4 + \alpha_5 H + \alpha_6 H^2\right)} \tag{3.12}$$

where $[M_{(\mathrm{o})}]$ and $[M_{(\mathrm{a})}]$ are the rare earth concentration in organic and aqueous solutions, respectively; H is the acidity in terms of hydrogen ion concentration and α_1–α_5 are coefficients to be determined.

Chemistry models are developed based on the data of extraction tests and mass balances under specified conditions. It is the expression of the solvent extraction distribution constant. It is often used to simulate the steady-state solvent extraction process.

Equilibrium models are the mathematic models of the equilibrium distribution constant. It correlates the solvent extraction reaction enthalpy, solute distribution ratio, as well as activity coefficients of each component in two phases.

3.2.3 Separation Factor

Under the same solvent extraction conditions, the ratio of the distribution ratios of two solutes $M1$ and $M2$ is defined as their separator factor (β):

$$\beta_{M1/M2} = \frac{D_{M1}}{D_{M2}} \tag{3.13}$$

where $\beta_{M1/M2}$ is the separation factor of solute M1 and M2, D_{MI} and D_{M2} are the distribution ratios of solute M1 and M2. The higher the separation factor is, the easier the separation of M1 from M2. If the separation factor is 1, it means the solutes M1 and M2 have the same distribution ratios. They cannot be separated by solvent extraction under the conditions where they have the same distribution ratios.

3.2.4 Extraction Percentage and Extraction Ratio

In practice, extraction percentage (Eq. 3.14) is often used for designing separation schemes. It is the percentage of the valuable solute extracted into the organic phase out of the total valuable solute in the system.

$$E\ \% = \frac{\text{Solute M1 in Organic Phase}}{\text{Total Solute M1 in Organic and Aqueous Phases}} = \frac{D}{1+D} \times 100 \tag{3.14}$$

The distribution ratio curve and extraction percentage curve as a function of a system variable can be constructed by experiment measurements. As shown in Fig. 3.8a, normally the distribution ratio curve is plotted on a logarithmic scale. The extraction percentage curve, Fig. 3.8b, is very useful in solvent extraction process design. The pH of the system is always used and denoted as $pH_{0.5}$ or $pH_{1/2}$ where extraction is 50 % or the distribution ratio is 1. The system variables can be acidity, extractant concentration in organic phase, free ligand concentration, or salt concentration in aqueous phase.

Different from the extraction percentage, extraction ratio is defined as the ratio of the solute M in organic phase to the solute M in aqueous phase.

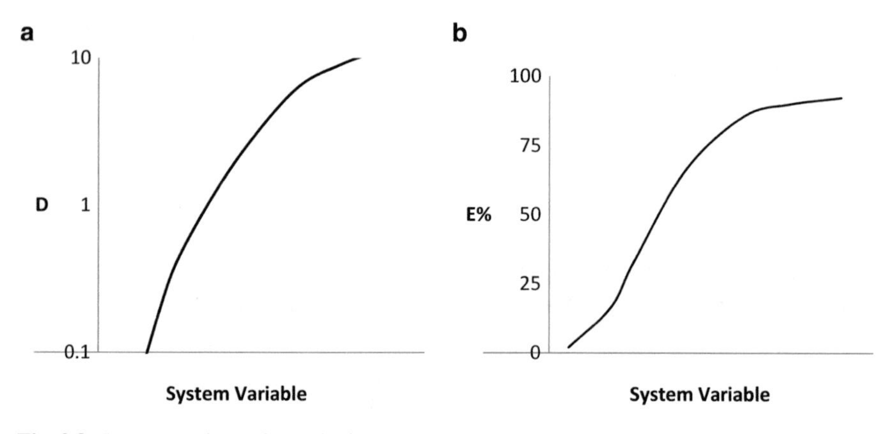

Fig. 3.8 Representations of (**a**) Distribution ratio curve and (**b**) Extraction percentage curve

$$q = \frac{\text{Solute M in Organic Phase}}{\text{Solute M in Aqueous Phases}} \tag{3.15}$$

When the organic phase equals to the aqueous phase, the extraction ratio (q) equals to the distribution ratio (D).

3.3 Solubility Rules in Solvent Extraction

Solvent extraction is a competing distribution process of a solute or solutes between the aqueous and organic solutions. It is not only related to the solute–solvent interaction but also the solvent–solvent interaction. An understanding of the general rules of these interactions is of great value in choosing a solvent extraction system for the separation of metals.

A solute will dissolve in a solvent with similar chemical properties or structure to itself. This is often referred to as "Like dissolves Like." For example, polar solute will best dissolve in polar solvent and nonpolar solute will best dissolve in nonpolar solvent.

3.3.1 Classification of Solvent

The van der Waals force and hydrogen bonding are the two cohesive forces between liquid molecules. The van de Waals force exists between any molecules. It is proportional to the product of the electric dipole moments of the molecules and is inversely proportional to the sixth power of the mean distance between them. A hydrogen bond is stronger than van der Waals force. The formation of a hydrogen bond requires that the liquid has an electronegative element B which can donate electrons and an A–H bond that can accept electrons. A solvent can be classified into one of the four different groups according to the interactions between molecules (Xu and Yuan 1987).

3.3.1.1 N Type of Solvent

N type of solvent is inert solvent which cannot form a hydrogen bond. van der Waals force is the cohesive force of the liquids. Examples are alkanes, benzene, carbon tetrachloride, carbon disulfide, and kerosene.

3.3.1.2 A Type of Solvent

A type of solvent can accept electrons such as chloroform. It can form a hydrogen bond with B type of solvent:

$$B + H - CCl_3 \leftrightarrow B\cdots H - CCl_3 \quad (3.16)$$

3.3.2 *B Type of Solvent*

B type of solvent can donate electrons. Examples include ethers, ketones, aldehydes, esters, and tertiary amines, etc. They can form a hydrogen bond with A type of solvent:

$$R_1R_2O + H - A \leftrightarrow R_1R_2O\cdots H - A \quad (3.17)$$

$$R_1R_2C = O + H - A \leftrightarrow R_1R_2C = O\cdots H - A \quad (3.18)$$

$$R_1R_2R_3N + H - A \leftrightarrow R_1R_2R_3N\cdots H - A \quad (3.19)$$

where, R_1, R_2, and R_3 are the hydrocarbon groups of the solvent.

3.3.2.1 AB Type of Solvent

AB type of solvent can donate and accept electrons. It includes three sub-types:

1. AB1 type that can form cross chain hydrogen bonds includes water, polyhydric alcohols, and polymeric carboxylic acids etc.
2. AB2 type that can form straight chain hydrogen bonds includes monohydric alcohols, carboxylic acids, and amines.
3. AB3 type that can form internal hydrogen bonds. Due to the formation of internal hydrogen bonds, AB3 type solvents lose their ability to accept electrons and become B type of solvents.

3.3.3 *Solvent Miscibility*

In many situations, a mixed solvent is required. The miscibility of different types of solvent is different depending on the increase in the number of hydrogen bonds or the intensity of hydrogen bonds after mixing. The miscibility of different types of solvents follows the general relationship as shown in Fig. 3.9 (Xu and Yuan, 1987).

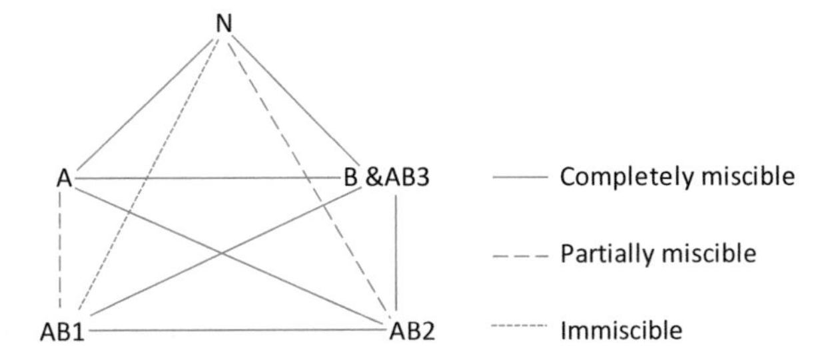

Fig. 3.9 Miscible relationship of solvents

N solvents are completely miscible with A and B (and AB3) solvents and partially miscible with AB2 solvents. N solvents are immiscible with AB1 solvents.

A solvents are completely miscible with B (and AB3) and AB2 solvents. A solvents are partially miscible with AB1 solvents.

B (and AB3) solvents are completely miscible with all other solvents.

AB1 solvents are completely miscible with AB2.

3.3.4 Water–Solvent Mutual Solubility

The aqueous–organic solvent extraction system is most applied in metal hydrometallurgy. If a solvent has similar chemical structure as water, the solvent and water will dissolve in each other to cause phase disengagement issues. A solvent less miscible or completely immiscible with water will favor extraction. Table 3.1 gives the mutual solubility of water and some organic solvents with low solubility in water. Low mutual solubility is very important in solvent extraction. It will not only favor extraction but also reduce organic and solute loss to raffinate.

3.4 Solute–Solvent Interactions

When a solute is dissolved in a solvent, it interacts with the solvent in terms of solvation. If the solvent is water, the solute–water interaction is called hydration. The solvation process will result in changes of some solute–solvent system properties such as energy, volume, fluidity, electrical conductivity, and/or spectroscopic properties. In order to dissolve the solute, a cavity must be formed in the solvent. The solute will fit in the cavity and is able to interact with its surrounding molecules by coordinate bonds to form a new entity called “solvated solute.” The solvated solute may further interact with its surrounding molecules.

Table 3.1 Mutual solubility of water and some organic solvents (Rydberg et al., 2004)

	Solvent in water		Water in solvent	
Solvent	Solubility, wt%	Temperature, °C	Solubility, wt%	Temperature, °C
c-Hexane	0.0055	25	0.0100	20
n-Hexane	0.00123	25	0.0111	20
n-Octane	6.6×10^{-7}	25	0.0095	20
n-Decane	5.2×10^{-8}	25	7.2×10^{-5}	25
n-Dodecane	3.7×10^{-9}	25	6.5×10^{-5}	25
Decalin (mixed isomers)	<0.02	25	0.0063	20
Benzene	0.179	25	0.0635	25
Toluene	0.0515	25	0.0334	25
Ethylbenzene	0.0152	25	0.043	25
p-Xylene	0.0156	25	0.0456	25
Chloroform	0.815	20	0.093	25
Carbon tetrachloride	0.077	25	0.0135	30
1,2-Dichloroethane	0.81	20	0.187	25
Trichloroethylene	0.137	25	0.32	25
Chlorobenzene	0.0488	30	0.0327	25
1,2-Dichlorobenzene	0.0156	25	0.309	25
Carbon disulfide	0.21	20	0.0142	25
1-Octanol	0.0538	25	–	
2-Ethyl-1-hexanol	0.07	20	2.6	20
Diisopropyl ether	1.2	20	0.57	20
Bis(2-chloroethyl) ether	1.02	20	0.1	20
Butyl acetate	0.68	20	1.2	20
Nitrobenzene	0.19	20	0.24	20
Benzonitrile	0.2	25	1.0	30
Quinoline	0.609	20	–	–
Tri-*n*-butyl phosphate	0.039	25	4.67	25

3.4.1 Metal in Aqueous Solution

Due to their ionic nature, metal salts are strong electrolytes. Their molecules can dissociate to ions partially or completely. When a metal ion is dissolved in water, the electronegative oxygen in the water molecule is attached to the positively charged metal ion to form a primary solvation shell of water molecules, the first hydration shell. Through hydrogen bonding, the coordinated water molecules will be associated with other water molecules in a second shell. The water molecules in the first shell exchange with molecules in the second shell and the bulk liquid. Take M^{z+} as an example, the hydration of metal ion can be expressed by the following equation:

$$M^{z+} + x\mathrm{H_2O} \rightarrow \left[M(\mathrm{H_2O})_x\right]^{z+} \tag{3.20}$$

where, x is the hydration number of the metal ion which can be determined by various experimental methods.

When anions exist, the M^{z+} ions are subjected to complexation. For example, in an aqueous solution containing ligand anion A^-, the following complexation can occur:

$$\begin{aligned} M^{z+} + A^- &= MA^{z-1} \\ MA^{z-1} + A^- &= MA_2^{z-2} \\ &\cdots \\ MA^{z-(n-1)} + A^- &= MA_n^{z-n} \end{aligned} \tag{3.21}$$

When $n > z$, a negatively charged metal complex is formed. The stability constant of the metal–anionic ligand complex, β_n, can be defined as:

$$\beta_n = \frac{[MA_n^{z-n}]}{[M^{z+}][A^-]^n} \tag{3.22}$$

In the case of rare earth (RE), it can form complexes with halogen, nitrate, sulfate, carbonates, etc. The rare earth–Cl complex is not stable at low Cl concentration. As well, the RE–NO_3^- complexes are not stable. At high NO_3^- concentration, they can form complexes with negative charges. However, rare earth ions can form stable complexes with SO_4^{2-} to increase the solubility of rare earth salts in sulfuric acid solution. The complexation between rare earth ion and carbonate can produce stable negatively charged complexes.

Also, RE ions will form complexes with phosphorous ligands like $H_2PO_4^-$, PO_4^{3-}, $P_2O_7^{4-}$, and $HP_3O_{10}^{4-}$. The major phosphorous RE complexion ions includes $RE_2P_2O_7^{2+}$, $REP_2O_7^-$, $RE(P_2O_7)_2^{5-}$, $REHP_3O_{10}^-$, $RE(HP_3O_{10})_2^{5-}$, and $RE(HP_3O_{10})_3^{9-}$. The formation of cationic, neutral, or anionic species strongly depends on pH (acidity) of the solution.

In order to extract the metal ions from aqueous phase to organic phase, the hydration of metal ions must be destroyed to increase the metal's hydrophobicity. One of the most employed methods is to use extractant to react with the metal ions to form neutral or less ionic species.

3.4.2 Metal in Organic Solution

Usually metal salts are not soluble in organic solvents. If the dielectric constant ε (also called relative permittivity) of a solvent is higher than 40, the solvent is highly polar and has great ability to stabilize charges. Metal salts can be dissociated into ions in the high polar solvents. When $\varepsilon < 5$, the solvent is not good for charged metal ions. However, metal ions can be rendered soluble by metal–extractant complexes. In the complexation between metal ion and extractant, the bonds between metal ion and water molecules are broken and the ionic charges of the metal ion are neutralized. The metal–extractant is a neutral species that can be dissolved in the organic solvent. Generally, there are three types of extractants:

1. Acidic extractants or cationic extractants which have ionizable hydrogen atoms that can be replaced by metal ions to generate a neutral species as indicated by Eq. (3.22). Examples of acidic extractants in rare earth solvent extraction include HEH/EPH, HDEHP, HBTMPP, naphthenic acid, and 1-(2-thienyl)-4,4, 4-trifluoro-1,3-butanedione. Acidic extractants are most effective for extracting cations by exchanging the cations for their protons.

$$M^{n+}_{(\mathrm{a})} + zHL_{(\mathrm{o})} = ML_{n\ (\mathrm{o})} + nH^{+}_{(\mathrm{a})} \tag{3.23}$$

   where M^{n+} denotes a metal ions with $+n$ charge, the subscripts (a) and (o) denote aqueous and organic phases, respectively, and HL is acidic extractant, L is hydrocarbon group of the HL.
2. Basic extractants or anionic extractants, which normally have high molecular weight such as primary, secondary, and tertiary amines and the quaternary compounds. They function by ion exchange mechanism to form ion-pairs with anionic species to reduce the effective ionic charge of the latter.

$$MA^{(z-n)}_{n(\mathrm{a})} + (n-z)R_3NHA_{(\mathrm{o})} = \left[(R_3NH)_{n-z}MA_n\right]_{(\mathrm{o})} + (n-z)A^{-} \tag{3.24}$$

   where, $n > z$ and $MA^{(z-n)}_{n(\mathrm{a})}$ is the anionic species of the metal in aqueous solution, and R is the alkyl group of the amine.
3. Neutral extractants or solvating extractants often extract uncharged metal complexes from aqueous solution. In this extraction, the metal species is coordinated with two different types of ligands, i.e., a water-soluble anion and an organic-soluble electron-donating functional group.

$$MX(\mathrm{H_2O})_{m(\mathrm{a})} + nS = MX(\mathrm{H_2O})_{m-n}S_{n\ (\mathrm{o})} + n\mathrm{H_2O} \tag{3.25}$$

   where, MX is metal ion pair and S is solvent.

3.5 Solvent Extraction System

There are many ways to classify a solvent extraction system. For example, a solvent extraction system can be classified according to the type of extractant used, the metal of interest, the nature of aqueous solution, etc. In this section, the classification according to extraction mechanism is discussed considering the extractant property, characteristics of metal, and nature of aqueous solution.

3.5.1 Acidic Solvent Extraction

Acidic solvent extraction utilizes weak organic acid as extractants to extract metals through cation exchanging as indicated by Eq. (3.23).

The weak organic acid extractants include:

1. Chelating extractants which have an acid functional group and a coordination functional group. They can form chelants with metal ions to enter organic phase.
2. Phosphonic acid extractants which normally exist as dimers in nonpolar solvent. They can form more stable chelants with metal ions by hydrogen bonds.
3. Carboxylic acid extractants.

The organic acid extractant can distribute in both aqueous phase and organic phase:

$$HL_{(a)} = HL_{(o)}\ D_{HL} = \frac{[HL_{(o)}]}{[HL_{(a)}]} \tag{3.26}$$

In the aqueous phase, HL is subjected to dissociation:

$$HL = H^+ + L^- \quad K_a = \frac{[H^+][L^-]}{[HL]} \tag{3.27}$$

where, K_a is the dissociation constant.

Therefore, the two phase dissociation constant of the extractant can be defined:

$$K_{dHL} = \frac{K_a}{D} = \frac{[H^+][L^-]}{[HL_{(o)}]} \tag{3.28}$$

Same for the extractant distribution in both phases, the metal–extractant complex also distributes in both phases.

$$ML_{n(a)} = ML_{n(o)} \quad D_{ML_n} = \frac{[ML_{n(o)}]}{[ML_{n(a)}]} \tag{3.29}$$

Based on Eq. (3.23), the extraction equilibrium constant K is:

$$K_{ex} = \frac{[ML_{n(o)}][H^+]^n}{[ML_{n(a)}][HL_{(o)}]^n} \tag{3.30}$$

Therefore,

$$D_{ML_n} = K_{ex}\frac{[HL_{(o)}]^n}{[H^+]^n} \tag{3.31}$$

The metal complex distribution ratio (D_{MLn}) is proportional to the nth power of free extractant concentration in organic phase [$HL_{(o)}$] and is inversely proportional to the nth power of hydrogen ion concentration in aqueous phase [H^+]. Normally, $D_{ML_n} \gg D_{HL} \gg 1$ in a solvent extraction system, the concentration of extractant in aqueous phase [$HL_{n(a)}$] and the concentration of metal–extractant complex in aqueous phase [$ML_{n(a)}$] can be neglected in the calculation of free extractant concentration in organic phase. Therefore:

$$\left[HL_{(o)}\right] = \left[HL_{(o)}\right]_i - n\left[ML_{n(o)}\right] \tag{3.32}$$

where, [$HL_{(o)}$]$_i$ is the initial extractant concentration in organic phase.

The logarithmic form of Eq. (3.31) can be expressed as:

$$\lg D_{ML_n} = \lg K_{ex} + n\lg\left[HL_{(o)}\right] + n\text{pH} \tag{3.33}$$

Equations (3.31) or (3.33) is the basic correlation of acidic solvent extraction systems.

For an acidic solvent extraction system, lg$D$–pH curve can be plotted by measuring the distribution ratios at various pH while maintaining the free extractant concentration unchanged. Solvent extraction equilibrium constant K_{ex} can be obtained from the lgD–pH curve.

According to Eqs. (3.31) or (3.33), the metal distribution ratio is subjected to the effects of pH, free extractant concentration, and the solvent extraction equilibrium constant. The solvent extraction equilibrium constant is related to many factors like metal ion species, extractant property, diluent property, etc.

3.5.2 *Ion Pair Solvent Extraction*

The extractants used in ion pair solvent extraction system are normally amine compounds. The feature of this type of solvent extraction system is that in aqueous phase metals react with inorganic ligands to form negatively charged complexes that can be extracted to organic phase by organic cations. The extraction process is usually referred to as liquid anion exchange. For example, uranyl ions react with sulfate to form negatively charged uranyl sulfate anions in sulfuric acid leaching solution. The tertiary amines accept extra protons at low pH to form cations in organic phase. The uranyl ions are extracted from the leaching solution in terms of anions into the organic solution.

$$UO^{2+}_{2(a)} + 2SO^{2-}_{4(a)} = UO_2(SO_4)^{2-}_{2(a)} \tag{3.34}$$

$$R_3N_{(o)} + H^+ = (R_3HN)^+_{(o)} \tag{3.35}$$

$$2(R_3HN)^+_{(o)} + UO_2(SO_4)^{2-}_{2(a)} = (R_3HN)_2UO_2SO_{4(o)} \tag{3.36}$$

The general ion pair solvent extraction is presented by Eq. (3.24). Its extraction constant can be expressed by the following equation:

$$K_{ex} = \frac{\left[(R_3NH)_{n-z}MA_{n(o)}\right]\left[A^-\right]^{n-z}}{\left[MA_{n(a)}^{(z-n)}\right]\left[R_3NHA_{(o)}\right]^{n-z}} \tag{3.37}$$

Assuming the aqueous phase contains all the stepwise metal–extractant complexes, the distribution ratio of the metal complexes is:

$$D_M = \frac{\left[(R_3NH)_{n-z}MA_{n(o)}\right]}{\sum\left[MA_{n(a)}^{(z-n)}\right]} \tag{3.38}$$

Therefore,

$$D_M = K_{ex}\frac{\left[MA_{n(a)}^{(z-n)}\right]\left[R_3NHA_{(o)}\right]^{n-z}}{\left[A^-\right]^{n-z}\sum\left[MA_{n(a)}^{(z-n)}\right]} \tag{3.39}$$

In a system with high concentration of ligand anion A^-, the metal–anionic ligand complex, MA_n^{z-n}, dominates in the aqueous phase where $\sum\left[MA_{n(a)}^{(z-n)}\right]$ can be approximated to $\left[MA_{n(a)}^{(z-n)}\right]$. Equation (3.39) becomes:

$$D_M = K_{ex}\frac{\left[R_3NHA_{(o)}\right]^{n-z}}{\left[A^-\right]^{n-z}} = K_{ex}\left(\frac{\left[R_3NHA_{(o)}\right]}{\left[A^-\right]}\right)^{n-z} \tag{3.40}$$

The metal distribution ratio is proportional to the $(n-z)$th power of free amine salt concentration in the organic phase and is inversely proportional to the $(n-z)$th power of free anionic ligand A^- concentration in the aqueous phase.

3.5.3 Solvating Solvent Extraction

The feature of the solvating solvent extraction is that a neutral extractant forms neutral metal–extractant complex with a neutral metal salt species. One of the major solvating extractants is organophosphine oxides with the functional group of:

$$-\overset{|}{\underset{|}{P}}=O$$

Other major solvating extractants are ketones, ethers, alcohols. They have the functional group of:

$$-\overset{|}{\underset{|}{C}}=O \quad \text{or} \quad -\overset{|}{\underset{|}{C}}-O-$$

Organophosphine oxides have a general formula of OPX_3 where $X =$ alkyl (R) or alkyl oxygen (R–O). Organophosphine oxides can form coordination complexes with water, acids, and metal salts. For example, it forms coordination complex with water by hydrogen bonding:

$$X_3P = O + H - O - H \leftrightarrow X_3P = O\cdots H - O\cdots H \tag{3.41}$$

A generic reaction with acid HA is:

$$X_3P = O + H - A \leftrightarrow X_3P = O\cdots H - A \tag{3.42}$$

The extraction of a metal salt by organophosphine oxide is to form a neutral complex through a coordinate bond between the metal and the oxygen in the P–O bond. The extraction capability is greater when the coordinate bond is stronger. The coordinate bond is affected by the electronegativity of the X groups in $X_3P=O$. $X_3P=O$ with more electronegative X groups has weaker extraction capability:

$$(RO)_3P = O < (RO)_2RP = O < (RO)R_2P = O < R_3P = O$$

A generic solvating extraction reaction equation is:

$$mX_3P = O_{(o)} + M^{n+}_{(a)} + nA^-_{(a)} \leftrightarrow (X_3P = O)_m \times MA_{n(o)} \tag{3.43}$$

The extraction equilibrium constant is:

$$K_{ex} = \frac{\left[(X_3P = O)_m \times MA_{n(o)}\right]}{\left[X_3P = O_{(o)}\right]^m \left[M^{n+}_{(a)}\right] \left[A^-_{(a)}\right]^n} \tag{3.44}$$

Assuming the metal ion is M^{n+} in aqueous phase, its distribution ratio is:

$$D_M = \frac{\left[(X_3P = O)_m \times MA_{n(o)}\right]}{\left[M^{n+}_{(a)}\right]} \tag{3.45}$$

Therefore,

$$D_M = K_{ex}\left[X_3P = O_{(o)}\right]^m \left[A^-_{(a)}\right]^n \tag{3.46}$$

Table 3.2 Examples of synergistic solvent extraction systems

Extractant combination	Denotation	Example	
		Aqueous	Organic
Chelating + solvating extractants	A + B	Eu^{3+}/H_2O–HNO_3	TTA + TBP in cyclohexane
Chelating + ion pair extractants	A + C	Th^{4+}/HCl, LiCl	TTA + TOA in benzene
Solvating + ion pair extractants	B + C	PuO_2^{2+}/H_2O–HNO_3	TBP + TBAN in kerosene
Chelating + chelating extractants	A1 + A2	RE^{3+}/H_2O–HNO_3	HA + TTA in benzene
Solvating + solvating extractants	B1 + B2	RE^{3+}/H_2O–HNO_3	TBPO + TOPO in kerosene
Ion Pair + ion pair extractants	C1 + C2	Pa^{5+}/H_2O–HCl	RCOR + ROH
Chelating + ion pair + solvating extractants	A + B + C	UO_2^{2+}/H_2O–H_2SO_4	D_2EHPA + TBP + R_3N in kerosene

where $[X_3P = O_{(o)}]$ is the free extractant concentration in organic phase and $[A^-_{(a)}]$ is the concentration of anionic ligand of the metal salt in aqueous phase.

3.5.4 Synergistic Solvent Extraction

A synergistic solvent extraction system utilizes two or more extractants. The metal distribution ratio is significantly higher than the combined distribution ratios when only one individual extractant exists. This enhanced extraction is called synergism. In contrary, reduced extraction is called antisynergism. The above three solvent extraction systems: acidic solvent extraction, ion pair solvent extraction, and solvating solvent extraction, can be employed separately or combined. Examples of synergistic solvent extraction systems are shown in Table 3.2 (Xu and Yuan 1987). A mixed metal–extractant complex is formed in the synergistic system. The mixed complexation follows the same rules as single-extractant systems. In practice, the chelating and solvating system with good synergism is more commonly investigated than others.

3.6 Kinetics of Solvent Extraction

The kinetics of solvent extraction is a function of the diffusion of various species and the reactions between the various species. It is the mathematic correlation between the solvent extraction rate and the concentration of species under consideration. The kinetic behavior of a solvent extraction system will dictate the size of the solvent extraction plant and the volume of reagent.

Metal solvent extraction involves three steps: (1) diffusion of reactants to the reaction zone, (2) reactions between reactants, and (3) diffusion of products away from the reaction zone. The slowest step is the rate determining step. The solvent extraction can be controlled by either diffusion or reaction. The diffusion depends on the interfacial area and the concentration of reactant or product species. The reaction can take places in the bulk phase, at the interface, or in the interface region. For reactions in the bulk phase, the reaction rate depends on the reactant solubilities, distribution coefficients, ionization constants, and phase volumes. For reactions at the interface and in the interfacial region, the interfacial area and reactant activities are important.

3.6.1 Diffusion Controlled Solvent Extraction

Diffusion is the process where a solute moves from a region of high concentration to one of low concentration. In a diffusion controlled solvent extraction system, the extraction rate depends on the diffusion of species under consideration.

Fick's first law of diffusion (Eq. (3.47)) relates the diffusive flux to the concentration at steady state.

$$J = -D\frac{\partial C}{\partial x} \tag{3.47}$$

where J, diffusion flux, is the amount of species flow through a small area in a small time interval. D is the diffusion coefficient of the species under consideration with unit of cm^2/s. C is the concentration of the species with unit of mol/cm^3, and x is its position with unit of cm. The negative sign ensures that the flux goes from the high concentration region to the low concentration region.

As shown in Fig. 3.10 for steady-state diffusion crossing a flat and thin diffusion film of thickness δ, only one dimension can be considered and Fick's first law can be simplified by replacing differentials with finite increments and assuming a linear concentration profile in the film of thickness δ.

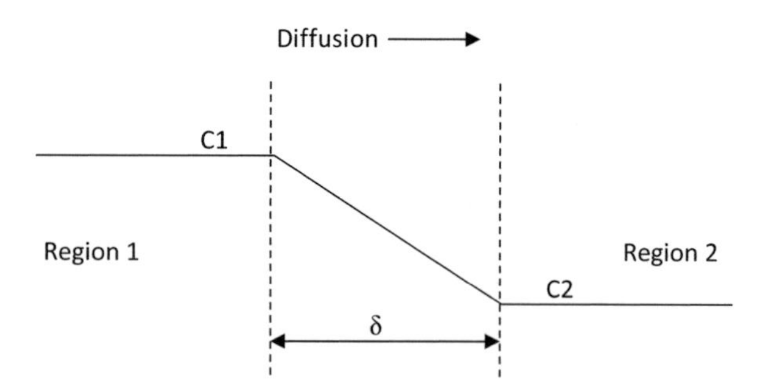

Fig. 3.10 Concentration profile of a solute diffusing across a film of thickness δ in contact with two regions of different concentration

$$J = -D(C_2 - C_1)/\delta \tag{3.48}$$

In a system where steady state cannot be assumed, the concentration change with time must be considered. Fick's second law of diffusion predicts how diffusion causes the concentration to change with time.

$$\frac{\partial C}{\partial t} = D\frac{\partial^2 C}{\partial x^2} \tag{3.49}$$

where, t is time.

3.6.2 *Reaction Controlled Solvent Extraction*

Assuming a chemical reaction occurs in the solvent extraction:

$$aA + bB \rightarrow cC + dD \tag{3.50}$$

The solvent exaction rate can be expressed as the formation rate of a product or the consumption rate of a reactant.

$$\frac{\mathrm{d}[C]}{\mathrm{d}t}, \quad \frac{\mathrm{d}[D]}{\mathrm{d}t}, \quad -\frac{\mathrm{d}[A]}{\mathrm{d}t}, \quad \text{or } -\frac{\mathrm{d}[B]}{\mathrm{d}t} \tag{3.51}$$

When the reaction is the controlling step, the reaction dictates the solvent extraction rate. The rate law of the solvent extraction system can be presented as a function of the system composition:

$$\frac{\mathrm{d}[C]}{\mathrm{d}t} = \frac{\mathrm{d}[D]}{\mathrm{d}t} = -\frac{\mathrm{d}[A]}{\mathrm{d}t} = -\frac{\mathrm{d}[B]}{\mathrm{d}t} = k[A]^a[B]^b[C]^c[D]^d \tag{3.52}$$

where k is the extraction rate constant. Powers a, b, c, and d are the reaction orders of corresponding species. For example, if $a=1$ and $b=2$, the reaction is a first order reaction for species A and second order reaction for species B. It is affected by variables such as concentrations of various species, temperature, solvent characteristics, etc.

The rate law can be developed by measuring concentration variations as a function of time or the initial rates as a function of the initial concentrations.

3.7 Factors Affecting Metal Solvent Extraction

Metal solvent extraction can be affected by many factors. The type of extractant normally dictates the extraction mechanism. The same factor can have different effects on metal extraction in different extractant systems. Factors that have significant effects on metal extraction are extractant, extractant concentration, aqueous phase acidity, metal concentration in aqueous phase, salting out agent, complexion agent, diluent, impurity in the feed solution, and synergistic extractant, etc. These affecting factors are discussed in detail based on extraction systems in Chap. 4.

References

Kislik, Vladimir S. (2012). Solvent extraction, classical and novel approaches. Elsevier Scientific Publishers.

Ritcey, G. M., Ashbrook, A. W. (1979). Solvent extraction: Principles and applications to process metallurgy, Vol. 1. Elsevier Scientific Publishers.

Rydberg, Jan., Cox, Michael., Usikas, laude., Choppin, Gregory. R. (2004) Solvent extraction principles and practice. Marcel Dekker.

Xu, G., Yuan, C., et al. (1987). *Solvent extraction of rare earth*. Beijing: China Science Press.

Zhang, C., & Zhang, X. (1995). *Rare earth extractive metallurgy: Principle and process*. Beijing: Metallurgy Industrial Press.

Chapter 4
Rare Earth Solvent Extraction Systems

As introduced in Chap. 3, a solvent extraction system can be classified according to extractants, metals of interest, the nature of aqueous solution, etc. In this chapter, rare earth solvent extraction systems are introduced according to the extractants. The selection of extractant is critical in developing a rare earth solvent extraction process. The rule of thumb in selecting an industrial extractants includes:

1. The extractant has at least one functional group that can form extractable complex with metal ions. Common functional groups contain O, N, P, and/or S elements.
2. The extractant has a relatively long hydrocarbon chain or contains benzene/substituted ring. This is to increase the solubility of the extractant and metal-extractant complex in organic phase.
3. The extractant has good selectivity towards the metal of interest.
4. The extractant has good chemical and thermal stability.
5. The extractant has low tendency to form emulsion in extraction and stripping operations. This requires the extractant to have low density, low viscosity, and large surface tension.

The commonly used rare earth extractants include 2-ethyl hexylphosphic mono-2-ethylhexyl ester (HEH/EPH, EHEHPA, P507), di-(2-ethylhexyl) phosphoric acid (D2EHPA, HDEHP, P204), di-(2,4,4-trimethylpentyl) phosphinic acid (HBTMPP), tributylphosphate(TBP), di-(1-methylheptyl) methyl-phosphonate (P350), Cyanex 923, naphthenic acid, and amines. These extractants can be used alone or together.

4.1 HEH/EHP Solvent Extraction System

One of the most widely used rare earth extractants is 2-ethyl hexylphosphic mono-2-ethylhexyl ester, often denoted by HEH (EHP) or EHEHPA. P507, PC-88A, and Ionquest 801 are the known trade names. HEH (EHP) and P507 are the two names

J. Zhang et al., *Separation Hydrometallurgy of Rare Earth Elements*,
DOI 10.1007/978-3-319-28235-0_4

frequently shown in publications. P507 is a phosphonic acid extractant. It is a colorless or yellowish transparent liquid. Its chemical formula is $(C_8H_{17})_2HPO_3$ with the following structure:

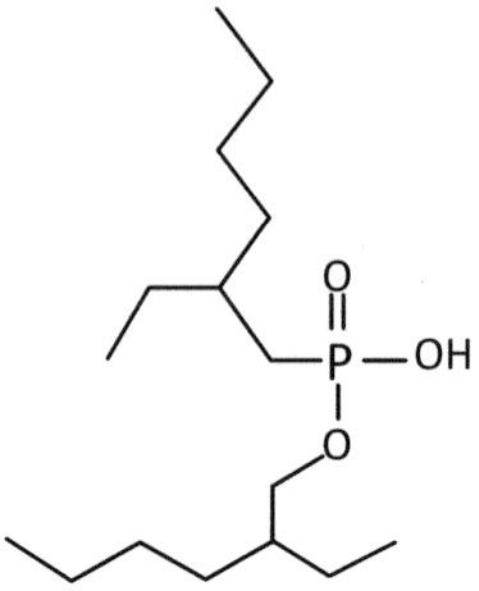

In nonpolar diluent, P507 exists as dimer which can be denoted by $(HL)_{2(o)}$. P507 has been used to extract rare earth ions in different inorganic acid solutions. It is an effective extractant for individual rare earth element separation.

4.1.1 *HEH/EHP-HNO$_3$ System*

Depending on the acidity of the solution, P507 acts in two different ways: (1) chelating extractant and (2) solvating extractant.

In low-acidic environment, P507 acts as chelating extractant. The extraction of RE^{3+} follows cation-exchanging mechanism:

$$RE^{3+}_{(a)} + 3(HL)_{2(o)} = RE(HL_2)_{3(o)} + 3H^{+}_{(a)} \tag{4.1}$$

The extraction equilibrium constant is

$$K = \frac{\left[RE(HL_2)_{3(o)}\right] \cdot \left[H^{+}_{(a)}\right]^3}{\left[RE^{3+}_{(a)}\right] \cdot \left[(HL)_{2(o)}\right]^3} \tag{4.2}$$

The distribution ratio is

$$D = \frac{\left[RE(HL_2)_{3(o)}\right]}{\left[RE^{3+}_{(a)}\right]} = K_{ex} \frac{\left[(HL)_{2(o)}\right]^3}{\left[H^{+}_{(a)}\right]^3} \tag{4.3}$$

Therefore,

$$\mathrm{Log}\left(D/\left[(\mathrm{HL})_{2(\mathrm{o})}\right]^{3}\right) = \mathrm{Log}\,K_{\mathrm{ex}} - 3\mathrm{Log}\left[\mathrm{H}^{+}_{(\mathrm{a})}\right] \quad (4.4)$$

Shanghai Institute of Organic Chemistry—Chinese Academy of Sciences conducted systematic research on the extraction of yttrium and all lanthanide elements expect promethium by P507 in HNO_3 solution. The organic phase is 0.5 mol/L P507 diluted in n-dodecane. The aqueous phases with various acidities contain 0.01 mol/L RE^{3+} and 1.0 mol/L (H, Na)NO_3. As shown in Fig. 4.1, the slope

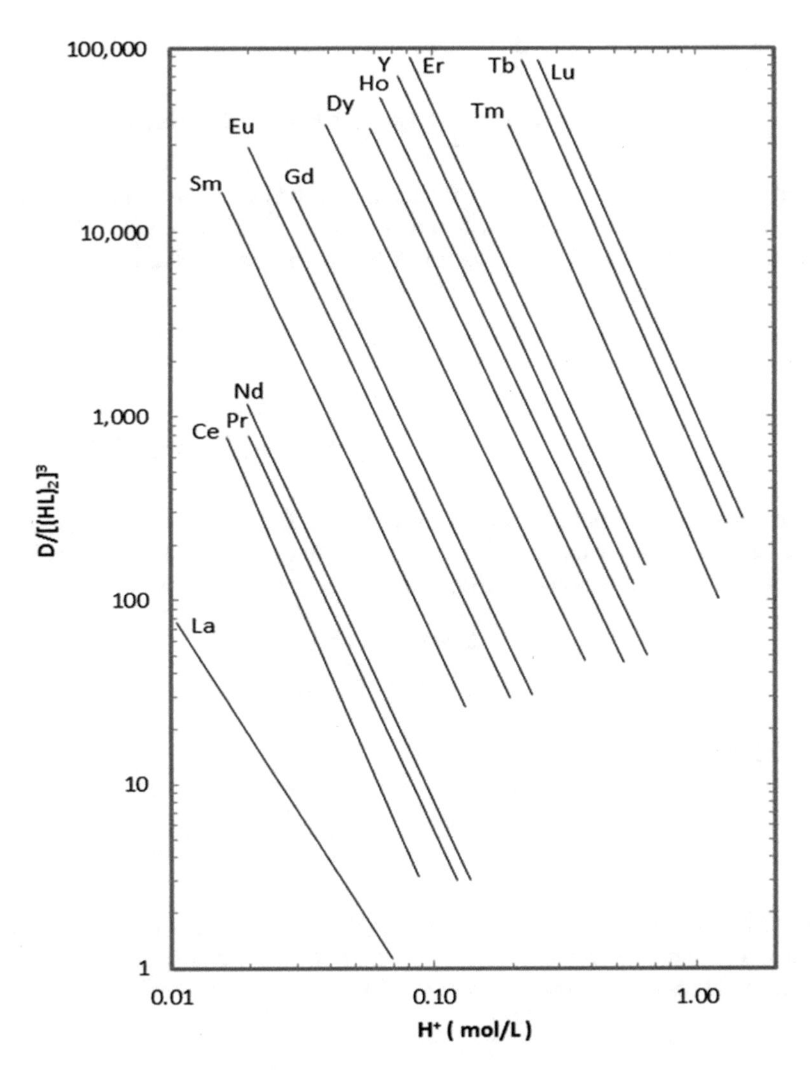

Fig. 4.1 Rare earth extraction using P507 at various acidities in HNO_3 solution at 25 °C and 1 mol/L (Na, H)NO_3

of the $\mathrm{Log}\left(D/\left[(\mathrm{HL})_{2(\mathrm{o})}\right]^3\right)$~$\mathrm{Log}\left[\mathrm{H}^+_{(\mathrm{a})}\right]$ curve of each rare earth element except La is about −3. This is in agreement with Eq. (4.4).

When HNO_3 is 0.05 mol/L, the average separation factor of rare earth ($\beta_{z+1/z}$) by P507 is 3.04 in HNO_3-H_2O-P507-kerosene system. Here z is the atomic number of the rare earth element.

In high-acidic environment, P507 acts as solvating extractant to form neutral metal salt-extractant complexes:

$$\mathrm{RE}^{3+}_{(\mathrm{a})} + 3\mathrm{NO}^-_{(\mathrm{a})} + 2(\mathrm{HL})_{2(\mathrm{o})} = \mathrm{RE}(\mathrm{NO}_3)_3 \cdot 4\mathrm{HL}_{(\mathrm{o})} \tag{4.5}$$

With the increase in acidity, P507 will extract HNO_3 and the dimer gets dissociated:

$$\mathrm{H}^+ + \mathrm{NO}_3^- + \mathrm{H}_2\mathrm{O} + \frac{1}{2}(\mathrm{HL})_{2(\mathrm{o})} = \mathrm{HNO}_3 \cdot \mathrm{H}_2\mathrm{O} \cdot \mathrm{HL}_{(\mathrm{o})} \tag{4.6}$$

There is a transaction from cation-exchanging mechanism to solvating mechanism with the increase in HNO_3 concentration (Li et al. 1985). Both P507 single molecules and dimers are in the system. A general extraction reaction is expressed by Eq. (4.7):

$$\mathrm{RE}^{3+}_{(\mathrm{a})} + \mathrm{nNO}^-_{(\mathrm{a})} + \left(3 - \frac{1}{3}n\right)(\mathrm{HL})_{2(o)} = \mathrm{RE}(\mathrm{NO}_3) \cdot n(\mathrm{HL}_2)_{3-n} \cdot \frac{4}{3}n(\mathrm{HL})_{(\mathrm{o})} + (3-n)\mathrm{H}^+ \tag{4.7}$$

where $n = 0$, 2, 3. When the extraction is cation-exchanging mechanism, $n = 0$. n increases with the increase in acidity. When the extraction is solvating mechanism, $n = 3$. The extraction equilibrium of solvating extraction is

$$K_{\mathrm{ex}} = \frac{[\mathrm{RE}(\mathrm{NO}_3)_3 \cdot 4\mathrm{HL}_{(\mathrm{o})}]}{[\mathrm{RE}^{3+}_{(\mathrm{a})}] \cdot [\mathrm{NO}^-_{(\mathrm{a})}]^3 \cdot [(\mathrm{HL})_{2(\mathrm{o})}]^2} \tag{4.8}$$

The distribution ratio of the rare earth is

$$D = \frac{[\mathrm{RE}(\mathrm{NO}_3)_3 \cdot 4\mathrm{HL}_{(\mathrm{o})}]}{[\mathrm{RE}^{3+}_{(\mathrm{a})}]} = K_{\mathrm{ex}} \cdot [\mathrm{NO}^-_{3(\mathrm{a})}]^3 \cdot [(\mathrm{HL})_{2(\mathrm{o})}]^2 \tag{4.9}$$

where $NO^-_{3(a)}$ is the nitrate in solution and $(HL)_{2(o)}$ is the free dimer of P507 in organic solution. According to Eqs. (4.5) and (4.6), the free dimer of P507 is the difference of initial and reacted dimer of P507:

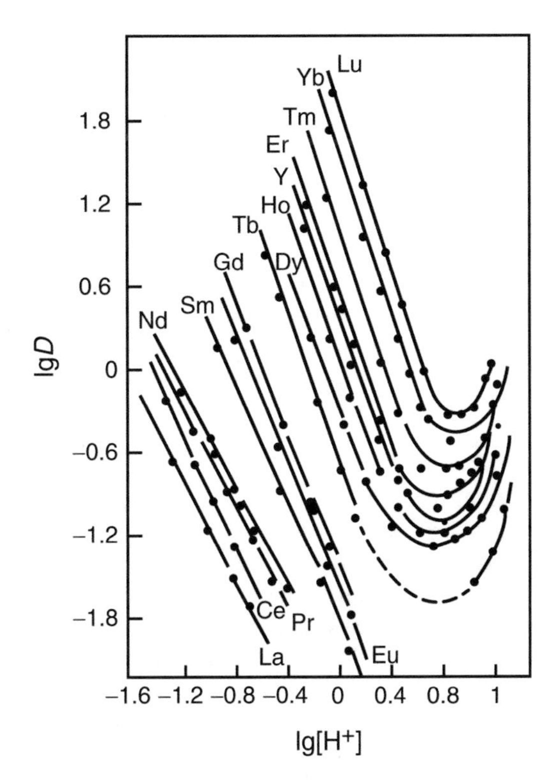

Fig. 4.2 Effects of equilibrium aqueous acidity on RE^{3+} extraction by P507 (Li et al. 1985)

$$[(HL)_{2(o)}] = \frac{1}{2}[P507]_i - 2[RE(NO_3)_3 \cdot 4HL_{(o)}] - \frac{1}{2}[HNO_3 \cdot H_2O \cdot HL_{(o)}] \quad (4.10)$$

The free dimer of P507 will depend on the initial P507 concentration $[P507]_i$, the rare earth equilibrium concentration in aqueous phase $[RE^{3+}_{(a)}]$, and the acidity of aqueous solution $[H^+_{(a)}]$ at equilibrium. Figure 4.2 shows the $\lg D \sim \lg[H^+]$ relationship of RE^{3+} in P507-HNO_3-$RE(NO_3)_2$ system. The distribution ratio decreases quickly to its lowest value with the increase of $[H^+]$ in aqueous solution. However, with the continuous increase of $[H^+]$, the distribution ratio of all heavy rare earth including yttrium starts to increase but no light rare earth (La-Gd) can be extracted in the HNO_3 concentration range of 2–11 mol/L.

There are no apparent effects of temperature on the extraction equilibrium of La, Sm, Eu, and Gd. The equilibrium constants of Ce, Pr, and Nd increase slightly when temperature is increased from 10 °C to 50 °C. The extraction equilibrium constants of the other rare earth elements decrease with the increase in temperature (Li et al. 1985). While the distribution ratio of light rare earths (La, Ce, Pr, Nd) is almost unchanged, the distribution ratio of heavy rare earth including yttrium becomes smaller at higher temperatures. The separation ratio is also slightly decreased. Therefore, industrial rare earth solvent extraction is normally operated in room temperature (Li 2011). In addition to the acidity, temperature, rare earth

Table 4.1 Extraction equilibrium constant of Yb^{3+} by P507 in different diluents

Diluent	Dielectric constant	Equilibrium constant of extraction	
		Cation-exchanging mechanism	Solvating mechanism
Kerosene	1.8	405	8.5×10^{-2}
n-Hexane	1.9	300	6.3×10^{-3}
Carbon tetrachloride	2.2	27.5	2.8×10^{-3}
Dimethylbenzene	2.4	22.6	2.6×10^{-3}
2-Ethyl hexanol	7.7	10.4	6.6×10^{-3}
Benzene	2.3	7.0	1.8×10^{-3}
Dichloromethane	9.1	5.0	2.5×10^{-3}
Chloroform	4.8	0.9	1.5×10^{-3}

concentration in aqueous phase, and the diluent of organic phase also have important impact on the rare earth separation by P507. In order to enhance the extraction capacity and selectivity of P507, Li et al. (1985) investigated the effects of diluent on the separation of Yb^{3+}. As shown in Table 4.1, the effects of diluent on equilibrium constant are different depending on the extraction mechanism. However, the extraction equilibrium constant in kerosene is always the largest.

4.1.2 P507-HCl System

In HCl acid solution, the rare earth extraction by P507 follows cation-exchanging mechanism (see Eq. 4.1). According to Eq. (4.3), the distribution ratio depends on the concentration of free dimer of P507 and the acidity of the aqueous phase at equilibrium.

The concentration of free dimer of P507 and the concentration of H^+ at equilibrium in HCl system can be expressed by the following Eqs. (4.11) and (4.12):

$$[(HL)_{2(o)}] = \frac{1}{2}[P507]_i - 3[RE(HL_2)_{3(o)}] = \frac{1}{2}[P507]_i - 3([RE^{3+}_{(a)}]_i - x) \tag{4.11}$$

$$[H^+_{(a)}] = [H^+_{(a)}]_i + 3\left([RE^{3+}_{(a)}]_i - x\right) \tag{4.12}$$

where $[P507]_i$ is the initial P507 concentration in organic phase, $[RE^{3+}_{(a)}]_i$ the initial RE^{3+} concentration in aqueous phase, x the equilibrium RE^{3+} concentration in aqueous phase, and $[H^+_{(a)}]_i$ the initial H^+ concentration in aqueous phase.

Therefore, the distribution ratio D can be expressed by Eq. (4.13):

$$D = K_{ex}\left(\frac{1}{2}[\text{P507}]_i - 3([\text{RE}^{3+}_{(a)}]_i - x)\right)^3 / \left([\text{H}^+_{(a)}]_i + 3([\text{RE}^{3+}_{(a)}]_i - x)\right)^3 \quad (4.13)$$

According to Eq. (4.13), the distribution ratio becomes smaller with the increase in initial acidity and initial rare earth concentration. Li et al. (1987a) conducted detailed investigation on the extraction equilibrium of individual rare earth element in P507-kerosene-HCl-$RECl_3$ system. The investigation was conducted at the conditions of 25 ± 1 °C, O/A = 1/1, and $[\text{RE}^{3+}_{(a)}]_i$=0.05−0.3 mol/L in 1 mol/L P507-sulfonated kerosene. The equilibrium data of individual rare earth in P507-kerosene-HCl-$RECl_3$ system are shown in Table 4.2. Based on the investigation, Li et al. (1987a) proposed a semiempirical model to correlate the distribution ratio (D) with P507 concentration (Z), acidity (H), and aqueous rare earth concentration (X) at equilibrium:

$$D = c1 \cdot Z^{c2} \cdot H^{c3} \cdot X^{c4} \quad (4.14)$$

Based on the data shown in Table 4.2, the values of coefficients $c1$, $c2$, $c3$, and $c4$ are calculated and shown in Table 4.3. The semiempirical model is very accurate to predict the P507-kerosene-HCl-$RECl_3$ extraction system.

The study on kinetics of rare earth extraction in P507-kerosene-HCl-$RECl_3$ system indicates that RE^{3+} extraction by P507 is a first-order reversible reaction Eq. (4.15) (Sun 1994):

$$\text{RE}_{(a)} \underset{k_b}{\overset{k_f}{\rightleftarrows}} \text{RE}_{(o)} \quad (4.15)$$

$$\frac{V}{A}\frac{d[\text{RE}_{(o)}]}{dt} = k_f[\text{RE}_{(a)}] - k_b[\text{RE}_{(o)}] \quad (4.16)$$

where k_f and k_b are the rate coefficients of forward reaction and reverse reaction, V the volume of organic phase (mL), and A the interphase area (cm^2).

At the beginning of extraction $[\text{RE}_{(O)}]_i = 0$, the following integral equations can be derived from Eq. (4.16):

$$\frac{[\text{RE}_{(a)}]_i - [\text{RE}_{(a)}]_e}{[\text{RE}_{(a)}]_i}\ln\frac{[\text{RE}_{(a)}]_i - [\text{RE}_{(a)}]_e}{[\text{RE}_{(a)}]_t - [\text{RE}_{(a)}]_e} = \frac{A}{V}k_f t \quad (4.17)$$

$$\frac{[\text{RE}_{(a)}]_e}{[\text{RE}_{(a)}]_i}\ln\frac{[\text{RE}_{(a)}]_i - [\text{RE}_{(a)}]_e}{[\text{RE}_{(a)}]_t - [\text{RE}_{(a)}]_e} = \frac{A}{V}k_b t \quad (4.18)$$

where $[\text{RE}_{(a)}]_i$, $[\text{RE}_{(a)}]_t$, and $[\text{RE}_{(a)}]_e$ are the aqueous rare earth ion concentration at beginning, at time t, and at equilibrium, respectively. t is extraction time.

Table 4.2 Equilibrium data of individual RE in P507-kerosene-HCl-RECl$_3$ system (Li et al. 1987a, 1987b)

$[H^+_{(a)}]_i$ (mol/L)	x (mol/L)	D	$[H^+_{(a)}]_i$ (mol/L)	x (mol/L)	D	$[H^+_{(a)}]_i$ (mol/L)	x (mol/L)	D
La			Ce			Pr		
0.0001	0.0363	0.3774	0.01	0.0305	0.6656	0.01	0.0259	0.9305
0.0001	0.0844	0.1848	0.01	0.0763	0.3106	0.01	0.0725	0.3793
0.0001	0.1345	0.1152	0.01	0.1232	0.2183	0.01	0.1673	0.1955
0.0001	0.1835	0.0899	0.01	0.1708	0.1710	0.01	0.2648	0.1329
0.0001	0.2329	0.0734	0.01	0.2672	0.1228	0.03	0.0307	0.6287
0.0001	0.2833	0.0589	0.03	0.0350	0.4285	0.03	0.0776	0.2899
0.01	0.0395	0.2658	0.03	0.0785	0.2739	0.03	0.1736	0.1521
0.01	0.0881	0.1351	0.03	0.1280	0.1719	0.03	0.2730	0.0989
0.01	0.1367	0.0973	0.03	0.1752	0.1416	0.05	0.0349	0.4335
0.01	0.1856	0.0781	0.03	0.2728	0.0997	0.05	0.0809	0.2358
0.01	0.3846	0.0541	0.05	0.0355	0.4113	0.05	0.1299	0.1547
0.03	0.0897	0.1148	0.05	0.0826	0.2107	0.05	0.2272	0.1004
0.03	0.1391	0.0784	0.05	0.1305	0.1494	0.07	0.0385	0.2987
0.03	0.1883	0.0621	0.05	0.1780	0.1236	0.07	0.0864	0.1574
0.03	0.2881	0.0413	0.07	0.2753	0.0897	0.07	0.1343	0.1168
0.04	0.0444	0.1329	0.07	0.0400	0.2500	0.07	0.2334	0.0711
0.04	0.0919	0.0881	0.07	0.0865	0.1595	0.10	0.0415	0.2048
0.04	0.1413	0.0616	0.07	0.1351	0.1103	0.10	0.0901	0.1099
0.04	0.1917	0.0459	0.07	0.1838	0.0911	0.10	0.1890	0.0582
0.04	0.2883	0.0388	0.07	0.2811	0.0672	0.10	0.2888	0.0388
Nd			Sm			Eu		
0.01	0.0239	1.0921	0.01	0.0115	3.3522	0.01	0.0405	1.4691
0.01	0.0683	0.4641	0.01	0.0480	1.0829	0.01	0.0849	0.7668
0.01	0.1170	0.2821	0.01	0.0944	0.5890	0.01	0.1318	0.5175
0.01	0.1680	0.1905	0.01	0.1864	0.3412	0.01	0.2326	0.2898
0.01	0.2637	0.1377	0.01	0.2371	0.2653	0.10	0.0164	2.0488
0.05	0.0201	0.7182	0.05	0.0156	2.2258	0.10	0.0546	0.8315
0.05	0.0757	0.3205	0.05	0.0556	0.7986	0.10	0.1000	0.5000
0.05	0.1722	0.1614	0.05	0.1035	0.4493	0.10	0.1473	0.3578
0.10	0.0389	0.2853	0.05	0.1997	0.5219	0.10	0.2459	0.2200
0.10	0.1330	0.1278	0.05	0.2471	0.2141	0.20	0.0266	0.8797
0.10	0.1807	0.1068	0.10	0.0219	1.2831	0.20	0.0683	0.4641
0.10	0.2320	0.0776	0.10	0.0623	0.6051	0.20	0.1142	0.3135
0.10	0.2807	0.0688	0.10	0.1105	0.3575	0.20	0.2592	0.1574
0.30	0.0478	0.0456	0.10	0.2073	0.2060	0.30	0.1259	0.1924
0.30	0.1464	0.0246	0.10	0.2572	0.1688	0.30	0.1739	0.1501
0.30	0.0189	0.0566	0.30	0.0408	0.2315	0.30	0.2703	0.1099
0.30	0.2957	0.0145	0.30	0.0842	0.1876	0.40	0.0424	0.1792
0.50	0.0482	0.0373	0.30	0.1334	0.1244	0.40	0.0880	0.3664

(continued)

Table 4.2 (continued)

$[H^+_{(a)}]_i$ (mol/L)	x (mol/L)	D	$[H^+_{(a)}]_i$ (mol/L)	x (mol/L)	D	$[H^+_{(a)}]_i$ (mol/L)	x (mol/L)	D
0.50	0.0976	0.0246	0.30	0.1814	0.1025	0.40	0.1336	0.1228
0.50	0.1968	0.0163	0.30	0.2782	0.0784	0.40	0.2314	0.0804
Gd			Tb			Dy		
0.10	0.0125	3.0000	0.10	0.0049	9.2041	0.10	0.0221	3.5249
0.10	0.0483	1.0704	0.10	0.0319	2.1348	0.10	0.0575	1.6087
0.10	0.0924	0.6234	0.10	0.0735	1.0408	0.10	0.1037	0.9286
0.10	0.1390	0.4388	0.10	0.1183	0.6906	0.10	0.1494	0.6734
0.10	0.1839	0.3594	0.10	0.1667	0.4997	0.10	0.1997	0.5023
0.10	0.2360	0.2712	0.10	0.2142	0.4006	0.30	0.0106	3.7170
0.30	0.0304	0.6447	0.30	0.0182	1.7473	0.30	0.0409	1.4449
0.30	0.0709	0.4104	0.30	0.0538	0.8587	0.30	0.0802	0.8703
0.30	0.1176	0.2755	0.30	0.0978	0.5337	0.30	0.1252	0.5974
0.30	0.1632	0.2255	0.30	0.1430	0.3986	0.30	0.2206	0.3599
0.30	0.2102	0.1893	0.30	0.2414	0.2428	0.50	0.0221	1.2624
0.30	0.2908	0.1503	0.50	0.0311	0.6077	0.50	0.0573	0.7452
0.50	0.0424	0.1802	0.50	0.0716	0.3966	0.50	0.0984	0.5244
0.50	0.1347	0.1136	0.50	0.1166	0.2864	0.50	0.1414	0.4144
0.50	0.1808	0.1062	0.50	0.1653	0.2099	0.50	0.2369	0.2664
0.50	0.2301	0.0865	0.50	0.2594	0.1565	0.70	0.0313	0.5335
0.50	0.2791	0.0749	0.70	0.0395	0.2658	0.70	0.0710	0.4085
0.70	0.0455	0.0980	0.70	0.0841	0.1891	0.70	0.1131	0.3263
0.70	0.0928	0.0776	0.70	0.1289	0.1637	0.70	0.1586	0.2598
0.70	0.2377	0.0517	0.70	0.2257	0.1077	0.70	0.2517	0.1919
Ho			Er			Tm		
0.10	0.0070	6.1429	0.10	0.0121	7.2645	0.50	0.0226	3.4248
0.10	0.0173	4.7803	0.10	0.0422	2.5545	0.50	0.0589	1.5467
0.10	0.0962	1.0790	0.10	0.1731	0.7331	0.50	0.0891	1.0182
0.10	0.1909	0.5715	0.50	0.0023	7.6783	0.50	0.1413	0.7693
0.50	0.0173	1.8902	0.50	0.0113	3.4230	0.50	0.1835	0.6349
0.50	0.0500	1.0000	0.50	0.0388	1.5121	1.00	0.0239	1.0962
0.50	0.1338	0.4948	0.50	0.0800	0.8746	1.00	0.0568	0.7921
0.50	0.2258	0.3286	0.50	0.2098	0.4299	1.00	0.0932	0.6094
0.75	0.0691	0.4472	1.00	0.0113	0.7699	1.00	0.1349	0.4826
0.75	0.1147	0.3078	1.00	0.0301	0.6611	1.00	0.1743	0.4343
0.75	0.2055	0.2165	1.00	0.1099	0.3649	1.00	0.2206	0.3599
0.75	0.2551	0.1760	1.00	0.2469	0.2151	2.00	0.0384	0.3051
1.00	0.0364	0.3736	1.50	0.0381	0.3123	2.00	0.0798	0.2531
1.00	0.0786	0.2735	1.50	0.0824	0.2136	2.00	0.1661	0.2041
1.00	0.1688	0.1848	1.50	0.1269	0.1820	2.00	0.2108	0.1860
1.00	0.2628	0.1416	1.50	0.2074	0.1219	2.00	0.2571	0.1669
1.25	0.0854	0.1710	2.00	0.0199	0.0500	2.50	0.0422	0.1848

(continued)

Table 4.2 (continued)

$[H^+_{(a)}]_i$ (mol/L)	x (mol/L)	D	$[H^+_{(a)}]_i$ (mol/L)	x (mol/L)	D	$[H^+_{(a)}]_i$ (mol/L)	x (mol/L)	D
1.25	0.1786	0.1747	2.00	0.0894	0.1186	2.50	0.0859	0.1644
1.25	0.2260	0.1062	2.00	0.1852	0.0799	2.50	0.1774	0.1274
1.25	0.2744	0.0933	2.00	0.2834	0.0586	2.50	0.2235	0.1186
Yb			Lu			Y		
0.50	0.0440	2.4091	0.50	0.0033	14.1515	0.10	0.0470	2.1915
0.50	0.0795	1.5157	0.50	0.0099	9.1010	0.10	0.0857	1.3337
0.50	0.1239	1.0178	0.50	0.0734	1.7248	0.10	0.1772	0.6930
0.50	0.1707	0.7576	0.50	0.1646	0.8226	0.50	0.0160	2.1250
1.00	0.0369	1.7100	1.00	0.0099	4.5556	0.50	0.0445	1.2472
1.00	0.0681	1.2026	1.00	0.0281	2.5587	0.50	0.0828	0.8116
1.00	0.1092	0.8315	1.00	0.0889	1.0222	0.50	0.1214	0.6474
1.00	0.1934	0.5512	1.00	0.1853	0.6190	0.50	0.2083	0.4402
2.00	0.0301	0.6611	2.00	0.0264	0.8939	1.00	0.0333	0.5015
2.00	0.0642	0.5574	2.00	0.0569	0.7575	1.00	0.0694	0.4409
2.00	0.1037	0.4465	2.00	0.0953	0.5740	1.00	0.1108	0.3538
2.00	0.1444	0.3850	2.00	0.1785	0.4006	1.00	0.1526	0.3106
2.00	0.2316	0.2953	2.00	0.2226	0.3477	1.00	0.2416	0.2417
2.50	0.0340	0.4735	2.50	0.0319	0.5674	1.00	0.1984	0.2601
2.50	0.0725	0.3793	2.50	0.0679	0.4728	1.50	0.0405	0.2346
2.50	0.1152	0.3021	2.50	0.1073	0.3979	1.50	0.0838	0.1933
2.50	0.2031	0.2309	2.50	0.1923	0.3001	1.50	0.1753	0.1409
3.00	0.0385	0.2987	3.00	0.0341	0.4665	2.00	0.0908	0.1013
3.00	0.0813	0.2300	3.00	0.0733	0.3643	2.00	0.1851	0.0801
3.00	0.1705	0.1730	3.00	0.1596	0.2531	2.00	0.2821	0.0635

According to Eqs. (4.17) or (4.18), $\ln\frac{[RE_{(a)}]_i-[RE_{(a)}]_e}{[RE_{(a)}]_t-[RE_{(a)}]_e}$ and time t have directly proportional relationship. When $\ln\frac{[RE_{(a)}]_i-[RE_{(a)}]_e}{[RE_{(a)}]_t-[RE_{(a)}]_e}$ is plotted against time t, a straight line passing through the origin should be produced for the first-order reaction. The kinetics data are measured and shown in Table 4.4 at the conditions O/A = 1/1, 25 °C, $[P507]_i = 0.20$ mol/L, pH = 3.0, and 0.20 mol/L NaCl. As shown in Fig. 4.3, the direct proportional relationship between rare earth concentration and extraction time can be obtained. This confirms the extraction is first-order reversible reaction.

The distribution ratio of each rare earth element becomes larger with the increase in extraction time. However, there is slight decrease in separation factors of adjacent rare earths with the increase of extraction time.

"Tetrad effects" is that the partition coefficients of REE complex against atomic number form four contiguous curves, each curve consisting of four elements: La–Ce–Pr–Nd, Pm–Sm–Eu–Gd, Gd–Tb–Dy–Ho, and Er–Tm–Yb–Lu. Figure 4.4 shows the distribution ratio of different rare earth elements in terms of LgD.

Table 4.3 Model coefficients of the P507-kerosene-HCl-$RECl_3$ system

$D = c1 \cdot Z^{c2} \cdot H^{c3} \cdot X^{c4}$					
RE	$c1$	$c2$	$c3$	$c4$	$\overline{r \cdot e}$, %
La	7.310	−12.61	0.0043	−1.007	2.0
Ce	7.110	−6.672	−0.082	−0.974	2.2
Pr	1.510	−5.223	−1.467	−0.799	2.4
Nd	2.110	−6.097	−1.071	−0.808	9.6
Sm	2.510	−1.384	−1.120	−0.777	2.9
Eu	9.310	0.3560	−2.169	−0.587	3.6
Gd	0.0211	0.6800	−2.477	−0.415	4.3
Tb	0.0776	1.1578	−2.644	−0.437	4.7
Dy	0.6760	1.9847	−3.161	−0.190	3.6
Ho	0.1616	0.6236	−2.140	−0.478	6.1
Er	0.4632	0.8341	−2.267	−0.364	4.5
Tm	0.7971	0.5160	−1.958	−0.272	4.5
Yb	3.1371	0.8793	−2.180	−0.255	3.8
Lu	4.3700	0.9095	−2.171	−0.269	2.9
Y	0.9776	1.1030	−2.496	−0.135	3.0

Clear tetrad effects can be observed at equilibrium and the beginning of the extraction. The distribution ratio becomes larger with the increase in atomic number in the same reaction time.

4.1.3 P507-H_2SO_4 System

Similar to the rare earth extraction in P507-HNO_3 system, P507 acts as a chelating extractant at low acidity and the extraction of rare earth follows cation-exchanging mechanism. However, P507 acts as solvating extractant at high acidity to form neutral complexes.

P507-octane extracting Er^{3+} in H_2SO_4 acid solution was studied by Li's team (Li et al. 1988; Le and Li 1990). At low acidity ($H_2SO_4 < 5$ mol/L), the extraction follows cation-exchanging mechanism but it is different with the extraction reaction in HNO_3 solution:

$$\mathrm{Er}^{3+}_{(a)} + \frac{5}{2}(\mathrm{HL})_{2(o)} = \mathrm{ErL} \cdot (\mathrm{HL}_2)_{2(o)} + 3\mathrm{H}^{+}_{(a)} \tag{4.19}$$

At high acidity ($H_2SO_4 > 5$ mol/L), the extraction of Er^{3+} follows solvating mechanism with the participation of sulfate ions:

$$\mathrm{Er}^{3+}_{(a)} + \frac{3}{2}\mathrm{SO}_4^{2-} + \frac{5}{2}(\mathrm{HL})_{2(o)} = \mathrm{Er(SO_4)}_{1.5} \cdot 4\mathrm{HL}_{(o)} \tag{4.20}$$

Table 4.4 Kinetics data of RE^{3+} extraction in P507-kerosene-HCl-$RECl_3$ system

RE	Kinetic data	Time (min)					
		0	20	40	60	80	Equilibrium
La	$[RE]_t$, mol/L	0.0196	0.0191	0.0188	0.0184	0.0180	0.0171
	D	0	0.025	0.044	0.065	0.086	0.146
Ce	$[RE]_t$	0.0208	0.0201	0.0196	0.0194	0.0190	0.0179
	D	0	0.036	0.061	0.074	0.094	0.159
	$\beta_{Ce/La}$	0	1.441	1.387	1.139	1.093	1.089
Pr	$[RE]_t$, mol/L	0.0203	0.0195	0.0190	0.0188	0.0185	0.0174
	D	0	0.043	0.069	0.082	0.096	0.163
	$B_{Pr/Ce}$	0	1.195	1.131	1.093	1.021	1.026
Nd	$[RE]_t$, mol/L	0.0203	0.0195	0.0189	0.0186	0.0184	0.0172
	D	0	0.043	0.075	0.089	0.103	0.179
	$B_{Nd/Pr}$	0	1.000	1.087	1.086	1.073	1.098
Sm	$[RE]_t$, mol/L	0.0203	0.0191	0.0184	0.0182	0.0177	0.0154
	D	0	0.056	0.103	0.118	0.148	0.315
	$B_{Sm/Nd}$	0	1.302	1.373	1.326	1.437	1.760
Eu	$[RE]_t$, mol/L	0.0214	0.0201	0.0193	0.0190	0.0185	0.0159
	D	0	0.065	0.104	0.125	0.154	0.343
	$B_{Eu/Sm}$	0	1.160	1.010	1.059	1.041	1.089
Gd	$[RE]_t$, mol/L	0.0220	0.0207	0.0203	0.0196	0.0193	0.0160
	D	0	0.063	0.082	0.121	0.135	0.370
	$B_{Gd/Eu}$	0	0.969	0.789	0.968	0.876	1.079
Tb	$[RE]_t$, mol/L	0.0220	0.0202	0.0196	0.0190	0.0186	0.0147
	D	0	0.088	0.121	0.156	0.178	0.492
	$\beta_{Tb/Gd}$	0	1.397	1.475	1.289	1.323	1.330
Dy	$[RE]_t$, mol/L	0.0202	0.0188	0.0178	0.0172	0.0166	0.0124
	D	0	0.076	0.133	0.172	0.214	0.619
	$\beta_{Dy/Tb}$	0	0.864	1.099	1.103	1.202	1.258
Ho	$[RE]_t$, mol/L	0.0197	0.0183	0.0172	0.0166	0.0160	0.0114
	D	0	0.078	0.145	0.186	0.230	0.729
	$\beta_{Ho/Dy}$	0	1.027	1.091	1.081	1.075	1.178
Er	$[RE]_t$, mol/L	0.0207	0.0188	0.0179	0.0173	0.0167	0.0119
	D	0	0.101	0.152	0.192	0.234	0.740
	$\beta_{Er/Ho}$	0	1.295	1.048	1.032	1.012	1.015
Tm	$[RE]_t$, mol/L	0.0204	0.0186	0.0179	0.0171	0.0165	0.0114
	D	0	0.096	0.139	0.194	0.237	0.792
	$B_{Tm/Er}$	0	0.951	0.914	1.010	1.013	1.071
Yb	$[RE]_t$, mol/L	0.0191	0.0172	0.0159	0.0153	0.0147	0.0096
	D	0	0.11	0.201	0.248	0.298	0.988
	$B_{Yb/Tm}$	0	1.145	1.455	1.279	1.258	1.247
Lu	$[RE]_t$, mol/L	0.0192	0.0167	0.0159	0.0152	0.0146	0.0095
	D	0	0.149	0.209	0.266	0.317	1.025
	$B_{Pr/Ce}$	0	1.355	1.040	1.073	1.064	1.037
Y	$[RE]_t$, mol/L	0.0209	0.0195	0.0188	0.0178	0.0172	0.0126
	D	0	0.073	0.114	0.173	0.214	0.660

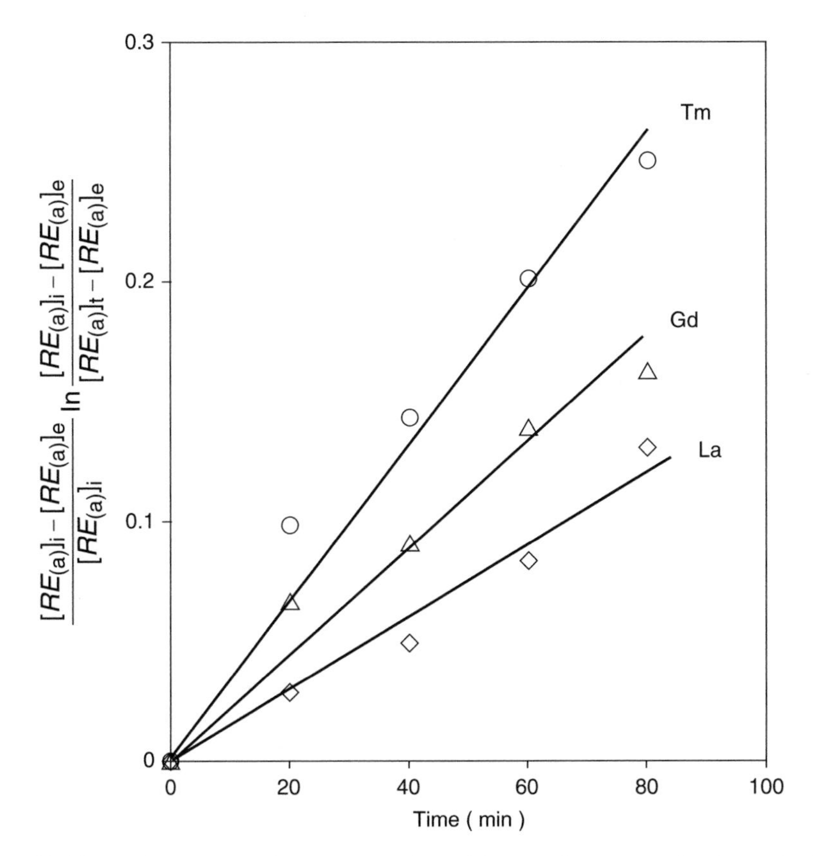

Fig. 4.3 First-order RE extraction reaction reflected by the direct proportional relationship between rare earth concentration and extraction time

The extraction of tetravalent Ce(IV) is similar to the extraction of trivalent RE^{3+}. In a low-acidic environment ($H_2SO_4 < 5$ mol/L), the extraction is cation-exchanging mechanism. In a high-acidic environment ($H_2SO_4 > 5$ mol/L), sulfate anions participate in the extraction to form neutral complexes by solvating mechanism (Li et al. 1984):

$$Ce^{4+}_{(a)} + 3(HL)_{2(o)} = CeL_2 \cdot (HL_2)_{2(o)} + 4H^{+}_{(a)} \tag{4.21}$$

$$Ce^{4+}_{(a)} + 3(HL)_{2(o)} = CeL_2 \cdot (HL_2)_{2(o)} + 4H^{+}_{(a)} \tag{4.22}$$

Figure 4.5 shows the effects of H_2SO_4 concentration on the extraction of Ce^{4+}, Th^{4+}, and RE^{3+} including Sc^{3+} and Y^{3+} by 0.94 mol/L P507 in kerosene. When H_2SO_4 is higher than 0.75 mol/L, the extraction of Ce^{4+} is higher than Th^{4+}. The extractability in P507 follows the order of $Sc^{3+} > Ce^{4+} > Th^{4+} > RE^{3+}$. The difference of extractability can be utilized to achieve the separation of Sc^{3+}, Th^{4+}, and Ce^{4+} from the other RE^{3+} ions using P507-kerosene-H_2SO_4 system.

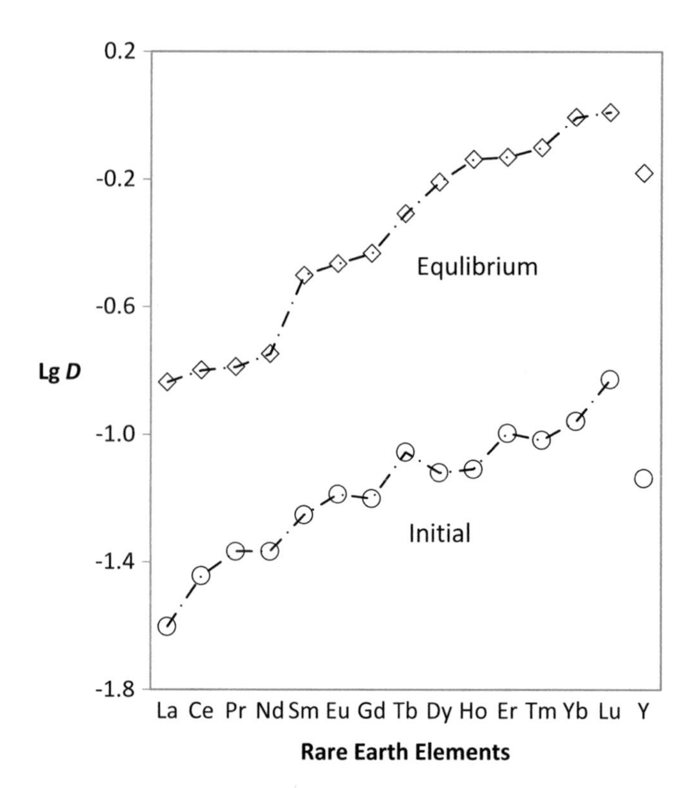

Fig. 4.4 Distribution ratios of rare earth elements in P507-kerosene-HCl-$RECl_3$ system

4.1.4 Saponification

When P507 acts as chelating extractant, the rare earth extraction is cation-exchanging reaction. While the rare earth-extractant complex forms, H^+ is produced leading to the increase of acidity. As discussed in Sects. 4.1.1–4.1.3, the rare earth distribution ratio and percentage of extraction decrease with the increase in acidity of the aqueous solution. To overcome the issues related to acidity increase and extraction reduction, saponification is used before extraction. As shown by Eq. (4.23), saponification is to substitute the exchangeable H^+ of the extractant with Na^+, NH_4^+, Ca^{2+}, or other cations. The available saponification agents include NH_4OH, NH_4HCO_3, NaOH, Na_2CO_3, and $Ca(OH)_2$. NaOH and Na_2CO_3 are strong alkalis with a lot of heat generation during saponification that can cause equipment failure and other operation issues. The commonly used saponification agents are NH_4OH and NH_4HCO_3. The disadvantage of ammonium-based saponification is

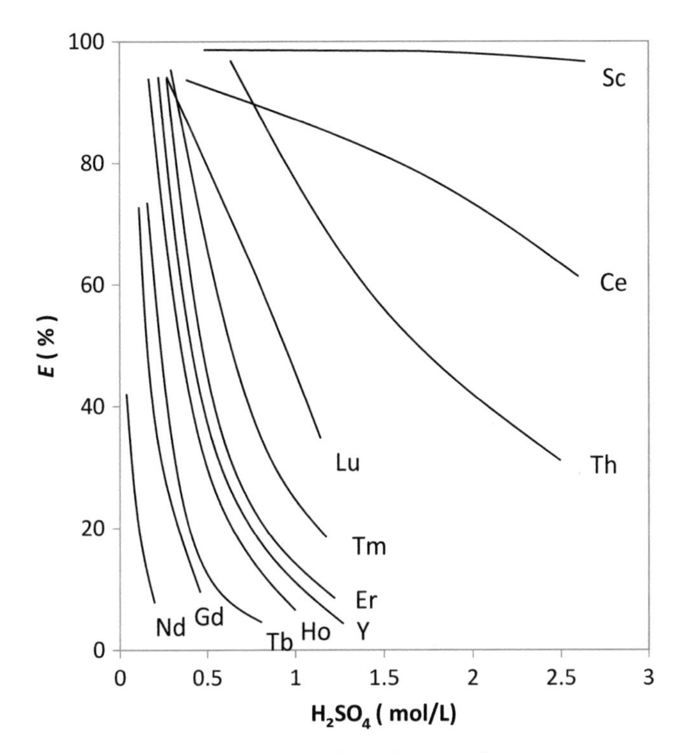

Fig. 4.5 Effects of H_2SO_4 on extraction of RE^{3+}, Ce^{4+}, and Th^{4+} by 0.94 mol/L P507-kerosene

the generation of large amount of waste water containing NH_4^+. The dimer of P507 will dissociate to single molecules after saponification to form a surfactant:

$$NH_4OH + HL_{(o)} \rightarrow NH_4L_{(o)} + H_2O \tag{4.23}$$

Reactions in the saponified P507 system include

$$RE^{3+}_{(a)} + 3NH_4L_{(o)} = REL_{3(o)} + 3NH^+_{4(a)} \tag{4.24}$$

$$RE^{3+}_{(a)} + 3(HL)_{2(O)} = RE(HL_2)_{3(o)} + 3H^+_{(a)} \tag{4.25}$$

$$NH_4L_{(o)} + H^+_{(a)} = HL_{(o)} + NH^+_{4(a)} \tag{4.26}$$

As shown in Fig. 4.6 (Liu and Wang 2014), the extraction of Nd^{3+} increases from 50 to 98 % at the initial aqueous pH of 3.3 when P507 saponification increases from 0 to 40 %. However, to increase the P507 saponification to 50 % will cause emulsion and difficulty of phase disengagement resulting in poor extraction. Also when excessive ammonia is added, saponified P507 will form gel-like precipitates with RE^{3+} to negatively affect the extraction.

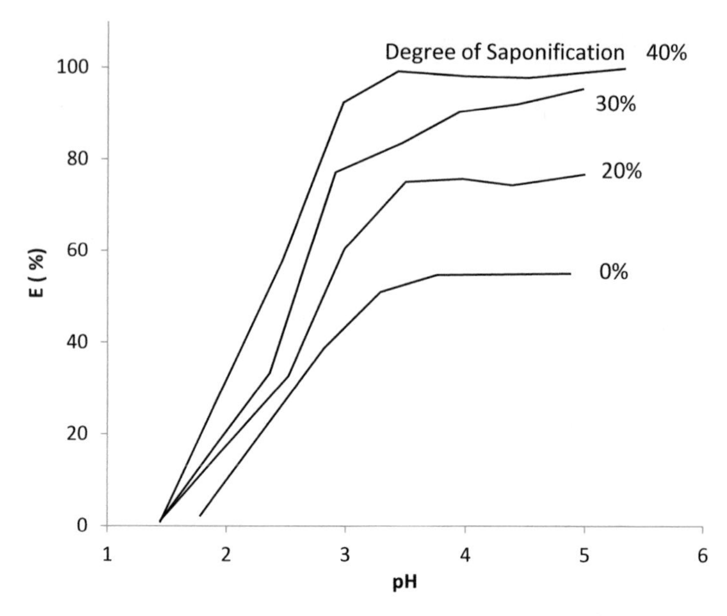

Fig. 4.6 Effects of initial aqueous pH and degree of saponification on Nd^{3+} extraction

Table 4.5 Effects of saponification and H^+ on Dy/Tb separation (Shen et al. 1985)

$[H^+]$, mol/L	1.5 mol/L P507-kerosene (no saponification)	1.5 mol/L P507-kerosene (43 % saponification)
0.6	1.87	2.00
0.7	1.81	1.95
0.8	1.78	1.96

Saponification can also improve the separation factor of rare earth elements. Table 4.5 shows the comparison of Dy/Tb separation factor in P507 and saponified P507. Utilizing the improved Dy/Tb separation in saponified P507, Shen et al. (1985) realized the Dy/Tb separation of the organic solution containing Sm-Lu and Y to two groups: Sm-Dy group with 100 % Tb recovery and Dy-Lu and Y group. The schematic separation process is shown in Fig. 4.7. In this process, the saponified P507 is used to realize Sm/Nd separation of a mixed rare earth feed containing mainly light rare earth elements. The aqueous phase containing LaCePrNd is further separated to pure La product and mixed Ce-Pr-Nd product. The organic phase containing Sm-Lu and Y is further separated to Sm-Dy group and Dy-Lu and Y group.

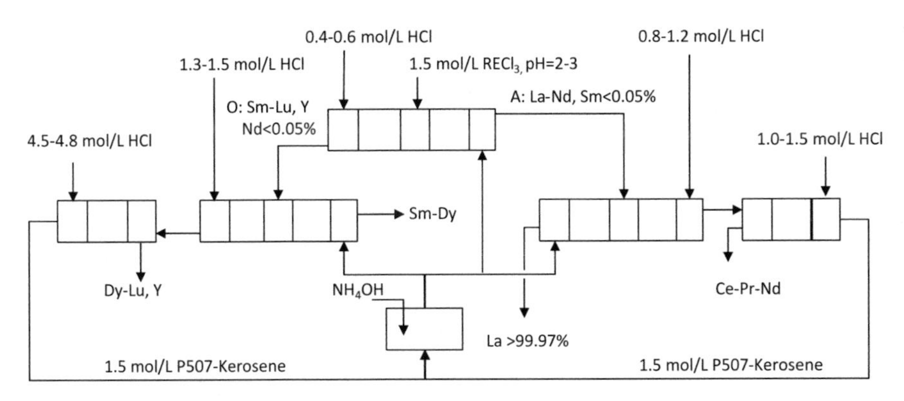

Fig. 4.7 Saponified P507-kerosene separation process of rare earth elements

Table 4.6 Composition of $RECl_3$ feeds and targeted product of separation (wt%)

Feed	La_2O_3	CeO_2	Pr_6O_{11}	Nd_2O_3	Sm_2O_3	Eu_2O_3	Gd_2O_3	Tb_4O_7	Dy_2O_3	$(Ho\text{-}Lu)_2O_3$	Y_2O_3
1	32.0	50.0	4.2	12.5	0.8	0.11	0.15	<0.10	<0.10	<0.10	0.10
2	25.0	50.0	5.5	15.5	1.5	0.20	0.50	<0.10	<0.10	<0.10	0.20
3	22.5	46.0	5.4	17.4	3.0	<0.10	1.90	0.30	0.70	<0.12	2.70
Product	99.99	99.99	99	99.99	99.9	99.99	99.99	99.95	99.9		99.99

4.1.5 *Full Rare Earth Separation Using P507*

Solvent extraction is well accepted as the most efficient commercial technology to separate individual rare earth elements or groups of rare earth elements. However, the rare earth solvent extraction processes are normally complicated due to the chemical similarity of rare earth elements. It often needs hundreds of stages of mixing and settling. The configuration of a solvent extraction process needs to consider the specific feed composition and the targeted products. P507 has been commercially used in many rare earth solvent extraction separation processes. All processes are different in one way or the other due to the variation of feed composition and targeted products.

Depending on the deposits, the rare earth feeds for separation are normally different but they all can be separated to pure individual or groups of rare earth elements. Table 4.6 shows three rare earth feeds with different compositions but with the same targeted full rare earth element separation. Figure 4.8 shows the five different separation processes to produce the targeted products (Ding and Chen 2001). In practice, most of the solvent extraction processes are designed to separate a group of a few rare earth elements instead of a full set of rare earth elements.

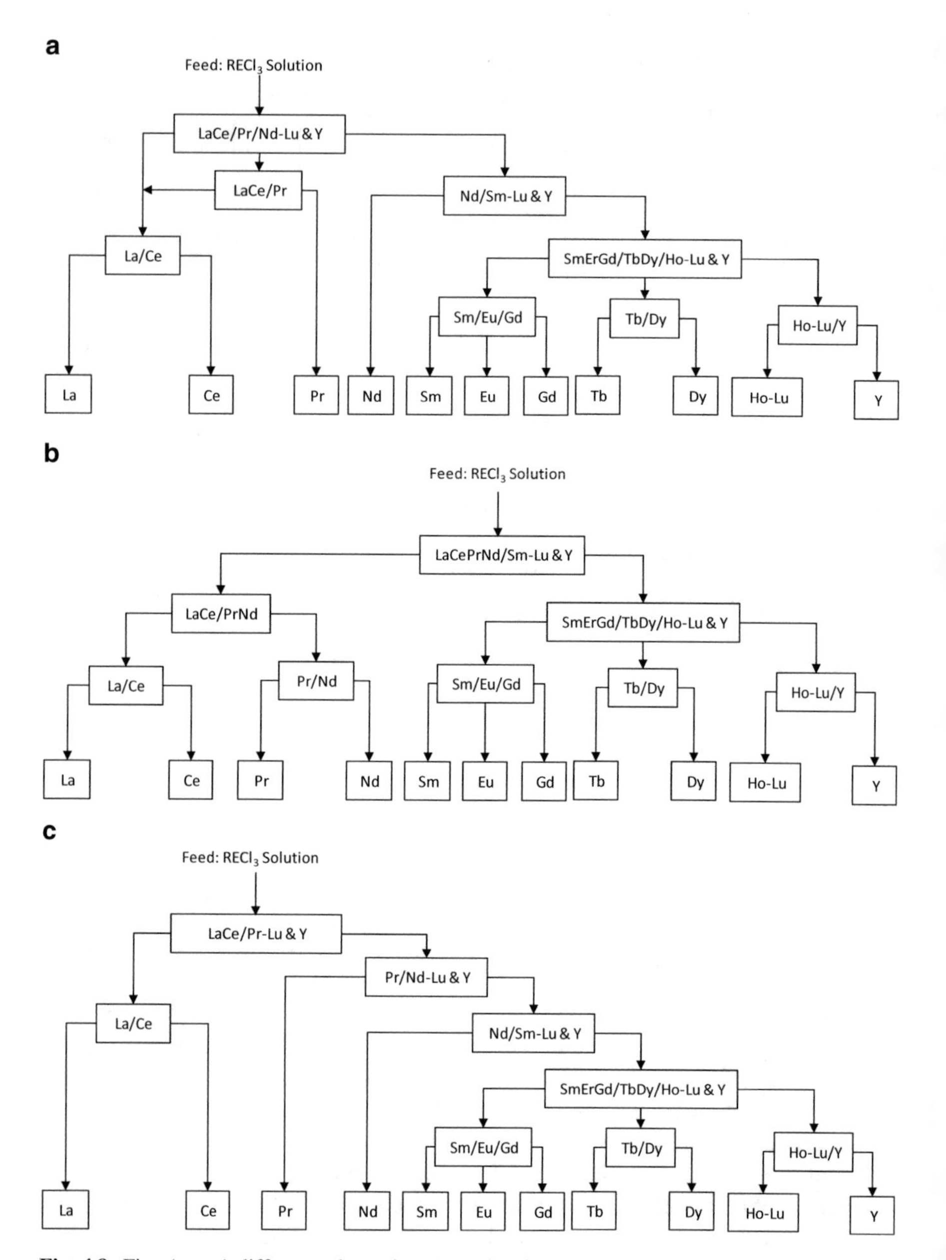

Fig. 4.8 Five (a to e) different schematic rare earth solvent extraction separation processes

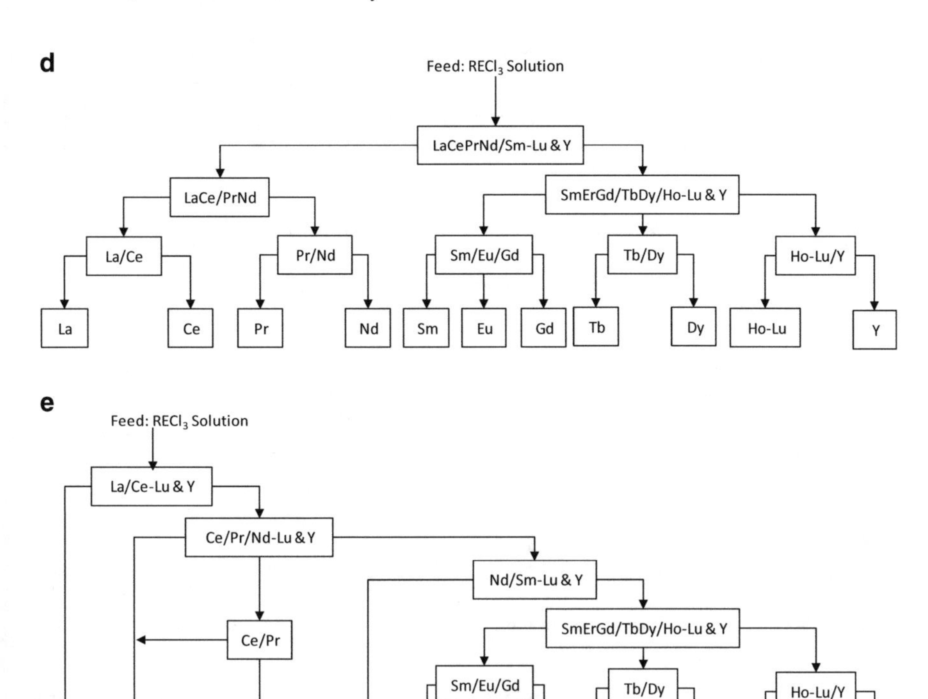

Fig. 4.8 (continued)

Table 4.7 Light rare earth feed and targeted product composition

Composition	La_2O_3, %	CeO_2, %	Pr_6O_{11}, %	Nd_2O_3, %
Feed	26.3	52.1	5.3	16.3
Targeted product	99.99	99.9–99.99	99.9–99.99	99.9–99.99

4.1.6 Light Rare Earth Element Separation Using P507

Bastnasite and monazite are the two most important industrial rare earth minerals. They contain mainly light rare earth elements. Light rare earth separation is the major process for these kinds of deposits. Table 4.7 shows the composition of a light rare earth feed for separation and the composition of targeted products (Deng, et al. 2003). Table 4.8 shows the separator factors when 1.5 mol/L P507 with 36 % saponification in sulfonated kerosene is used to separate the aqueous solution containing 1.5 mol/L REO at pH 3–4. There are A and B two separation options as shown in Fig. 4.9. Separation process A is the traditional cascade separation process. Separation process B is an optimized separation process with the application of fuzzy separation technology. In process B fuzzy separation is used to

Table 4.8 Separation factors in 1.5 mol/L P507-sulfonated kerosene at aqueous pH 3–4

$\beta_{Nd/La}$	$\beta_{Ce/La}$	$\beta_{Nd/Pr}$	$\beta_{Pr/Ce}$
12.0	4.0	1.5	2.0

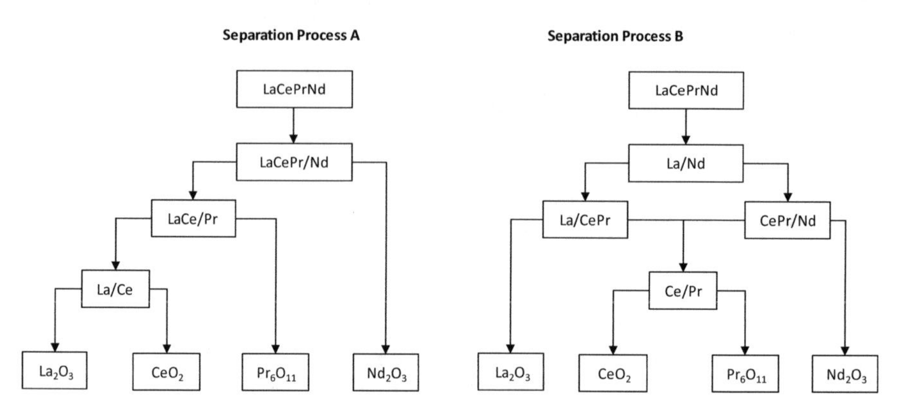

Fig. 4.9 Separation processes of mixed light rare earth

separate the mixed rare earths into two groups by utilizing the largest separation factor between La and Nd to realize the complete separation of La and Nd as the preliminary separation without worrying about the separation of middle elements Ce and Pr. The application of fuzzy separation technology can significantly reduce the reagent consumption and the storage volume of organic solution and rare earth solution. Based on the processing of 1000 *t*/year mixed rare earth oxide, the major process parameters of the two process options are compared and shown in Table 4.9 (Deng, et al. 2003).

4.1.7 Light and Heavy Rare Earth Element Separation Using P507

It is common that heavy rare earth elements only account for a small portion in a rare earth deposit. In order to separate the individual heavy rare earth elements, light/heavy rare earth separation to two groups can be performed as the first step of separation. Li et al. (1982) conducted detailed research on the separation of light/heavy rare earth starting from Tb/Gd separation in HNO_3 solution. The rare earth composition for separation is shown in Table 4.10. The REO extraction isotherm curve at 25 ± 1 °C in 1.0 mol/L P507-sulfonated kerosene is shown in Fig. 4.10. Changing the O/A ratio from 1 to 2 and 4 can increase the separation factor of Tb/Gd from 2.4 to 4.5 and 5.3, respectively. The Tb/Gd separation can be done at O/A = 3–4. Li et al. (1982) obtained good Tb/Gd separation using 10 stages of

Table 4.9 Major process parameters of the mixed light rare earth separation process

Parameter	Process A (traditional process)	Process B with fuzzy separation
REO (*t*/year)	1000	1000
Feed concentration (mol/L)	1.5	1.5
Organic phase	1.5 mol/L P507-sulfonated kerosene	1.5 mol/L P507-sulfonated kerosene
Mixing time (min)	5	5
Mixer-settler ratio	1/2.5	1/2.5
Separation V_{mix} (L) × stage	Pr/Nd: 538 × 130	La/Nd: 260 × 26
	Ce/Pr: 272 × 120	La/Ce: 288 × 32
	La/Ce: 365 × 54	Pr/Nd: 470 × 120
		Ce/Pr: 211 × 100
Organic storage (m^3)	214.0	163.6
Rare earth storage (*t*)	50.0	37.8
HCl consumption (*t*/*t* REO)	5.391	3.186
NH_4OH consumption (*t*/*t* REO)	0.783	0.474
Total volume (m^3) and stages	428 × 304	327 × 278

Table 4.10 Rare earth composition of feed solution (wt%)

Feed	La_2O_3	Nd_2O_3	Sm_2O_3	Gd_2O_3	Tb_4O_7	Ho_2O_3	Er_2O_3	Tm_2O_3	Yb_2O_3	Lu_2O_3	Y_2O_3
1	12.0	12.5	5.0	12.1	2.2	2.8	8.5	0.7	6.5	1.5	21.5
2	12.0	15.5	4.7	11.5	2.5	2.8	8.2	0.4	6.8	1.6	21.2

extraction at O/A = 4 and 8–10 stages of stripping at O/A = 8. In raffinate, Tb_4O_7, Dy_2O_3, and Er_2O_3 are less than 0.12 %, 0.1 %, and 0.18 %, respectively. No yttrium is detected in the raffinate. In stripping solution, the Gd_2O_3 is less than 0.28 % and no Sm and Nd are detected. After separation, the recovery and purity of the light rare earth and heavy rare earth are both higher than 99 %.

4.1.8 Heavy Rare Earth Separation Using P507

Fractional extraction is to introduce the feed from a middle stage while the barren organic solution and clean scrubbing solution are fed from each end of an extraction circuit. P507-kerosene system is commonly used in fractional extraction to separate heavy rare earth. To separate the feed shown in Table 4.11, fractional solvent extraction can be used in three steps: (1) Ho-Er group separation and Er-Tm separation, (2) Tb-Dy group separation and Dy-Ho separation, and (3) Gd-Tb group separation and Tb-Dy separation (Li 2011).

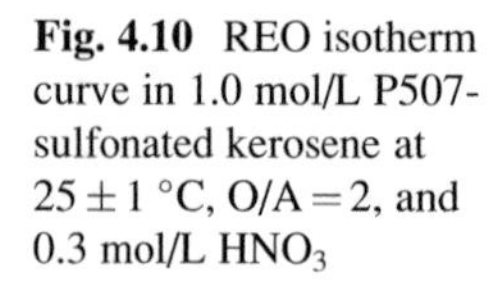

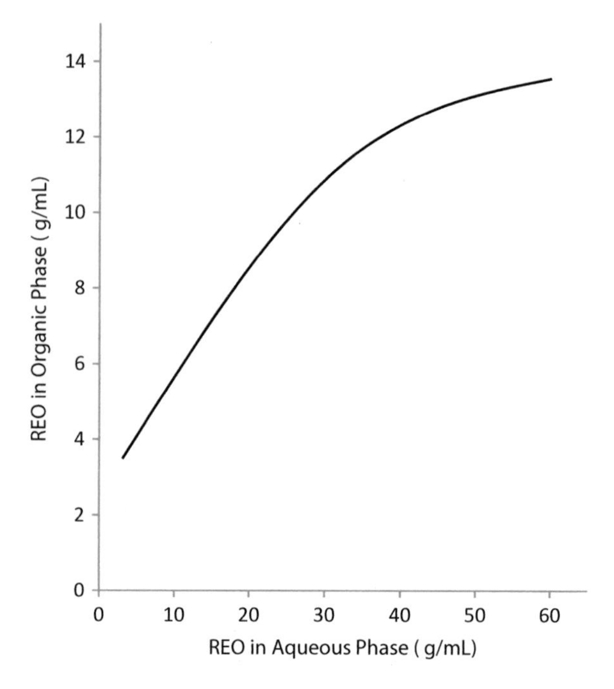

Fig. 4.10 REO isotherm curve in 1.0 mol/L P507-sulfonated kerosene at 25 ± 1 °C, O/A = 2, and 0.3 mol/L HNO_3

Table 4.11 Feed composition for heavy rare earth separation

RE	Eu	Gd	Tb	Dy	Ho	Er	Tm	Yb	Lu	Y
Composition (%)	<0.3	8.8	3.0	~5.0	2.3	14.3	1.5	6.7	1.4	2–10

As shown in Fig. 4.11, in STEP1 aqueous feed containing 1.0 mol/L RE^{3+} with the composition shown in Table 4.11 at pH 2 is separated to Eu-Ho group in raffinate and Er-Tm&Y group in organic solution. Er product with purity >95 % is produced at over 95 % recovery from the loaded organic solution after multiple stages of scrubbing and stripping. The raffinate containing Eu-Ho is fed to STEP2 separation where it is prepared to a feed solution containing 1.0 mol/L RE^{3+} at pH 2. After multiple stages of extraction, scrubbing, and stripping, a Dy product with purity >99.9 % is produced at over 90 % recovery. In STEP3, the raffinate from STEP 2 containing EuGdTb is prepared to a feed solution with 0.8 mol/L RE^{3+} at pH 2. A Tb product with >99.9 % purity is produced at over 90 % recovery in STEP 3. Also, a Gd-rich product (>80 %) and Dy-rich product (>80 %) are produced.

In solvent extraction, the rare earth in raffinate becomes less concentrated after the separation of the easy extractable rare earth in previous steps. Low-concentration feeds will cause low separation efficiency of the following steps and also lower the equipment capacity. In industrial practice, evaporation is normally adapted to concentrate the difficult extractable rare earth in feeds.

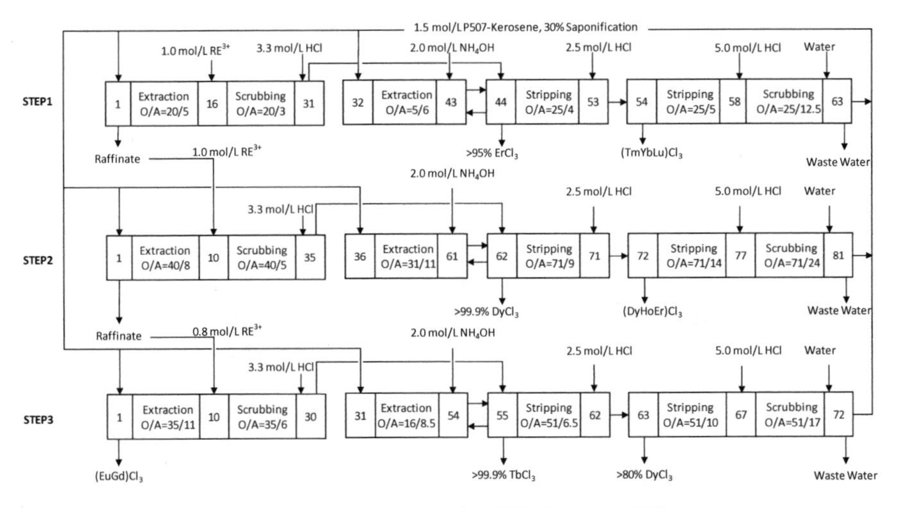

Fig. 4.11 Heavy rare earth separation process using P507-kerosene-HCl system

Table 4.12 Typical composition of ion-adsorption deposit and target products (wt%)

Composition	La_2O_3	CeO_2	Pr_6O_{11}	Nd_2O_3	Sm_2O_3	Eu_2O_3	Gd_2O_3	Tb_4O_7
Deposit-1	26.5	2.4	6.0	20.0	4.0	0.8	4.0	0.6
Deposit-2	28.0	1.6	6.8	23.9	4.8	0.9	4.5	0.6
Targeted product	99.9	99.9	99.0	99.9	99.9	99.99	99.99	99.95
Composition	Dy_2O_3	Ho_2O_3	Er_2O_3	Tm_2O_3	Yb_2O_3	Lu_2O_3	Y_2O_3	
Deposit-1	4.0	0.8	1.8	0.3	1.2	0.1	27.5	
Deposit-2	3.4	0.6	1.7	0.2	1.5	0.2	21.2	
Targeted product	99.9	–	99.9	–	–	–	99.999	

4.1.9 Ion-Adsorptive Rare Earth Separation Using P507

With the increasing demand for middle and heavy rare earth elements, many research and development works are done on the solvent extraction of ion-adsorptive deposit. One of the major ion-adsorptive rare earth deposits is rich in Eu content with moderate Y content. The distribution of light, middle, and heavy rare earth elements is relatively even. Table 4.12 shows the typical rare earth composition of these types of deposits and the targeted products (Yan, et al. 1999; Zhong, et al. 2001). In these kinds of deposits, light rare earth accounts for approximately 50 % with low Ce and Pr. Middle rare earth accounts for about 10 % with high Eu content up to 0.8 %. Heavy rare earth accounts for about 40 %. Based on the composition, the separation must consider all light, middle, and heavy rare earths. It is common to separate rare earth elements to light, middle, and heavy groups by utilizing the tetrad effects. The point of separation is normally selected to have large separation factors. But it is also important to consider the rare earth distribution in the feed, the targeted products, process and equipment, as well as operational factors.

Yan et al. (1999) made detailed comparison of three processes for the separation of rare earth shown in Table 4.12. In Fig. 4.12, Process a utilizes the large separation factor of Nd/Sm and two-outlet process to get light and heavy two groups of rare earth concentrates. The light rare earth is further separated using three-outlet process to get >99.9 % La, >99.9 % Ce, >99.0 % Pr, and >99.9 % Nd. The heavy rare earth in organic phase is directly fed to -Gd/Tb/Dy- three-outlet separation process to get SmEuGd concentrate, Tb-rich concentrate, and Y-rich concentrate containing Dy. Three products Sm >99.9 %, Gd > 99.9 %, and Eu-rich (>50 %) concentrate can be produced using three-outlet process. The Y-rich concentrate is further separated into high-purity Y >99.999 % and barren Y heavy rare earth product in naphthenic acid system. Based on 2000 t REO/year processing capacity and other design criteria as shown in Table 4.13, the detailed process parameters of each separation circuit of Process a are designed and shown in Table 4.14.

Nd/Sm separation is also first used in the Process b followed by La/Ce/Pr/Nd separation and Dy/Ho separation. The separation of La/Ce/Pr/Nd is the same as that in Process a to product La, Ce, Pr, and Nd products. Due to the large separation factor between Ho and Dy and low Ho content in the feed, the Sm-Dy group only contains <0.01 % Ho and no other heavy rare earths while the Y-rich concentrate only contains <0.5 % Dy. Pure Dy >99.9 % can be separated from the Sm-Dy group by three-outlet process while SmEuGd and Tb-rich concentrates are produced. The SmEuGd concentrate can be further separated to pure Sm and Gd products and Eu-rich product. The Y-rich concentrate is further processed to produce pure Y and Er products. In comparison with Process a, the Er processing work is reduced. According to the design criteria shown in Table 4.13, the detailed process parameters of each separating circuit of Process b are designed and shown in Table 4.15.

Three-outlet process is utilized in Process c in the first step to produce light rare earth group, Sm-Dy group with ~40 % Ho-Lu, and Y-rich group. The light rare earth separation is the same as Process a and Process b. Dy/Ho separation is done on the Sm-Dy concentrate to remove the Ho-Lu which is then combined with the Y-rich concentrate. The following processing of the Sm-Dy group without Ho-Lu and the Y-rich concentrate is the same as that in Process b. In comparison with Process b, the Dy/Ho separation is simplified in Process c. According to the design criteria, the process parameters of each separation circuit of Process c are designed and shown in Table 4.16.

The Tb-rich concentrate and Eu-rich concentrate produced in the above processes can be further processed to produce pure Tb >99.95 % and Eu > 99.99 %.

4.1.10 Ce^{4+} Separation by P507-H_2SO_4 System

In the H_2SO_4 leaching solution of roasted rare earth concentrate, the majority of cerium is in its tetravalent state. Table 4.17 shows the typical composition of H_2SO_4

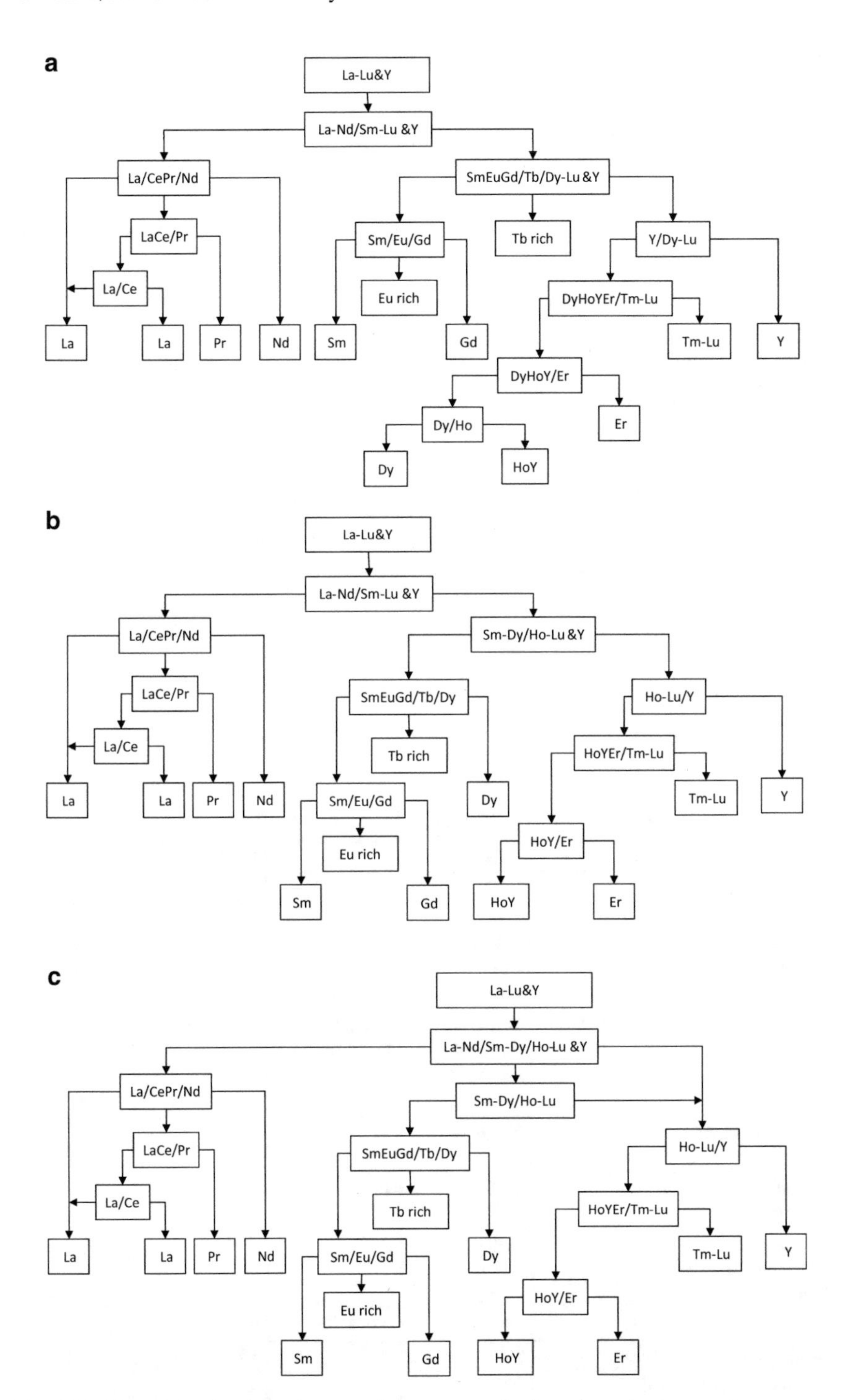

Fig. 4.12 Three (a to c) schematic separation processes for ion-adsorption rare earth

Table 4.13 Design criteria of ion-adsorption rare earth separation

Processing capacity	2000 t REO/year
Operation days	300
Organic solution	1.5 mol/L P507-sulfonated kerosene, 36 % saponification
Rare earth loading capacity	0.18 mol/L
Scrubbing HCl solution	3.6 mol/L
Stripping HCl solution	4.5 mol/L
Mixer-settler ratio	1/2.5
Mixing O/A ratio	Organic flow/aqueous flow
Settling O/A ratio	1.2/1
Scale-up factor of stages	75–90 %
Separator factors	$\beta_{Ce/La}=6.0$, $\beta_{Pr/Ce}=2.0$, $\beta_{Nd/La}=1.5$, $\beta_{Sm/Nd}=8.0$, $\beta_{Eu/Sm}=2.0$, $\beta_{Gd/Eu}=1.5$, $\beta_{Tb/Gd}=2.0$, $\beta_{Dy/Tb}=3.0$, $\beta_{Ho/Dy}=2.0$, $\beta_{Y/Ho}=1.5$, $\beta_{Er/Ho}=1.5$, $\beta_{Tm/Er}=2.5$

Table 4.14 Process parameters of each separation circuit of ion-adsorption separation process A

	Separation circuit				
Items	Nd/Sm	La/CePr/Nd	LaCe/Pr	La/Ce	SmEuGd/Tb/Dy
Feed (mol/min)	31.0	15.3	4.70	3.10	15.7
RE^{3+} in feed (mol/L)	1.70	1.50	1.30	1.20	0.14
Feed composition x_{RE} (%)	Sm: 51.0 Nd: 49.0	Nd:36.0 CePr:15.0 La:49.0	Ce:35.0 Pr:14.0 La:51.0	Ce:21.0 La:79.0	Dy:84.4 Tb:0.95 Sm-Gd:14.6
Product purity P_{RE}(%)	Sm:99.99 Nd:99.99	Nd:99.95 CePr:49.0 La:99.95	LaCe:99.95 Pr:99.95	Ce:99.95 La:99.95	Dy-:99.95 Tb:45.0 Sm-Gd:99.95
Extraction recovery (%)		Nd:99.8			Sm-Gd:99.95
Normalized extraction, S	0.65	1.75	1.10	0.50	1.50
Normalized scrubbing, W	0.14	1.39	0.75	0.29	1.65
$V_O/V_F/V_W/V_S$	112/18.2/2/3.62	149/10.2/17.7/4.24	28.7/3.61/2.94/1.25	8.61/2.58/0.75/0.50	131/112/17.3/10.9
V_{mixing} (L)	600	700	150	50	1300
Number of stages	48	80	60	36	80

(continued)

Table 4.14 (continued)

Items	Separation circuit				
	Nd/Sm	La/CePr/Nd	LaCe/Pr	La/Ce	SmEuGd/ Tb/Dy
Organic holding (m^3)	65.7	126	20.1	3.98	238
Rare earth holding (*t*)	8.98	16.5	2.5	0.42	34.9
HCl consumption (m^3/day)	2.08	13.2	2.6	0.79	22.0
NH_3 consumption (*t*/day)	1.48	1.97	0.38	0.11	1.73
	Separation circuit				
Items	Sm/Eu/Gd	Y/non-Y[a]	Er/Tm	Dy-Y/Er	Dy/HoY
Feed (mol/min)	2.30	13.1	1.97	1.62	1.22
RE^{3+} in feed (mol/L)	1.30	1.00	1.00	1.00	1.00
Feed composition x_{RE} (%)	Gd:44.5 Eu:9.2 Sm:46.3	Y:85.0 Dy-:15.0	Dy-Er:82.2 Tm-:17.8	Dy-,Y:75.0 Er:25.0	Dy:77.0 HoY:23.0
Product purity P_{RE} (%)	Gd:99.9 Eu:50 Sm:99.95	Y:99.999 Dy-:99.5	Dy-Er:99.999 Tm-:99.95	Dy-,Y:99.5 Er:99.95	Dy:99.99 HoY:99.5
Extraction recovery (%)	Sm:99.95				
Normalized extraction, *S*	2.80	1.60	0.95	1.20	1.40
Normalized scrubbing, *W*	2.45	1.45	0.77	0.95	1.17
$V_O/V_F/V_W/V_S$[b]	35.8/1.77/ 3.76/0.68	105/13.1/ 15.8/1.79	10.4/1.97/ 1.01/0.29	10.8/1.62/ 1.03/0.34	9.49/1.22/ 0.95/0.23
V_{mix}(L)[c]	200	600	80	80	70
Number of stages	80	100	70	90	82
Organic holding (m^3)	36.0	127	12.2	16.1	12.9
Rare earth holding (*t*)	5.70	10.8	1.81	2.37	1.50
HCl consumption (m^3/day)	3.30	10.1	0.98	1.04	0.89
NH_3 consumption (*t*/day)	0.47	1.54	0.14	0.14	0.13

[a]The separation of Y/non-Y is done by naphthenic acid-HCl system

[b]V_O, V_F, V_W, and V_S are flow rates of organic solution, feed solution, scrubbing solution, and stripping solution (L/min)

[c]V_{mix} is the mixer volume

Table 4.15 Process parameters of each separation circuit of ion-adsorption separation process b

	Separation circuit[a]				
Items	Dy/Ho	SmEuGd/ Tb/Dy	Y/non-Y[b]	HoYEr/Tm	HoY/Er
Feed (mol/min)	15.7	3.46	12.3	0.98	0.59
RE^{3+} in feed (mol/L)	0.140	1.25	1.00	1.00	1.00
Feed composition x_{RE} (%)	Ho-:78.0 -Dy:22.0	Dy:28.9 Tb:4.3 Sm-Gd:66.8	Y:92.0 Ho-:8.0	HoYEr:60.0 Tm:40.0	HoY:41.0 Er:59.0
Product purity P_{RE}(%)	Ho-:99.5 -Dy:99.99	Dy:99.95 Tb:50.0 Sm-Gd:99.95	Y:99.9999 Dy-:99.5	HoYEr:99.99 Tm:99.5	HoY:99.5 Er:99.95
Extraction recovery (%)		Sm-Gd:99.95			
Normalized extraction, *S*	1.05	0.80	1.40	1.10	3.00
Normalized scrubbing, *W*	1.27	0.55	1.32	0.70	2.41
$V_O/V_F/V_W/V_S$	91.6/112/ 13.3/10.2	15.4/2.77/ 1.28/0.71	86.1/12.3/ 13.5/0.89	5.99/0.98/ 0.46/0.33	9.83/0.59/ 0.95/0.29
V_{mix} (L)	1100	100	500	40	60
Number of stages	80	80	100	70	90
Organic holding (m^3)	200	17.9	105	6.12	12.0
Rare earth holding (*t*)	19.5	2.64	9.09	0.87	1.72
HCl consumption (m^3/day)	18.5	1.55	8.29	0.62	0.94
NH_3 consumption (*t*/day)	1.21	0.20	1.26	0.08	0.13

[a]The separation circuits of Nd/Sm, La/CePr/Nd, LaCe/Pr, La/Ce, and SmEuGd/Tb/Dy in Process B are the same as those in Process A

[b]The separation of Y/non-Y is done by naphthenic acid-HCl system

leaching solution of a roasted Baiyun Obo rare earth concentrate (Li 2011). In order to selectively extract Ce^{4+}, P507 can be used at acidity higher than 1.5 mol/L H_2SO_4 where RE^{3+} is not extractable in H_2SO_4 system. At 3.0 mol/L H_2SO_4, iron is the major impurity that can be extracted with the extraction of Ce^{4+}.

As shown in Fig. 4.13, the Ce^{4+} separation process consists of total 20 stages of extraction, scrubbing, and stripping. The organic solution contains 1.0 mol/L P507 in sulfonated kerosene. The aqueous solution contains 55–60 g/L REO with 25–30 g/L $\sum CeO_2$. The scrubbing solution contains 0.5 mol/L H_2SO_4 and 5 g/L Al_2O_3. The purpose of using H_2SO_4 and $Al_2(SO_4)_3$ in the scrubbing solution is to stabilize the F^- in the solution in terms of $(AlF_6)^{3-}$. Otherwise, F will be extracted in terms of $(CeF_2)^{2+}$ and form CeF_3 precipitate during stripping to cause emulsion

Table 4.16 Process parameters of each separation circuit of ion-adsorption separation process c

Items	Separation circuit[a]	
	La-Nd/Sm-Dy/HoY-Lu	Sm-Dy/Ho-
Feed (mol/min)	31	5.72
RE^{3+} in feed (mol/L)	1.70	1.45
Feed composition x_{RE} (%)	Sm-Dy:11.1 -Nd:49.3 Ho-:39.6	Sm-Dy:60.0 Ho-:40.0
Product purity P_{RE}(%)	Sm-Dy:60.0 -Nd:99.99, Ho-:99.5	Sm-Dy:99.99 Ho-:99.5
Extraction recovery (%)	Sm-Dy:99.99	
Normalized extraction, *S*	0.80	1.10
Normalized scrubbing, *W*	0.48	0.70
$V_O/V_F/V_W/V_S$	138/18.2/9.92/8.27	35.0/3.94/2.67/1.91
V_{mix} (L)	850	250
Number of stages	54	80
Organic holding (m^3)	104	45.0
Rare earth holding (*t*)	11.2	5.6
HCl consumption (m^3/day)	14.4	3.6
NH_3 consumption (*t*/day)	1.82	0.46

[a]La/CePr/Nd, LaCe/Pr, La/Ce, and Sm/Eu/Gd circuits are the same as those in Process a. SmEuGd/Tb/Dy, Y/non-Y, HoYEr/Tm-, and HoY/Er circuits are the same as those in Process b

Table 4.17 Composition of H_2SO_4 leaching solution (wt%)

Leachate	REO	CeO_2	$Ce^{4+}/\sum Ce$	CaO	Fe_2O_3	ThO_2	Al_2O_3	P_2O_5	F	H^+
1	59.80	27.95	88.01	1.50	0.51	0.20	0.21	1.65	2.01	2.45
2	57.42	28.59	87.16	1.46	0.53	0.20	0.28	1.72	2.14	2.60
3	59.40	28.91	89.62	1.31	0.48	0.12	0.20	1.42	2.01	2.80

issue. The stripping of Ce^{4+} is done using 2.0 mol/L HCl solution. High-concentration HCl solution (6.0 mol/L) with the addition of $H_2C_2O_4$ and NH_4HCO_4 is used to strip the Fe and Th out of the organic solution before recycling. H_2O_2 and $KMnO_4$ are used in different stages of the process to control the oxidation state of Ce.

4.2 HDEHP Solvent Extraction System

Di-(2-ethylhexyl) phosphoric acid is another often used phosphorous acid extractant in rare earth separation, often denoted by D2EHPA, DEHPA, and HDEHP. The trade name P204 is frequently used in publications. It is a colorless

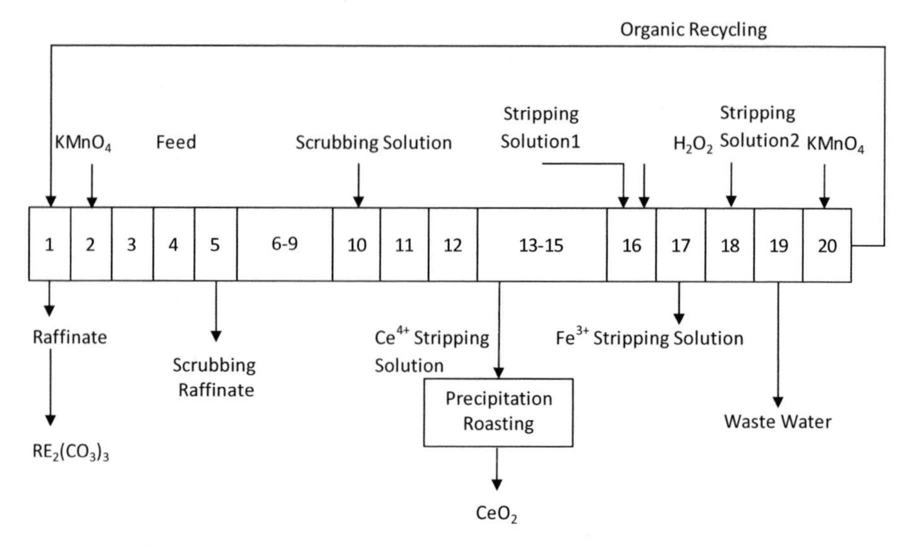

Fig. 4.13 Ce^{4+} separation process by P507-H_2SO_4 system (Li 2011)

or yellowish transparent vicious oily liquid. Its chemical formula is $(C_8H_{17}O)_2PO_2H$ with the following structure:

Similar to HEH/EHP, HDEHP exists as dimer in nonpolar diluent and can be denoted by $(HL)_{2(o)}$. It is an effective extractant for group rare earth separation and individual rare earth element separation. In comparison with HEH/EHP, it is a stronger acid with stronger extraction ability for rare earth elements.

4.2.1 HDEHP Extracting Rare Earths

In 1950s, the use of HDEHP was studied by Peppard and co-workers (1957; 1958a, b) for rare earth fractional solvent extraction in HCl solution. They found that the HDEHP exists largely as a dimer in the organic phase and each trivalent RE^{3+} ion in the organic phase is associated with three of these dimers. Based on this observation, a reaction between HDEHP and RE^{3+} is concluded as follows:

$$RE^{3+}_{(a)} + 3(HL)_{2(o)} = RE(HL_2)_{3(o)} + 3H^+_{(a)} \tag{4.27}$$

The extraction equilibrium constant is

$$K_{ex} = \frac{\left[RE(HL_2)_{3(o)}\right] \cdot \left[H^+_{(a)}\right]^3}{\left[RE^{3+}_{(a)}\right] \cdot \left[(HL)_{2(o)}\right]^3} \tag{4.28}$$

The distribution ratio is

$$D = \frac{\left[RE(HL_2)_{3(o)}\right]}{\left[RE^{3+}_{(a)}\right]} = K_{ex} \frac{\left[(HL)_{2(o)}\right]^3}{\left[H^+_{(a)}\right]^3} \tag{4.29}$$

Therefore,

$$\lg D = \lg K_{ex} + 3\lg\left[(HL)_{2(o)}\right] - 3\lg\left[H^+_{(a)}\right] \tag{4.30}$$

The separation factor between two adjacent rare earths is

$$\beta_{z+1/z} = \frac{D_{z+1}}{D_z} = \frac{Kz + 1}{K_z} \tag{4.31}$$

The distribution ratio depends on the free HDEHP and the concentration of hydrogen ions. Due to the fact that hydrogen ion concentration increases with the increase of extraction, the distribution ratio varies during the different stages of a solvent extraction system operating with HDEHP. To avoid significant change of distribution ratio during the extraction, relatively high acid concentration in the aqueous feed is normally used. Also, relatively low organic loading is adopted by increasing the HDEHP concentration in organic phase to avoid excessive depression of the distribution ratios. At the same extraction conditions, the distribution ratio increases with the increase in atomic number or the decrease of ion radium with Y between Ho and Er. However, the separation factors do not change significantly. Table 4.18 shows the separation factors of rare earth elements in HDEHP-HCl system (Song et al. 2009).

In H_2SO_4 solution at low acidity, the RE^{3+} extraction is cation-exchanging reaction as shown by Eq. (4.27). At high SO_4^{2-} concentration, RE^{3+} and SO_4^{2-} form sulfate complexes. The extraction of RE^{3+} is controlled by the dissociation of the sulfate-RE complex:

Table 4.18 Separation factors of rare earths in HDEHP-HCl system

β	Ce	Pr	Nd	Sm	Eu	Gd	Tb	Dy	Ho	Er	Y
La	2.17	2.07	3.99	2.76	7.60	8.37	9.03	12.07	16.37	19.78	18.10
Ce		0.95	1.84	2.66	3.50	3.86	4.16	5.56	7.55	9.12	8.34
Pr			1.93	2.79	3.68	4.05	4.37	5.84	7.92	9.57	8.76
Nd				1.44	1.90	2.10	2.26	3.02	4.10	4.96	4.54
Sm					1.32	1.45	1.57	2.09	2.84	3.43	3.14
Eu						1.10	1.19	1.59	2.15	2.60	2.38
Gd							1.08	1.44	1.95	2.36	2.16
Tb								1.34	1.81	2.19	2.00
Dy									1.36	1.64	1.50
Ho										1.21	1.11
Er											0.91

$$RE^{3+} + nSO_4^{2-} = RE(SO_4)_n^{3-2n} \tag{4.32}$$

In the leaching solution of H_2SO_4 roasted bastnasite, over 99 % of Ce is tetravalent. The solution also contains F^- which exists with Ce^{4+} as $(CeF_2)^{2+}$ complexes. It is extracted to organic phase by the following reaction (Long et al. 2000):

$$(CeF_2)^{2+}_{(a)} + (HL)_{2(o)} = CeF_2L_{2(o)} + 2H^+_{(a)} \tag{4.33}$$

In industrial practice, sometimes grey white or yellow white precipitates are generated in the HDEHP extraction system due to supersaturation:

$$nRE^{3+}_{(a)} + nRE(HL_2)_{3(o)} = (REL_3)_n + 3nH^+_{(a)} \tag{4.34}$$

The precipitate does not dissolve in diluent but dissolves in excessive HDEHP. Therefore, the precipitate can be avoided at higher extractant concentration or O/A ratio, lower rare earth feed concentration, and higher aqueous acidity. Where F^- is removed before H_2SO_4 leaching, HDEHP can be used to separate the Ce^{4+} with the rest of rare earths to produce high-purity CeO_2 product (99–99.99 %) and rich La rare earth product (Zhou et al. 1998).

In phosphoric acid solution, the rare earth extraction by HDEHP can be expressed by Eq. (4.35):

$$RE^{3+}_{(a)} + H_3PO_4 + 2(HL)_{2(o)} = RE(H_2PO_4)L_2(HL)_{2(o)} + 3H^+_{(a)} \tag{4.35}$$

At high acidity, HDEHP exists as single molecules in organic phase. The P=O bond of HDEHP acts as electron accepting group to extract $H_2PO_4^-$ by solvation mechanism. At low acidity HDEHP exists as dimer in organic phase. The rare earth extraction follows cation-exchange mechanism. At moderate acidity, single molecule and dimer of HDEHP exist together. There is a transaction between solvating and cation-exchanging mechanisms. Due to the partial dissociation of H_3PO_4, some $H_2PO_4^-$ will form complexes with RE^{3+} which are extracted to organic phase. The extraction of $H_2PO_4^-$ to organic phase in turn leads to more H_3PO_4 dissociation and the generation of more H^+(Wang et al. 2009a, b).

4.2.2 Factors Affecting Rare Earth Extraction with HDEHP

4.2.2.1 Aqueous-Phase Acidity

According to Eqs. (4.29) and (4.30), the rare earth distribution ratio is reversely proportional to the third power of the concentration of hydrogen ions at constant-free HDEHP concentration. As shown in Fig. 4.14, at low-acidity plot lg$D$ against lg[H^+], a straight line with slope −3 can be obtained with interception of $\lg K + 2\lg[(HL)_2]$ for each rare earth element (Xu 2003; Li 2011). At higher acidity, the distribution ratio becomes smaller for the same rare earth element in a system with constant extractant concentration. The effects of acidity on distribution ratio can be utilized to determine the aqueous acidity in order to separate a group or an individual rare earth from the other by extraction or stripping.

According to Eq. (4.27), acidity will increase with the increase of extraction due to the generation of hydrogen ions. The distribution ratios and separation factors are all affected. In order to minimize the negative effects of acidity changes on rare earth extraction, saponification is commonly used.

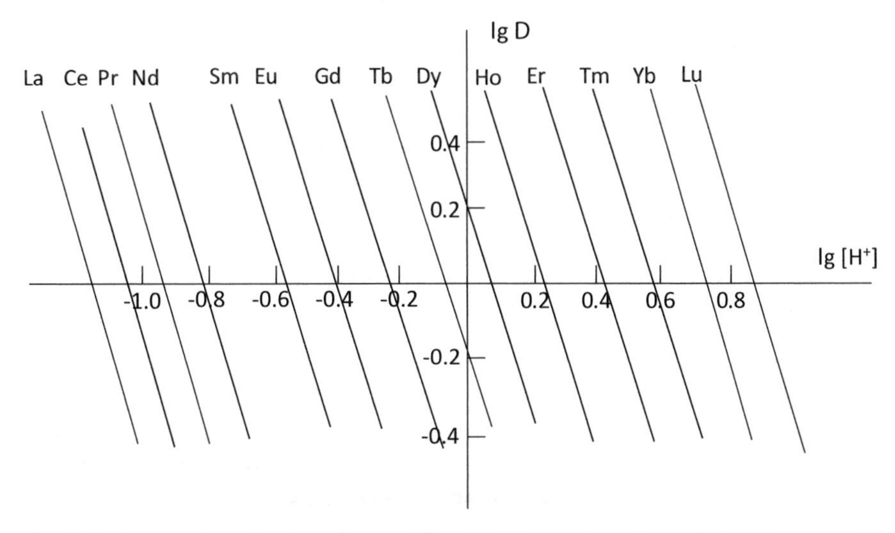

Fig. 4.14 Rare earth distribution ratio vs. acidity. 1.0 mol/L HDEHP- $C_6H_5CH_3$-HCl system with 0.05 mol/L $RECl_3$

4.2.2.2 Saponification

Saponification can increase the loading capacity of HDEHP. Commonly used saponification agents include ammonium hydroxide, ammonia gas, sodium hydroxide, sodium carbonate, and ammonium bicarbonate. The following Eqs. (4.36) and (4.37) show the reactions between HDEHP and ammonium hydroxide:

$$(HL)_2 + NH_4OH = NH_4HL_2 + H_2O \tag{4.36}$$

$$NH_4HL_2 + NH_4OH = 2NH_4L + H_2O \tag{4.37}$$

Rare earth extraction by saponified HDEHP produces NH_4^+ instead of H^+. Therefore, the acidity of the extraction system does not increase and the loading capacity of the HDEHP is enhanced. However, excessive saponification will cause emulsion issues. Figure 4.15 shows the HDEHP loading capacity as the function of degree of saponification. The loading capacity of the HDEHP increases about 50 % at 30 % saponification. However, there is phase disengagement problem at 40 % saponification due to the formation of emulsion.

4.2.2.3 Diluent

The effect of diluent on HDEHP is similar to that on HEH/EPH. The extraction ability of HDEHP will increase with the decrease of relative dialectical constant of the diluent. Table 4.19 shows the extraction equilibrium constant of various rare earth elements at 25 °C. The extracting ability of HDEHP in different diluent follows the order n-C_6H_{14} > c-C_6H_{14} > CCl_4 > $C_6H_5CH_3$ > C_6H_6 > $CHCl_3$ (Chai 1998).

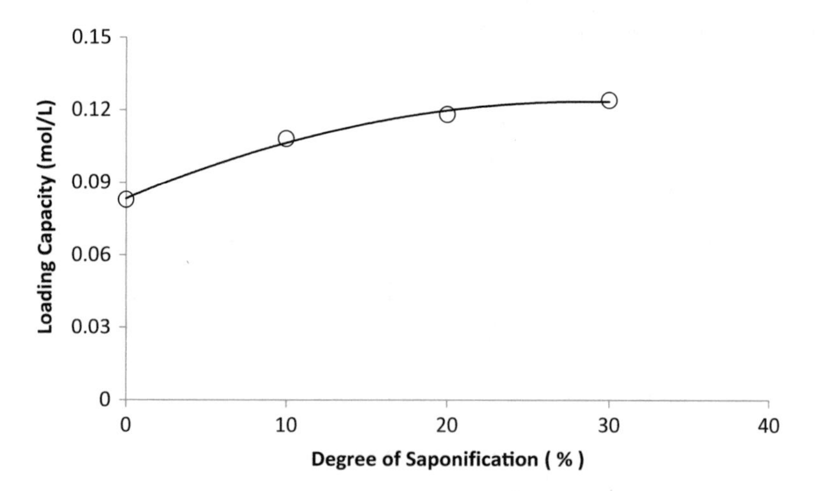

Fig. 4.15 Effects of saponification on HDEHP loading capacity (Li 2011). Feed: 0.9814 mol/L, pH = 4–5; organic phase: 1 mol/L HDEHP; O/A = 1/1

4.2.2.4 Rare Earth Ion Radius

For rare earth ions with same charges, smaller ions form more stable rare earth-extractant complexes with higher distribution ratio. Due to lanthanide contraction, the radius of rare earth ions becomes smaller with the increase in atomic number. Therefore, as shown in Table 4.19, rare earth with higher atomic number has larger extraction equilibrium constant when HDEHP is dissolved in the same diluent.

4.2.2.5 Feed Rare Earth Concentration

Rare earth distribution ratio decreases with the increase in feed rare earth concentration when other conditions are maintained the same. As shown in Fig. 4.16, the distribution ratios of Sm and Nd decrease with the increase in feed rare earth concentration (Zhang et al. 1993).

Table 4.19 Extraction equilibrium constant of rare earths by HDEHP in different diluents

		Lg*K* (25 °C)					
Diluent	ε	Pr	Nd	Sm	Dy	Ho	Yb
n-C_6H_{14}	1.88	0.885	1.545	1.799	3.622	4.130	5.245
c-C_6H_{12}	2.02	0.440	1.000	1.441	3.296	3.584	4.995
CCl_4	2.24	−0.116	0.300	0.622	2.502	2.635	4.216
$C_6H_5CH_3$	2.38	−1.249	−0.987	−0.423	1.257	1.503	2.713
C_6H_6	2.28	−1.881	−1.440	−0.868	0.700	0.993	2.562
$CHCl_3$	4.81	−2.220	−1.503	−1.270	0.370	0.560	1.700

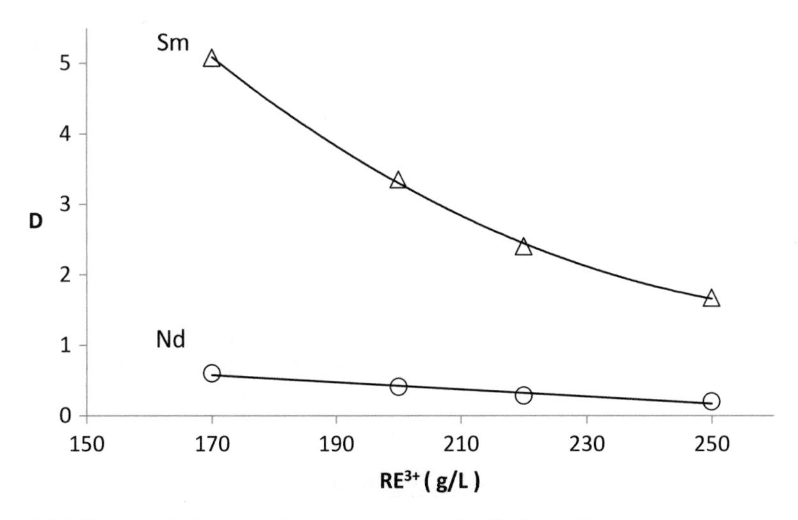

Fig. 4.16 Effects of feed rare earth concentration on distribution ratios

4.2.2.6 Temperature

The rare earth extraction in HDEHP is exothermal. Higher temperature does not favor extraction. Rare earth distribution ratio and extraction speed all decrease with the increase in temperature.

4.2.3 Group Separation of Mixed Rare Earth

HDEHP forms stable complexes with heavy rare earths, especially Tm, Yb, and Lu. It is difficult to strip the stable complexes from the organic solution. Therefore, it is normally used for group rare earth separation and individual light rare earth separation. Figure 4.17 shows a schematic solvent extraction process of group rare earth element separation (Chen 1993; Li 2005). The feed contains 1.0–1.2 mol/L

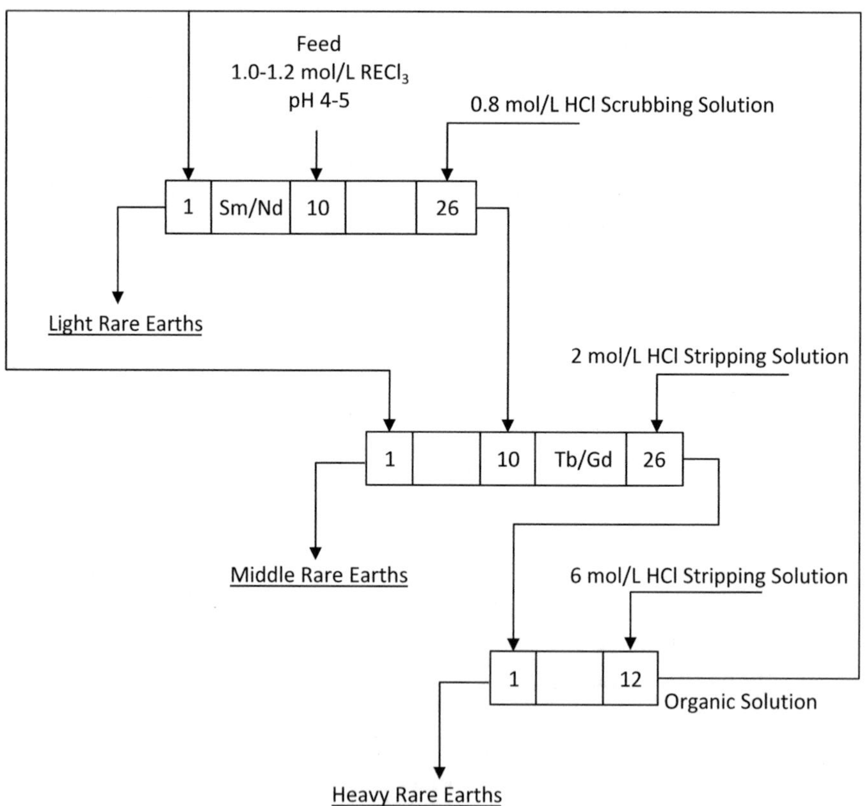

Fig. 4.17 Schematic process of group rare earth separation with HDEHP

$RECl_3$ with pH 4–5. The organic solution is 1.0 mol/L HDEHP-kerosene. The first step is group separation between Nd and Sm by 10 stages of extraction and 16 stages of scrubbing. The volume ratio of barren organic solution/feed solution/scrubbing solution is 2.5/1.0/0.5. The light rare earths will stay in the raffinate while the middle and heavy rare earths are in the pregnant organic solution. The second step is to use 2 mol/L HCl solution to strip the pregnant organic solution to realize the group separation between Tb and Gd. A small stream of barren organic solution is used to remove the entrained heavy rare earths in the stripping solution to purify the middle rare earths. The volume ratio of barren organic solution/pregnant organic solution/stripping solution is 0.25/2.0/0.25. The third step is to strip the heavy rare earth out of the organic solution. High-acidity stripping solution containing 6.0 mol/L HCl is used to strip the heavy rare earths out of the organic solution which will be recycled back to the process as barren organic solution.

In operation, it is important to control flow rates of aqueous solution and organic solution as well as the rare earth concentration in them in order to control the rare earth extraction percentage which is determined according to the feed composition and separation target.

According to the rare earth distribution in the feed, an alternative group separation process is adopted using mixer-settlers with internal aqueous recirculation for the middle-heavy rare earth separation. As shown in Fig. 4.18, high-acidity stripping solution with 7 mol/L HCl is introduced in the 34th stage. Two heavy rare earth concentrates are produced. Heavy rare earth concentrate rich in Y_2O_3 is produced from the 24th stage and heavy rare earth concentrate rich in Tb_2O_3 from the sixth stage. The group of middle rare earth is produced from the first stage (Deng et al. 1990). Due to the strong extraction capability of HDEHP to RE^{3+} and the high-level impurity in feed, saponification can cause emulsion issues. Therefore, normally it is not used in Sm/Nd group separation.

In the "third-generation rare earth separation method" developed by the General Research Institute of Nonferrous Metals in Beijing, China, purified rare earth H_2SO_4 solution is directly separated by HDEHP (Zhang et al. 1988). Following similar schematic processes shown in Figs. 4.17 and 4.18, Sm/Nd group separation is performed in H_2SO_4 solution with <50 g REO/L at pH 4 to 0.315 mol/L H_2SO_4.

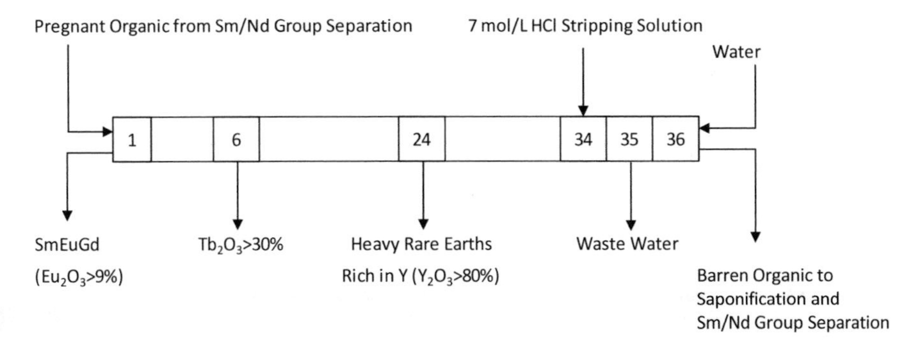

Fig. 4.18 Middle-heavy rare earth group separation process

0.25–1.0 mol/L H_2SO_4 solution is used as scrubbing solution. HDEHP concentration in organic solution is 1.0–1.5 mol/L. Light rare earth group and middle rare earth group can be produced directly from the H_2SO_4 solution by adjusting the acidity of stripping solution. Conversion of rare earth sulfate to rare earth chloride and HDEHP saponification are avoided in this process.

4.2.4 Individual Separation of Light Rare Earth

HEH/EPH has been widely used in light rare earth group separation and individual separation. However, HDEHP is still used in many light rare earth separation processes due to its low cost in comparison with HEH/EPH (Qiao et al. 2002). The group of light rare earths from Sm/Nd group separation can be further separated using HDEHP. Table 4.20 shows an example of light rare earth composition after Sm/Nd group separation. It contains over 50 % CeO_2 and less than 0.01 % Sm_2O_3. Figure 4.19 shows the schematic LaCe/PrNd separation process (Hou 2005). Using saponified HDEHP as extractant, the LaCe/PrNd in HCl solution can be well separated into two groups with ≥99.95 % purity.

Pr/Nd separation is relatively difficult due to their small separation factor. Saponification can improve the loading capacity of HDEHP, the distribution ratios of Pr and Nd, and the separation factor between Pr and Nd. The separation factor of Nd/Pr in saponified HDEHP is between 1.30 and 1.38. Where further Pr/Nd separation is required, the separation process shown in Fig. 4.20 with staged stripping can be used to produce pure Nd product with >98 % Nd_2O_3 where a small recycled organic stream is used to extract the Nd from the Pr stripping solution. It includes ten stages of Nd extraction, ten stages of Pr stripping, and ten stages of Nd stripping (Li 2011). Where low Sm_2O_3 impurity is required in the

Table 4.20 Composition of light rare earths from Sm/Nd group separation

La_2O_3, %	CeO_2, %	Pr_6O_{11}, %	Nd_2O_3, %	Sm_2O_3, %
28.17	51.61	4.29	15.93	<0.01

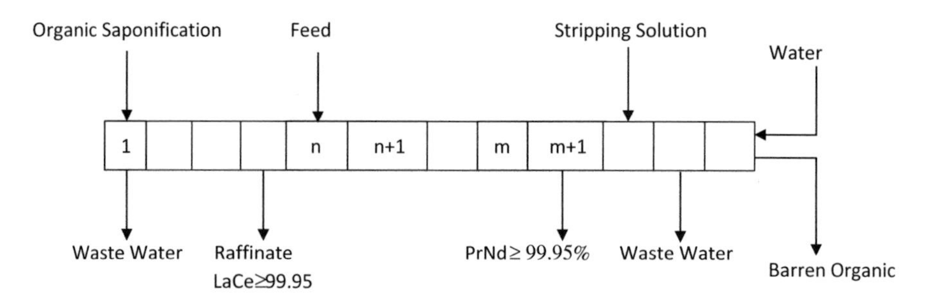

Fig. 4.19 LaCe/PrNd separation using 1.0 mol/L HDEHP-kerosene

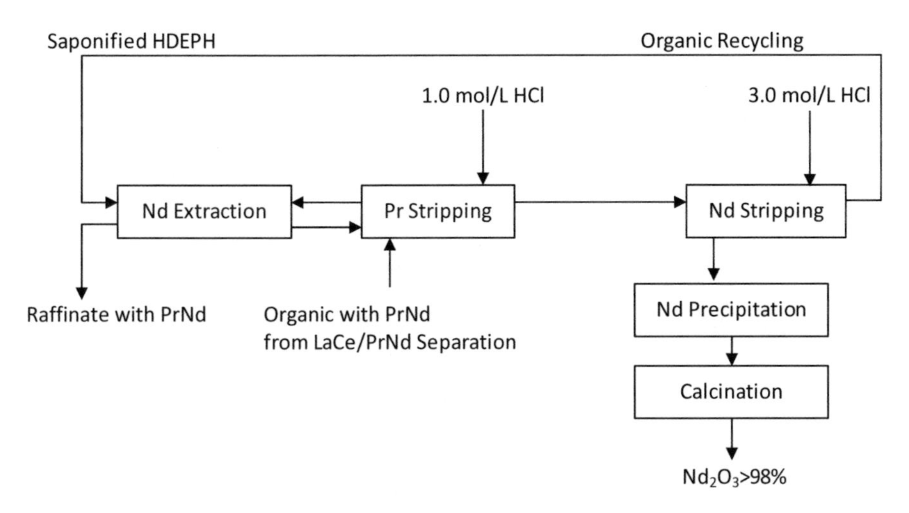

Fig. 4.20 Pr/Nd separation using 1.0 mol/L HDEHP-kerosene with 30–31 % saponification

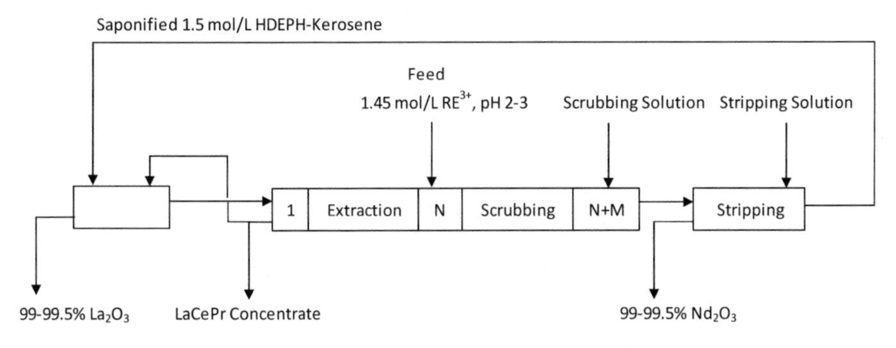

Fig. 4.21 La/CePr/Nd separation process using 1.5 mol/L HDEHP in kerosene

Nd_2O_3 product, Sm extraction is normally performed on the Nd stripping solution to remove Sm prior to Nd precipitation (Yang et al. 1998).

An alternative Pr/Nd separation process is shown in Fig. 4.21 using three-outlet separation technology without LaCe/PrNd group separation. The feature of this process is to add a rare earth saponification circuit to treat the LaCePr raffinate to realize partial La/CePr separation. A La product is produced with purity between 99.0 % and 99.5 %. The Sm/Nd group separation raffinate with the composition shown in Table 4.20 is the feed solution. The organic solution contains 1.5 mol/L HDEHP in kerosene with 31 % saponification (Zang and Wang 1995).

HDEHP is also used for Nd recovery and purification from secondary resources such as waste magnetic materials containing Nd (Fan et al. 2002). The Nd produced from the secondary resources normally contains high non-rare earth impurities such as Fe, Ca, and S. Totally 18 stages of extraction, scrubbing, and stripping are used in the Nd purification process shown in Fig. 4.22. The organic solution contains

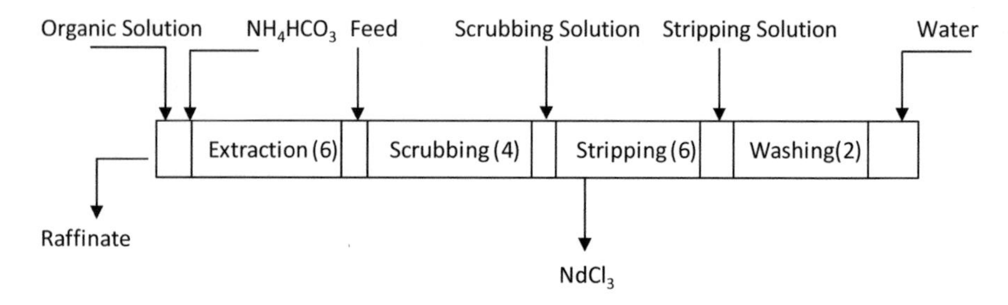

Fig. 4.22 Nd purification process using 1.5 mol/L HDEHP-kerosene

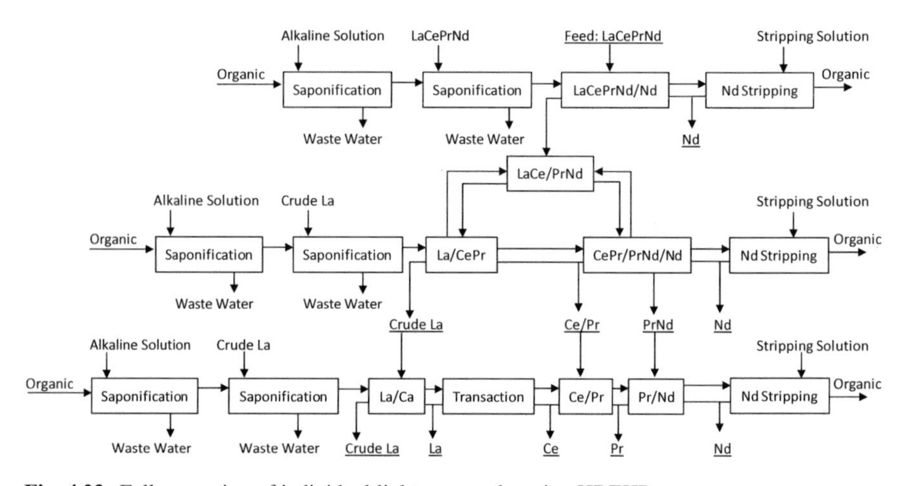

Fig. 4.23 Full separation of individual light rare earths using HDEHP

1.5 mol/L HDEHP in kerosene and the feed contains 0.6 mol/L RE^{3+}. HCl is used in scrubbing solution and stripping solution at 2 mol/L and 4 mol/L, respectively. The ratio of organic solution, NH_4HCO_3 solution, feed solution, scrubbing solution, stripping solution, and water is 15/3.75/2/1/1.1/0.25.

Yang et al. (2013) developed a full separation process to produce high-purity individual light rare earth elements. As shown in Fig. 4.23, the process incorporates many new technologies developed in rare earth separation in China including fuzzy separation, combined linkage extraction, replacement extraction, three-outlet technology, organic feeding, scrubbing-stripping common feeding, and rare earth saponification technology. In comparison with traditional light rare earth separation, significant reduction of reagent consumption and waste water generation can be achieved to reduce rare earth production cost.

A feed with low Ce is separated by the process shown in Fig. 4.23 to produce high-purity targeted products. Both feed and product composition are shown in Table 4.21. The process parameters are shown in Table 4.22.

Table 4.21 Composition of light rare earth feed for individual element separation

REO	La_2O_3	CeO_2	Pr_6O_{11}	Nd_2O_3	Sm_2O_3
Feed (wt%)	43–49	3.8–4.1	10–12	36–41	<0.01
Product (wt%)	>99.99	>99.99	>99.9	>99.9	

Table 4.22 Process parameters of individual light rare earth separation

		Number of stage		Flow ratio		
Step	Circuit	Extraction	Scrubbing	Organic	Feed	Scrubbing solution
1	LaCePr/Nd	10	50	70.37	14.66	5.50
2	LaCePr/CePrNd	20	20	73.66	12.81	3.74
	La/CePr	30	50	37.83	8.45	2.82
	CePr/PrNd/Nd	30	40		Organic feeding	5.41
3	Ca/La	Scavenging La	30	66.85	7.29	0.61
	La/Ce	60			Transaction	6.56
	Ce/Pr	35	25		1.16	6.17
	Pr/Nd	30	50		1.95	5.83

Table 4.23 Composition of light rare earth feed for individual element separation

(La-Nd)xOy, %	Sm_2O_3, %	Eu_2O_3, %	Gd_2O_3, %	(Tb-Lu, Y)$_mO_n$, %
<0.015	46–66	6–12	23–47	<0.005

4.2.5 *Individual Separation of Middle Rare Earths*

EHE/EPH has been widely used for individual middle rare earth separation since 1980s. However, due to its high cost, new research and development work using HDEHP never stopped. A process using external reflux and three-outlet technology was developed (Hao 1995; Hao et al. 1995) to separate the middle rare earths after Sm/Nd group separation. The feed composition is shown in Table 4.23. The organic solution is saponified HDEHP at 1.5 mol/L in kerosene.

As shown in Fig. 4.24, external reflux is used in the same stage as the scrubbing solution. Reflux is to send partial stripping solution containing rare earth to the stage where scrubbing solution is fed. The reflux can increase the product purity. Without reflux, the reactions expressed by Eqs. (4.37) and (4.38) are the major reactions occurring in the stripping:

$$\mathrm{Gd(HL_2)_{3(o)}} + 3\mathrm{H^+_{(a)}} = 3\mathrm{(HL)_{2(o)}} + \mathrm{Gd^{3+}_{(a)}} \tag{4.38}$$

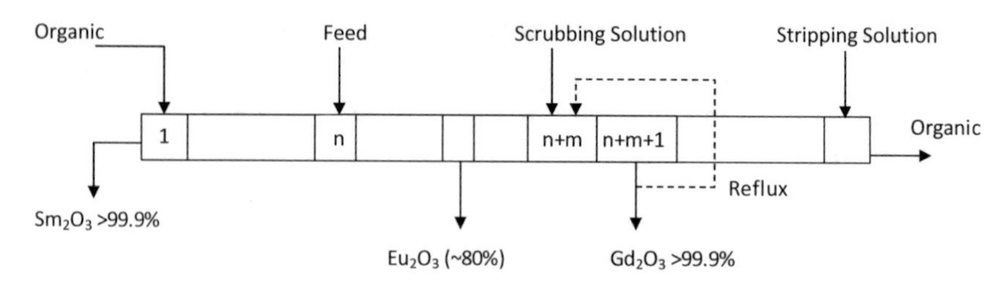

Fig. 4.24 Sm/Eu/Gd separation process with external reflux

$$Eu(HL_2)_{3(o)} + 3H^+_{(a)} = 3(HL)_{2(o)} + Eu^{3+}_{(a)} \tag{4.39}$$

With reflux of partial stripping solution containing Gd^{3+}, reaction Eq. (4.40) will occur leading to the replacement of Eu by Gd in extractant-metal complexes in the organic solution. The Eu to stripping stage is reduced consequently leading to the improvement of Gd purity. Due to less Eu to stripping, more Eu is scrubbed to Eu concentrate to increase its concentration:

$$Eu(HL_2)_{3(o)} + Gd^+_{(a)} = Gd(HL_2)_{3(o)} + Eu^{3+}_{(a)} \tag{4.40}$$

4.2.6 *Scandium Recovery*

Scandium is classified as one of the rare earth elements due to its chemical and physical similarity to lanthanides. However, scandium is distributed sparsely and occurs in trace amounts in many deposits. Scandium is produced as by-product from the extraction of other elements. The current world scandium production is about 2000 kg/year in the form of Sc_2O_3. About 50 % is produced in Baiyun Obo Mine in China. Scandium can render special properties of functional materials. However, the lack of reliable supply has limited its commercial application. Therefore, the research and development of scandium extraction process is of great interest.

He et al. (1991) conducted detailed research on the extraction mechanism of scandium by HDEHP from HCl acid solution. At low acidity ($HCl < 5$ mol/L), the extraction of scandium with HDEHP is cation-exchanging mechanism:

$$Sc^{3+}_{(a)} + 3(HL)_{2(o)} = Sc(HL_2)_{3(o)} + 3H^+_{(a)} \tag{4.41}$$

At high acidity ($HCl > 6$ mol/L), there is salvation and competing HCl extraction.

Due to the low concentration of scandium, the relative impurity concentration is high. Therefore, the scandium concentrating and separation with other impurities

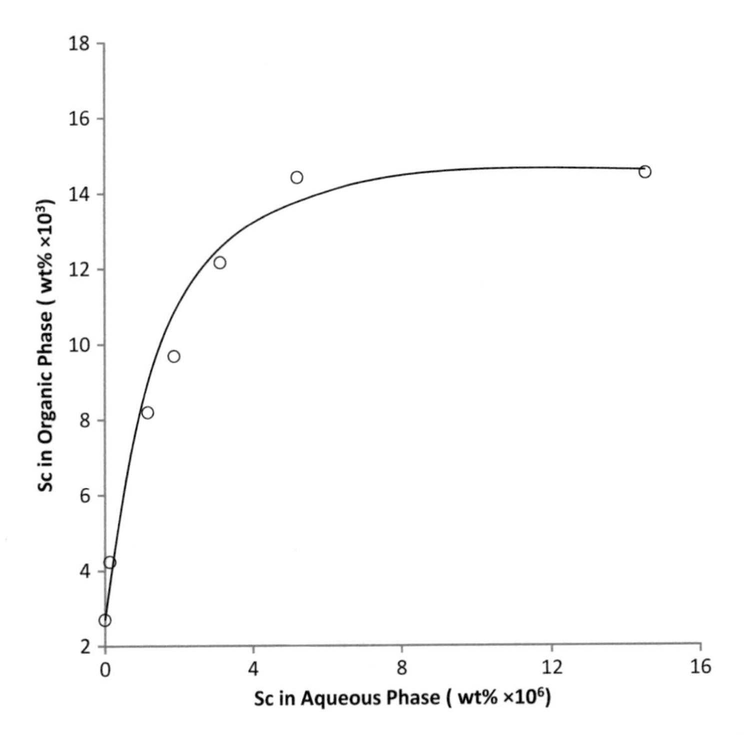

Fig. 4.25 Sc extraction isotherm in 25 % HDEHP + modifier +60 % kerosene at 25 °C and O/A = 0.05

are important. Major coexisting elements include Fe, Ti, Mn, Cr, and V. In 0.05 mol/L HDEHP-kerosene system, the separation factors of Sc to other elements are normally large in the order of 1000 (Wang 1998). At HCl concentration of 5 mol/L, separation factor of Sc/Ti is about 100 and Sc/Fe higher than 1000. Over 99.5 % scandium can be extracted with only about 9 % Ti forming stable complex with H_2O_2 in 0.7–1.8 mol/L H_2SO_4 solution. HDEHP does not extract the Ti-H_2O_2 complexes. Therefore, the extracted Ti can be scrubbed by 4 % H_2O_2-1 mol/L H_2SO_4 solution. Due to the low extraction of Fe with Sc, low-alkaline solution containing 0.2 mol/L NaOH can be used to strip the Fe out of the organic phase. Eventually, scandium can be stripped by 3.5 mol/L NaOH solution. Over 96 % scandium can be stripped. Total scandium recovery is about 95.5 %.

As shown in Fig. 4.25, Tian et al. (1998) measured the scandium extraction isotherms at room temperature and O/A = 0.05/1. It shows good distribution ratio of Sc. Good separation of Sc with Fe and Mn is achieved at aqueous acidity of 4.0 mol/ HCl and O/A ratio = 0.05. At the above conditions, the distribution ratio, separation factor, and percentage of extraction are measured and shown in Table 4.24.

Table 4.24 Distribution ratio, separation factor, and percentage of extraction of Sc, Fe, and Mn

Element	Distribution ratio	Separation factor	Percentage of extraction
Sc	2800		99.30
Fe	0.312	$\beta_{Sc/Fe} = 8970$	1.53
Mn	0.320	$\beta_{Sc/Mn} = 8750$	1.56

4.2.7 *Th and Other Non-rare Earth Impurity Separation*

It is common that the rare earth solution for extraction separation contains thorium and other non-rare earth impurities. Precipitation method is normally used to remove these impurities so that extraction can be used for further separation. In addition to rare earth extraction, HDEHP is also used for impurity removal.

In HNO_3 solution, the extraction of Th^{4+} is considered as follows (Gao et al. 2012):

$$Th^{4+}_{(a)} + 2NO^{-}_{3(a)} + (HL)_{2(o)} = Th(NO_3)_2L_{2(o)} + 2H^{+}_{(a)} \tag{4.42}$$

In HCl solution, the extraction of Th^{4+} follows cation-exchange mechanism:

$$Th^{4+}_{(a)} + 4(HL)_{2(o)} = Th(HL_2)_{4(o)} + 4H^{+}_{(a)} \tag{4.43}$$

At 6 mol/L HCl and O/A = 4/1, the extraction percentage of Th^{4+} increases rapidly with the increase of HDEHP concentration. However, the increase of Th^{4+} extraction becomes slower with the increase of HDEHP concentration higher than 5 %. The percentage of Th^{4+} extraction does not increase with the increase in HDEHP concentration higher than 20 % where Th extraction reaches 97 % (Ding et al. 2007). Solution with $\geq$0.5 mol/L Na_2CO_3 solution can be used to strip Th. Over 96 % Th can be recovered by Na_2CO_3 solution.

In rare earth operations where "third-generation rare earth separation method" is used, HDEHP is commonly used for rare earth group separation in H_2SO_4 solution. Wang et al. (2002) investigated the non-rare earth impurity removal during the rare earth separation in HDEHP-H_2SO_4 system. The removal of impurities is performed during scrubbing. Using scrubbing solutions with 0.2 mol/LH^+ at the O/A ratio of 3/1, Ca and Mg impurities are reduced significantly to 0.057 % and 0.023 % after six stages of countercurrent scrubbing.

4.3 HBTMPP Extraction System

Bis(2,4,4-trimethylpentyl) phosphinic acid is a colorless to light amber liquid with a molecular weight of 290. It is often denoted by HBTMPP or BTMPPA. Trade name Cyanex 272 is commonly seen in publications. It has the chemical formula of

$C_{16}H_{35}O_2P$ with the following structure. It is totally miscible with common aromatic and aliphatic diluents.

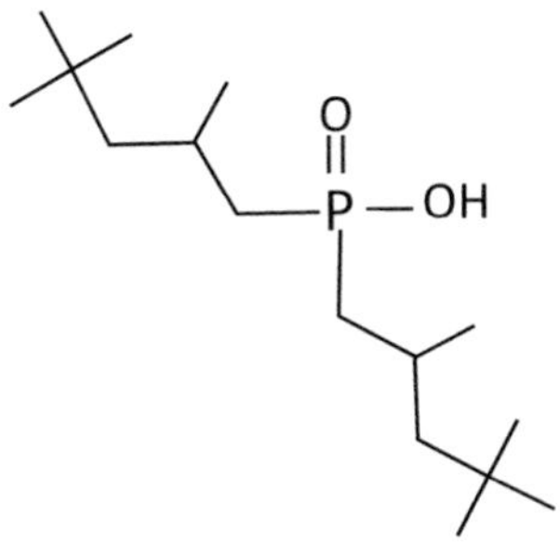

4.3.1 *HBTMPP Extracting Rare Earths*

HEH/EHP, HDEHP, and HBTMPP are acidic extractants. They all can be denoted as HL. In comparison with HEH/EHP and HDEHP, HBTMPP has higher acid-ionization constant. Therefore, extraction and stripping of metals can be done at a relatively low acidity. Zhang and Li (1993) conducted investigation on the extraction of 15 rare earth ions in purified HBTMPP-octane-HCl system. The rare earth extraction follows cation-exchanging mechanism. Equations 4.27–4.31 also apply to the HBTMPP-RE^{3+} extraction. The rare earth distribution ratio decreases with the increase in aqueous acidity. Their extraction follows the order of Le < Ce < Pr < Nd < Sm < Eu < Gd < Tb < Dy < Ho < Y < Er < Tm < Yb < Lu. They measured the separation factors as shown in Table 4.25 at the conditions of pH = 2.8, RE^{3+} concentration = 4.0×10^{-4} mol/L, and HCl concentration = 2.0×10^{-2} mol/L, O/A = 1/1, and 25 ± 1 °C. The average separation factor of adjacent rare earths $\beta_{z+1/z}$ is 3.24 which is higher than that in HEH/EHP and HDEHP systems. It indicates that HBTMPP has better selectivity towards different rare earth ions. However, the actual separation is affected by many conditions such as aqueous acidity, organic composition, aqueous composition, temperature, O/A ratio. Liao et al. (2007) investigated the extraction of Nd and Y in HCl solution by purified HBTMPP in n-heptane at the O/A = 1/2. The organic solution contains 0.102 mol/L HBTMPP in n-heptane. The aqueous solution contains 0.01 mol/L $NdCl_3$ and 0.01 mol YCl_3 with 1.0 mol/L NaCl to maintain the ionic strength of the solution. The separator factor between Y and Nd is 12–26 at pH 1–3 and temperature 25–35 °C. With multiple stages of extraction, Y and Nd can be well separated.

Wang et al. (1995) compared the extraction of rare earth in HBTMPP-HCl system and HBTMPP-HNO_3 system. The HBTMPP concentration in kerosene is 0.838 mol/L. The rare earth extraction in HNO_3 solution follows the same distribution rule as in HCl solution. The extractability of HBTMPP in HNO_3 solution is higher than in HCl solution. The stripping curves of heavy rare earths (Ho, Er, Tm,

Table 4.25 Separation factors of rare earths in HBTMPP-HCl system

β	Ce	Pr	Nd	Sm	Eu	Gd	Tb	Dy	Ho	Y	Er	Tm	Yb	Lu
La	11.7	39.7	51.4	679	910	1052	3389	6515	14525	23388	28056	74655	179077	236073
Ce		3.39	4.39	58.0	77.8	89.9	290	557	1241	1998	2397	6379	15301	20171
Pr			1.30	17.1	23.0	26.5	85.4	164	366	590	707	1882	4515	5952
Nd				13.2	17.7	20.5	66.0	127	283	455	546	1453	3486	4596
Sm					1.34	1.55	4.99	9.59	21.4	34.4	41.3	110	264	347
Eu						1.16	3.72	7.16	16.0	25.7	30.8	8.0	197	259
Gd							3.22	6.19	13.8	22.2	26.7	71.0	170	224
Tb								1.92	4.29	6.90	8.28	22.0	52.8	69.7
Dy									2.23	3.59	4.31	11.4	27.5	36.2
Ho										1.61	1.93	5.14	12.3	16.2
Y											1.20	3.19	7.66	10.1
Er												2.66	6.38	8.41
Tm													2.40	3.16
Yb														1.32

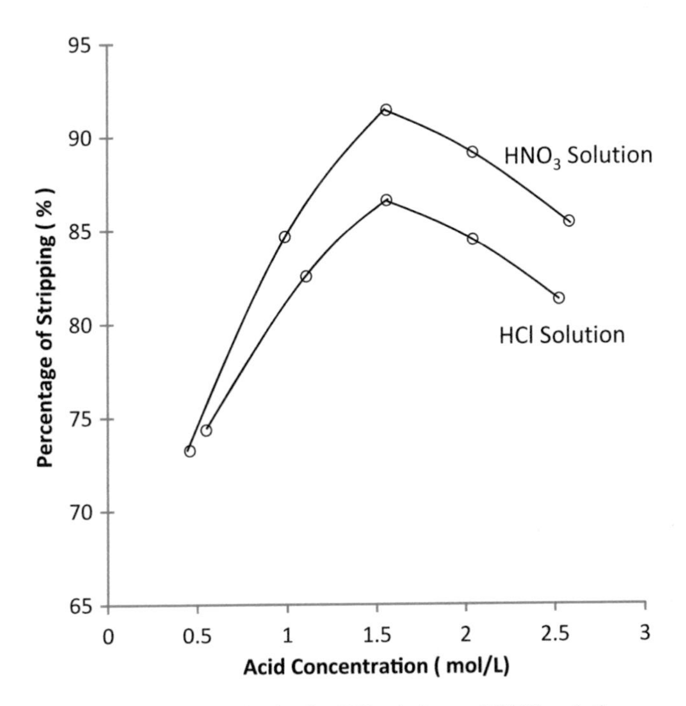

Fig. 4.26 Mixed heavy rare earth stripping by HCl solution and HNO_3 solution

Yb, Lu, and Y) by HCl solution and HNO_3 solution are shown in Fig. 4.26. The percentage of stripping reaches maximum at 1.5 mol/L acid concentration. HNO_3 has higher stripping capability than HCl. Kinetics study conducted on Er and Y indicates that the extraction process using HBTMPP in HCl solution may be governed by a mixed mechanism of diffusion and chemical reaction (Lu et al. 1998a, b; Zhang et al. 1999a; Xiong et al. 2004; 2006a, b).

Using 60 % saponified HBTMPP by sodium hydroxide, El-Hefny et al. (2010) investigated the extraction of Sm^{3+} in HCl solution. Based on slope analysis, the extraction is considered taking place as follows:

$$2Na^+_{(a)} + 2(HL)_{2(o)} = 2NaL \cdot 2HL_{(o)} + 2H^+_{(a)} \tag{4.44}$$

$$Sm(OH)^{2+}_{(a)} + 2NaL \cdot 2HL_{(o)} = Sm(OH)L_2 \cdot 2HL_{(o)} + 2Na^+_{(a)} \tag{4.45}$$

4.3.2 *HBTMPP Extracting Sc^{3+} and Fe^{3+}*

Ma and Li (1992) investigated the extraction and Sc^{3+} and Fe^{3+} using HBTMPP in HCl solution. The same as other rare earth ions, at low acidity the extraction of Sc^{3+} follows cation-exchange mechanism. Equation (4.41) can be applied to the

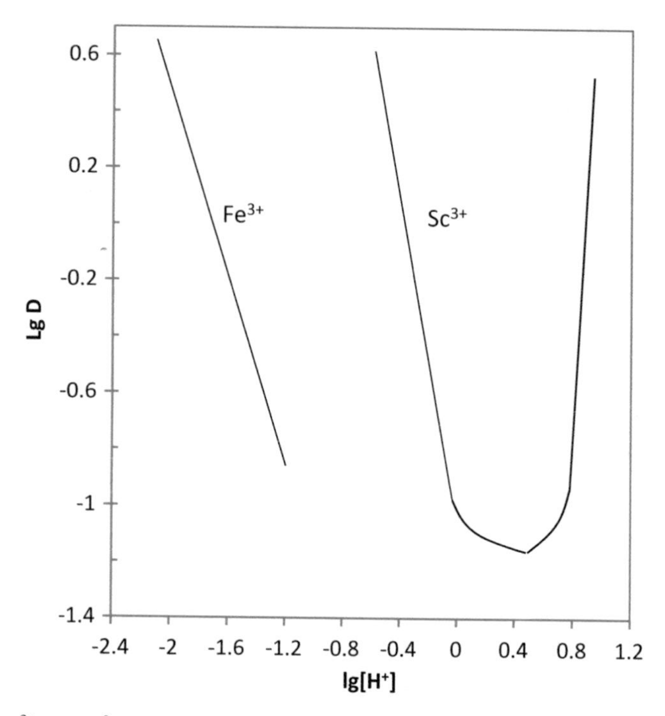

Fig. 4.27 Sc^{3+} and Fe^{3+} distribution ratio as a function of acidity

extraction by HBTMPP. Purified HBTMPP is diluted in n-octane to make 0.0196 mol/L organic solution. The aqueous solution contains 18 ppm of Sc^{3+} and 24 ppm of Fe^{3+}. At various acidity and O/A = 1/1, the lg$D$-lg[H^+] relation is shown in Fig. 4.27. When HCl is between 3.5 mol/L and 4.0 mol/L, there is no Sc^{3+} extraction. This can be utilized by stripping. At the conditions of Sc^{3+} extraction, there is no extraction of Fe^{3+}. This indicates that good separation between Sc^{3+} and Fe^{3+} can be realized in a relatively large acidity range.

4.3.3 HBTMPP Extracting Th^{4+}

The extraction mechanism of Th^{4+} by HBTMPP in HCl acid solution was investigated by Li and Li (1995). The effects of acidity on Th^{4+} extraction at different HBTMPP concentrations and ionic strengths are shown in Fig. 4.28, where the concentration of Th^{4+} is 90 ppm and O/A = 1/1.

In the system where ionic strength is not controlled, the lg$D$-pH lines have the slope of 3. The extraction equilibrium can be represented by the following equation:

$$Th^{4+}_{(a)} + Cl^-_{(a)} + 3(HL)_{2(o)} = ThClL_3 \cdot (HL)_{3(o)} + 3H^+_{(a)} \tag{4.46}$$

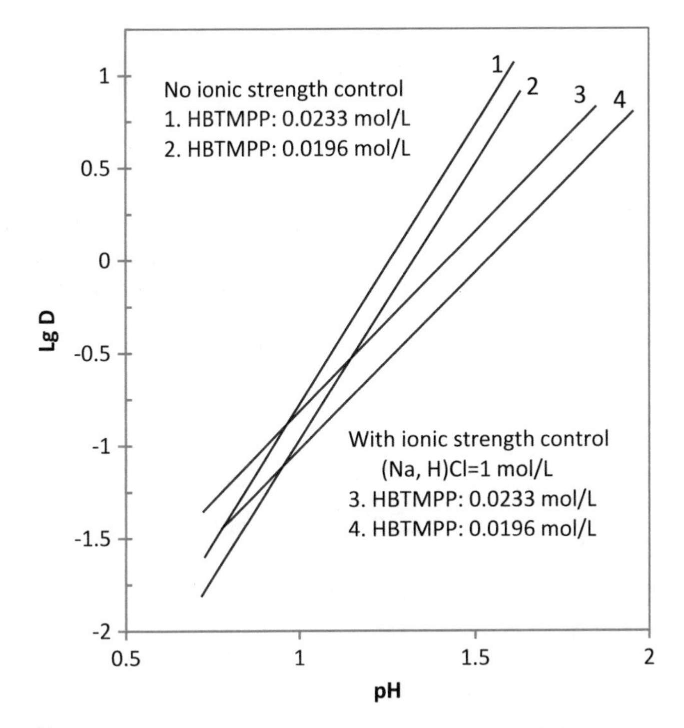

Fig. 4.28 Th^{4+} distribution ratio as a function of aqueous pH at equilibrium

When the ionic strength is controlled at 1 mol/L, the slope of lg*D*-pH lines is 2. The Th^{4+} extraction is deemed as follows:

$$Th^{4+}_{(a)} + 2Cl^{-}_{(a)} + 3(HL)_{2(o)} = ThCl_2L_3 \cdot (HL)_{3(o)} + 2H^{+}_{(a)} \quad (4.47)$$

In either case, Cl^- participates the Th^{4+} extraction in HCl solution by HBTMPP. Slope analysis on the lg*D*-lg[Cl^-] relationship indicates that one or two Cl^- ions participated in the extraction reaction. The investigation on temperature effects on equilibrium shows that the Th^{4+} extraction by HBTMPP is endothermic. High temperature favors the extraction of Th^{4+}.

4.4 TBP Solvent Extraction System

4.4.1 TBP Extracting Rare Earth

TBP is a colorless and odorless neutral organophosphate extractant with the chemical formula of $(CH_3CH_2CH_2CH_2O)_3PO$ with the structure shown below.

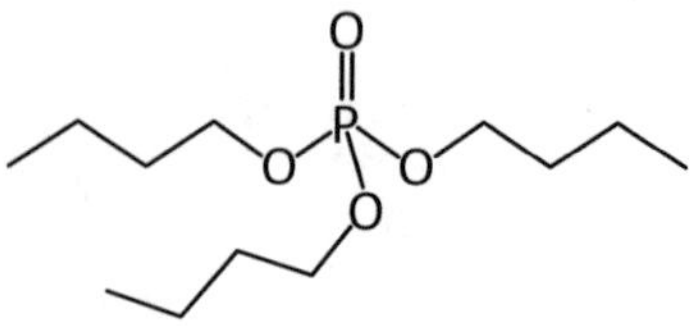

It forms neutral complex with rare earth ions. The extraction follows ion pair solvent extraction mechanism. In the extraction of rare earth nitrate salts, neutral complexes of $RE(NO_3)_3 \cdot 3TBP$ are formed through coordinate bonds (Eq. 4.48):

$$RE(NO_3)_{3(a)} + 3TBP_{(o)} = RE(NO_3)_3 3TBP_{(o)} \tag{4.48}$$

The extraction equilibrium constant is

$$K_{ex} = \frac{\left[RE(NO_3)_3 3TBP_{(o)}\right]}{\left[RE(NO_3)_{3(a)}\right] \times \left[TBP_{(o)}\right]^3} \tag{4.49}$$

Assuming the following dissociation equilibrium exists

$$RE(NO_3)_{3(a)} = RE^{3+}_{(a)} + 3NO^-_{3\ (a)} \tag{4.50}$$

The dissociation equilibrium constant is

$$K_a = \frac{\left[RE^{3+}_{(a)}\right] \times \left[NO^-_{3\ (a)}\right]^3}{\left[RE(NO_3)_{3(a)}\right]} \tag{4.51}$$

Therefore, the distribution ratio of the rare earth is

$$D = \frac{[9RE(NO_3) \cdot 3TBP_{(o)}]}{\left[RE^{3+}_{(a)}\right]} = \frac{K_{ex}}{K_a} \cdot \left[NO^-_{3(a)}\right]^3 \cdot \left[TBP_{(o)}\right]^3 \tag{4.52}$$

According to Eq. (4.47), the rare earth distribution ratio depends on the free TBP concentration and the nitrate ligand concentration. At low HNO_3 concentration, the concentration of NO_3^- increases with the increase of HNO_3 concentration. The distribution ratio of rare earth will increase. However, with the continuous increase of HNO_3 concentration, the concentration of TBP-HNO_3 complex will increase resulting in the decrease of the free TBP and then the rare earth distribution ratio. At high HNO_3 concentration, the rare earth distribution ratio increases due to the salting effects of water where H^+ and NO^- form hydration ions $H^+(H_2O)_x$ and $NO_3^-(H_2O)_y$. The consumption of free water molecules leads to the increase of effective rare earth concentration in the aqueous phase and then the distribution ratio.

Table 4.26 Separation factors of adjacent rare earth elements in 50 % (V/V) TBP-diluent

Aqueous REO g/L	Ce/ La	Pr/ Ce	Nd/ Pr	Sm/ Nd	Gd/ Sm	Dy/ Gd	Ho/ Dy	Er/ Ho	Yb/ Er
460	1.7	1.75	1.50	2.26	1.01	1.45	0.92	0.96	0.81
430	n/a	n/a	n/a	n/a	n/a	1.20	0.96	0.65	n/a
310	n/a	n/a	n/a	2.04	1.07	1.17	0.94	0.82	n/a
220	n/a	n/a	n/a	1.55	0.99	1.08	0.89	0.78	n/a
125	n/a	n/a	n/a	1.58	0.82	0.92	0.83	0.72	n/a
60	n/a	n/a	n/a	1.40	0.78	0.89	0.77	0.70	0.63

Table 4.27 Separation factors of lanthanide elements to yttrium in 50 % (V/V) TBP-diluent

Aqueous REO g/L	Ce/Y	Pr/Y	Nd/Y	Sm/Y	Gd/Y	Dy/Y	Ho/Y	Er/Y	Yb/Y
460	n/a	n/a	0.39	0.88	0.89	1.30	1.20	1.15	0.93
430	n/a	n/a	0.50	n/a	1.14	1.37	1.31	0.85	n/a
310	n/a	n/a	0.69	1.41	1.51	1.76	1.65	1.35	n/a
220	0.60	n/a	1.30	0.02	1.99	2.15	1.91	1.48	0.83
125	0.75	n/a	1.77	2.79	2.29	2.10	1.75	1.25	0.83
60	n/a	n/a	4.08	5.71	4.44	3.96	3.03	2.13	n/a

In the separation of lanthanide–HNO_3 aqueous solution using 50 % V/V TBP diluent, the separator factors of adjacent rare earth elements are shown in Table 4.26 (Wu 1988). At higher REO concentrations, the separator factors are higher. To maintain a higher REO aqueous concentration favors the separation. However, the separation factors of lanthanide elements to yttrium are higher at lower REO concentrations as shown in Table 4.27 (Wu 1988).

In general, the distribution ratio increases with the increase of charges of cations when TBP is used as rare earth extractant. The distribution ratio decreases with the increase in radius of cations that have the same charges. In the separation of rare earth elements using TBP, it is common to add $LiNO_3$ or NH_4NO_3 as salting agents to improve rare earth distribution ratio or separation factor of adjacent rare earth elements. For example as shown in Fig. 4.29, the extraction of yttrium increases with the increase in $LiNO_3$ concentration (Xu and Yuan 1987).

The effects of salting agents on rare earth solvent extraction are mainly due to the ion hydration that reduces the free water molecules. The reduction of free water molecules effectively improves the activity of rare earth ions by reducing their hydration or the hydrophilicity. At the same concentration, cations with higher positive charges have stronger salting-out effects. For cations with the same charge, smaller ones have stronger salting-out effects due to their stronger hydration tendency. As shown below, the salting effects of the cations follow the order:

$$Al^{3+} > Fe^{3+} > Mg^{2+} > Ca^{2+} > Li^{+} > Na^{+} > NH^{4+} > K^{+}$$

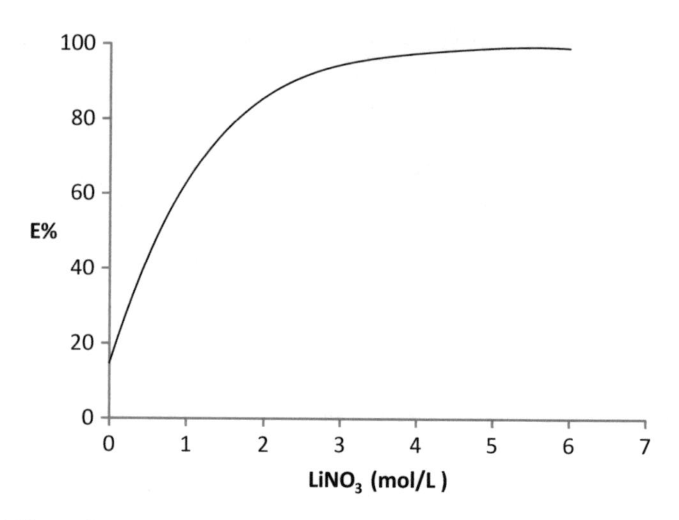

Fig. 4.29 Effects of $LiNO_3$ on yttrium extraction by 100 % TBP

However, many other factors such as effects on product and next stage separation have to be considered in the selection of salting agents.

As indicated by Table 4.26, the rare earth separation factors are not very high in TBP extraction system. Therefore, its application is limited. However, it can be used to concentrate La, Pr, and Nd. Using 50 % TBP-ShellsoA, three products of 98 % La, 90 % Pr, and 98 % Nd can be produced at the same time. La with purity of 99.99 % can be produced from La/Pr/Nd solution using 50 % TBP-Shellsol extraction system (Ma 1989; Wu 1988). Molycorp used to produce 99–99.99 % La_2O_3 through TBP-RE$(NO_3)_3$-HNO_3 separation system. This system has large separation factor but with high acid concentration (>10 mol/L HNO_3). La purification can also be done in TBP (kerosene)-NH_4SCN-HCl system to produce 99.9–99.99 % pure lanthanum product. However, this system is not commercially applied due to the safety concerns of ammonium thiocyanate (NH_4SCN) (Xu 2003; Li 2011).

4.4.2 *TBP Extracting Water and Acid*

While TBP extracts rare earth in an acidic aqueous solution, TBP extracts water and acid as well. As shown by Eq. (4.53), TBP reacts with H_2O through hydrogen bonding to form TBP-H_2O complex. One liter TBP can dissolve about 3.6 mol water:

$$TBP_{(o)} + H_2O = TBP \cdot H_2O_{(o)} \tag{4.53}$$

Table 4.28 TBP-inorganic acid complex species in 100 % TBP

Inorganic acid	H_2O equilibrium concentration, mol/L	TBP-inorganic acid complex
HNO_3	<4.0	$TBP \cdot HNO_3 \cdot H_2O$
HNO_3	>4.0	$TBP \cdot 2HNO_3 \cdot H_2O$
HCl	3.0–5.0	$(TBP)_2 \cdot HCl \cdot (H_2O)_6$
HCl	>6.0–8.0	$TBP \cdot HCl \cdot (H_2O)_3$
H_2SO_4	≤3.0	$TBP \cdot H_2SO_4 \cdot 2H_2O$
H_2SO_4	>3.0	$TBP \cdot H_2SO_4 \cdot H_2O$
HF	<3.0	$TBP \cdot HF$
HF	>3.0	$TBP \cdot (HF)_{2–4}$
HCNS	with HCl	$TBP \cdot HCNS$

TBP extracts not only organic acids but also forms complex with acids. Equation (4.54) shows the formation of TBP-HNO_3 complex:

$$TBP_{(o)} + HNO_3 = TBP \cdot HNO_{3(o)} \tag{4.54}$$

At high acid concentration (>4 M), TBP-2HNO_3 complex can form as follows:

$$TBP \cdot HNO_{3(o)} + HNO_3 = TBP \cdot 2HNO_{3(o)} \tag{4.55}$$

Table 4.28 lists the major extraction complex species of TBP-acid (Xu 2003; Li 2011). The extraction of acid by TBP follows the order of:

$$HCOOH{\sim}CH_3COOH > HClO_4 > HNO_3 > H_3PO_4 > HCl > H_2SO_4$$

This order is corresponding to the increase of hydration energy of the anionic ligands. For example, TBP has low extraction for SO_4^{2-} due to its high hydration energy (Xu and Yuan 1987).

4.4.3 TBP Extracting Uranium and Thorium

It is common that uranium and thorium are present in rare earth solution. For example in monazite processing, partial uranium and thorium are dissolved in nitric acid solution through the following reactions:

$$Na_2U_2O_7 + 6HNO_3 = 2UO_2(NO_3)_2 + 2NaNO_3 + 3H_2O \tag{4.56}$$

$$Th(OH)_4 + 4HNO_3 = Th(NO_3)_4 + 4H_2O \tag{4.57}$$

The extraction order of TBP for uranium, thorium, and rare earth is $UO_2^{2+} > Th^{4+} > RE^{3+}$ and the separation factors of uranium and thorium decrease with the

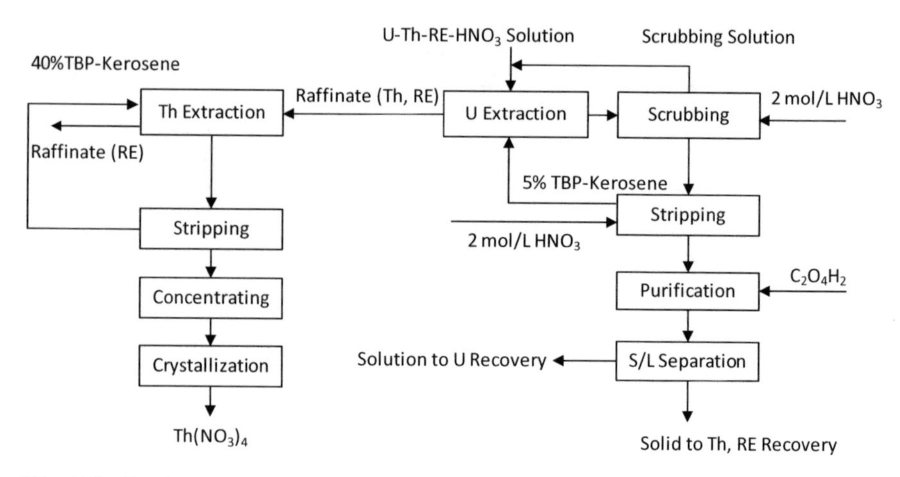

Fig. 4.30 Uranium, thorium, and rare earth separation process using TBP

increase of TBP concentration (Li 2011). Therefore, 5 % TBP-kerosene is used to extract uranium first:

$$UO^{2+}_{2(aq)} + 2NO^{-}_{3(aq)} + 2TBP_{(org)} = UO_2(NO_3)_2 \cdot 2TBP_{(org)} \tag{4.58}$$

And then, 40 % TBP-kerosene is used to extract thorium:

$$Th^{4+}_{(a)} + 4NO^{-}_{3(a)} + 2TBP_{(o)} = Th(NO_3)_4 \cdot 2TBP_{(o)} \tag{4.59}$$

Figure 4.30 shows the TBP solvent extraction process for the separation of uranium, thorium, and rare earth (Li 2011). The feed solution contains 3.5 mol/L HNO_3. The organic solution contains 5 % TBP diluted by kerosene. The organic/aqueous (O/A) ratio is 1/1. Ten stages of countercurrent extraction are used to extract the uranium from the solution. The organic solution containing uranium is scrubbed by 2 mol/L HNO_3 solution at O/A of 5/1 to remove entrained thorium and iron impurities. Uranium by-product is recovered after stripping and purification. The raffinate containing thorium and rare earth is fed to thorium extraction where 40 % TBO diluted by kerosene is used. The O/A ratio is 2–2.5/1. The organic containing thorium can be washed using a small amount of water followed by water stripping, concentrating, and crystallization to recover thorium by-product. The raffinate after thorium extraction containing the majority of rare earth is fed to rare earth recovery circuit.

The separation of uranium and thorium can also be done in HCl solution where the extraction follows the order of $Fe^{3+} > UO_2^{2+} > Th^{4+} > RE^{3+}$. In low-acidic solution, Fe^{3+} and UO_2^{2+} can be separated very well from Th^{4+} and RE^{3+}. Table 4.29 shows the distribution ratios and separation factors of uranium and thorium at various conditions (Li 2011).

Table 4.29 Distribution ratios and separation factors of uranium and thorium

TBP %	Salting agents	Distribution ratio		Separation factor $\beta_{U/Th}$
		D_U	D_{Th}	
5	HNO_3 (3 mol/L)	2020	48.8	42
5	NH_4NO_3 (2.5 mol/L)	2200	28.4	77
25	HCl (4 mol/L)	8330	1.6	5200
25	HCl (2 mol/L) + NH_4Cl (2 mol/L)	24700	2.7	9150
25	NH_4Cl (4 mol/L)	1640	n/a	

Aqueous solution: 1.01 g/L U and 1.45 g/L Th; O/A = 1/1

Table 4.30 Distribution ratio of Sc in TBP-HCl system

Extractant	HCl, mol/L	2	4	6	8	10	>11.5
100 % TBP at O/A = 1/1	D_{Sc}	0.02	0.08	1.9	41	125	>1100

The other TBP solvent extraction systems to separation thorium with rare earth include TBP-NH_4SCN-$CHCl_3$ system and TBP-NH_4SCN-CCl_4 (Wang et al. 1989). The former is mainly used to separate thorium and scandium and the latter is used to separate thorium with lanthanides.

4.4.4 TBP Extracting Scandium

TBP is used to extract scandium from HCl solution. Its distribution ratio increases with the increase in HCl acid concentration as shown in Table 4.30 (Xu 2003; Li 2011). TBP can also be used to extract scandium from its HNO_3 solution.

Isogawa et al. (2015) reported that TBP is used as a modifying agent for Versatic Acid 10 to improve the scandium separation from Sc^{3+}-Ti^{4+}-Zr^{4+} mixed solution. With the presences of TBP, the Sc^{3+} can be extracted at a higher pH 4.9. In stripping by 2 mol/L H_2SO_4 solution, the stripping of Sc^{3+} increased from 14 to 97 % when TBP is added to Versatic Acid 10.

4.5 P350 Solvent Extraction System

Di-(1-methylheptyl)methyl-phosphonate, often called by its trade name P350, is another important neutral organophosphate extractant commonly used in rare earth solvent extraction. It was first synthesized in China in 1964 for rare earth solvent extraction separation. It has a chemical formula of $CH_3PO(O(CH)CH_3C_6H_{13})_2$ and its structure is shown below.

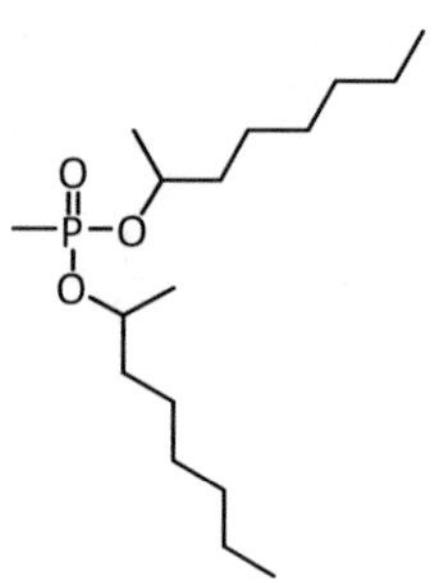

The same as TBP, P350 forms neutral extraction complex with rare earth ions through coordinate bonds. When P350 extracts rare earth nitrate, the following reaction is generally considered to occur:

$$RE(NO_3)_{3(a)} + 3P350_{(o)} = RE(NO_3)_3 \cdot 3P350_{(o)} \tag{4.60}$$

The extraction equilibrium constant is

$$K_{ex} = \frac{[RE(NO_3)_3 \cdot 3P350_{(o)}]}{[RE(NO_3)_{3(a)}] \times [P350_{(o)}]^3} \tag{4.61}$$

Assuming the same dissociation equilibrium (Eq. 4.50) exists with the same dissociation equilibrium constant (Eq. 4.51), the distribution ratio of the rare earth in P350 extraction system is

$$D = \frac{[RE(NO_3)_3 \cdot 3P350_{(o)}]}{[RE^{3+}_{(a)}]} = \frac{K_{ex}}{K_a} \cdot [NO_3^-]^3 \cdot [P350_{(o)}]^3 \tag{4.62}$$

where C_{P350} is the concentration of free P350 and $C_{NO_3^-}$ is the concentration of nitrate.

The same as TBP, as shown below, P350 also extracts water and acid while extracting rare earth:

$$P350_{(o)} + H_2O = P350 \cdot H_2O_{(o)} \tag{4.63}$$

$$P350_{(o)} + HNO_3 = P350 \cdot HNO_{3(o)} \tag{4.64}$$

The free P350 is the difference between initial P350 and reacted P350:

$$[P350_{(o)}] = [P350_{(o)}]_i - 3[RE(NO_3)_3 \cdot 3P350_{(o)}] - [P350 \cdot H_2O_{(o)}] - [P350 \cdot HNO_{3(o)}] \tag{4.65}$$

However, it was reported (Wang et al. 1986) that two P350 molecules formed complex with rare earth nitrate at pH 1.0 and rare earth concentration between 0.3 mol/kg and 0.9 mol/kg:

$$RE(NO_3)_{3(a)} + 2P350_{(o)} = RE(NO_3)_3 \cdot 2P350_{(o)} \tag{4.66}$$

This indicates that the rare earth extraction mechanism using P350 is affected by reaction conditions. Li and Cheng (1990) reported that at pH 2.5, P350 and rare earth react as follows:

$$RE(NO_3)_{3(a)} + nP350_{(o)} = RE(NO_3)_3 \cdot nP350_{(o)} \tag{4.67}$$

They found $n < 2$ when $RE(NO_3)_3$ concentration ≥ 1.0 mol/kg and $n = 2$ when $RE(NO_3)_3$ concentration < 1.0 mol/kg.

4.5.1 *Factors Affecting Rare Earth Extraction Using P350 (Xu 2003; Li 2011)*

4.5.1.1 Acid

Rare earth ions are dissolved in acidic solution. The acid and its concentration have important impact on the formation of rare earth-extractant complex and the rare earth distribution ratio. P350 has good extraction for rare earth ions in HNO_3 solution but almost no extraction for rare earth ions in HCl solution.

According to Eqs. (4.60) and (4.64), the rare earth ions and acid are competing for P350. The effects of acid concentration on the rare earth extraction are complicated. As shown in Fig. 4.31, the distribution ratios of Tb^{3+} and Yb^{3+} increase first with the increase in HNO_3 concentration and then decrease with the continuous

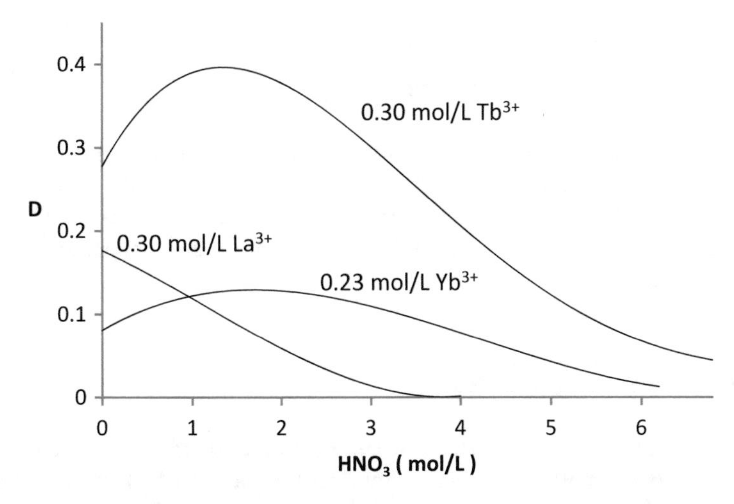

Fig. 4.31 Effects of HNO_3 on distribution ratio of rare earth in 1.52 mol/L P350-kerosene-HNO_3 system

Table 4.31 Effects of acid on distribution ratio, separator factor, and percentage of rare earth extraction

Feed aqueous solution		*D*					
RE, mol/L	Acid	RE	La	Pr	Sm	%E of RE	$\beta_{Pr/La}$
1.255	pH 5.5	0.390	0.284	0.499	–	43.9	1.76
1.285	pH 0.5	0.369	0.244	0.613	–	48.5	2.51
1.280	0.5 mol/L	0.324	0.204	0.448	0.658	39.3	2.20
1.238	1.0 mol/L	0.293	0.164	0.415	0.967	36.9	2.55
1.178	2.0 mol/L	0.201	0.125	0.248	0.512	28.7	1.98

increase in HNO_3 concentration. However, the distribution ratio of La^{3+} decreases quickly as the HNO_3 concentration increases in the same extraction system.

Table 4.31 shows the relationship between acid concentration and rare earth distribution ratio, separator factor, and percentage of extraction. The distribution ratio and percentage of extraction decrease with the increase of acid concentration. When acid concentration (pH = 5.5) is low, the rare earth distribution ratios and Pr/La separation factor are high. However, rare earth hydrolysis is easy to occur at low acid concentration resulting in slow phase disengagement (Li 2011). In practice, an optimal acid concentration is often determined by considering the overall rare earth separation efficiency.

Without salting agent, the separation factor $\beta_{Pr/La}$ increases with the increase of HNO_3 concentration when P350 is used to separate La in HNO_3 solution. With the presence of $LiNO_3$ salting agent, the largest $\beta_{Pr/La}$ is in the HNO_3 concentration range of 1.0–1.5 mol/L. With the presence of NH_4NO_3 salting agent, the largest $\beta_{Pr/La}$ is in the HNO_3 concentration range of 0.5–1.0 mol/L. Therefore, it is common to use P350 to separate La in the HNO_3 concentration range of 0.5–1.0 mol/L (Xu 2003; Li 2011).

4.5.1.2 Salting Agent (Xu and Yuan 1987)

In rare earth solvent extraction process using P350, it is often to add NH_4NO_3 or $LiNO_3$ as salting agent to enhance rare earth distribution ratio and separator factor of adjacent rare earth elements. As shown in Fig. 4.32, the addition of $LiNO_3$ in 70 % P350-sulfonated kerosene-HNO_3 system improves the separator factor of Pr/La significantly.

4.5.1.3 Rare Earth Concentration

It is hard to achieve good separation when the separation factor between two rare earth elements is small. As shown in Fig. 4.33, separation factors vary with the change of total rare earth concentration in aqueous solution. The separation factor between Pr and La increases first and then drops after reaching its maximum value

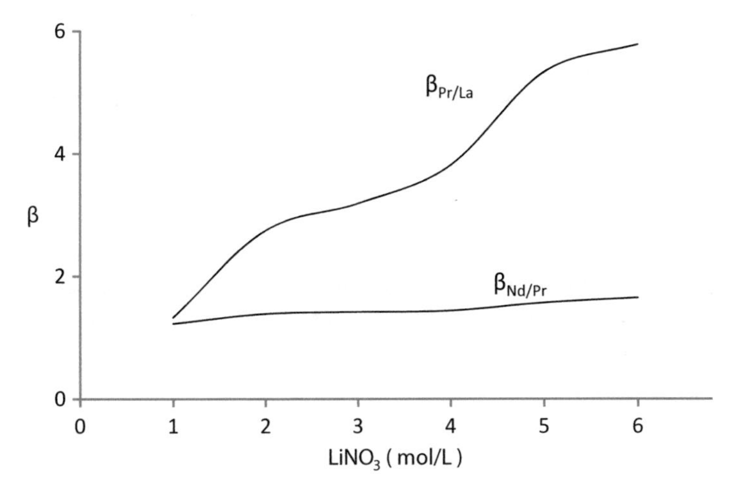

Fig. 4.32 Effects of salting agent $LiNO_3$ on separator factors of $\beta_{Pr/La}$ and $\beta_{Nd/Pr}$ in 70 % P350-sulfonated kerosene-0.5 mol/L HNO_3 system, O/A = 2.5/1

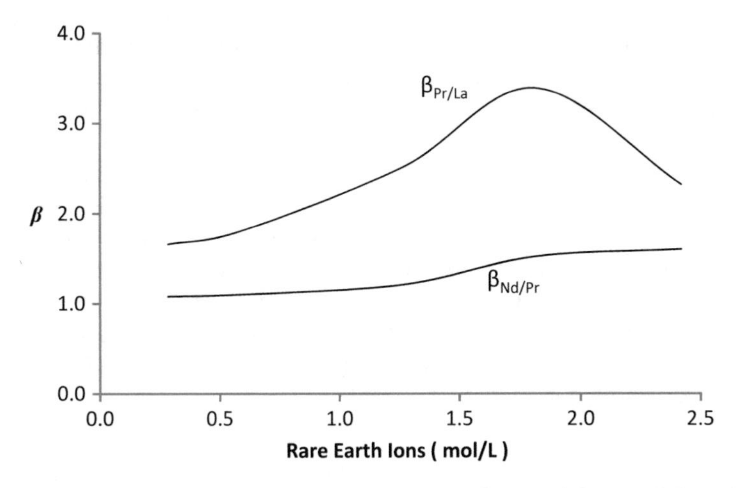

Fig. 4.33 Effects of rare earth concentration on separator factors of $\beta_{Pr/La}$ and $\beta_{Nd/Pr}$ in 50 % P350-sulfonated kerosene-0.5 mol/L HNO_3 system

with the increase of rare earth concentration. The separation factor between Nd and Pr increases continuously with the rare earth concentration. At low rare earth concentration, the separation factors are too low to render good separation. Therefore, the separation of rare earth elements by P350 is always performed in an appropriate rare earth concentration range. For example, good La/Pr separation by P350 can be achieved when the feed solution contains 1.1–1.3 mol/L RE^{3+} in the feed solution while good Pr/Nd separation can be done when the feed contains 2.3 mol/L RE^{3+}.

4.5.1.4 P350 Concentration

According to Eq. (4.62), the rare earth distribution ratio is in proportional to the third power of free P350 concentration in organic solution. When P350 concentration is lower than 50 %, rare earth distribution ratios are small for good separation. With the increase in P350 concentration, the rare earth distribution ratio goes up and favors separation. However, when P350 concentration is higher than 70 %, the rare earth separation factors become smaller and are not favorable for good separation. Also, high P350 concentration will increase the viscosity of the organic solution and will have negative effects on operation. Therefore, P350 is normally used in the concentration range of 50–70 %.

4.5.1.5 Other Factors

Other factors that have impact on the rare earth extraction by P350 include the diluent, the addition of other extractants, and temperatures.

As a rule of thumb, low-polarity diluent is normally used in the organic phase. The diluent has important impact on extractant and rare earth-extractant complex. For example, the La distribution ratio is related to the permittivity of the diluent in the P350 extraction system. Generally the rare earth distribution ratio becomes lower when a diluent with higher permittivity is used.

Since extraction is normally exothermic reaction, low temperature favors the formation of extractant-rare earth complex. Rare earth has larger distribution ratio. However, at low temperature there are issues of high organic phase viscosity, slow phase disengagement, and formation of emulsion. Therefore, rare earth solvent extraction is normally performed at ambient or slightly heated temperatures.

The addition of a second extractant to enhance the rare earth separation is often utilized in rare earth separation as synergetic solvent extraction. Synergetic solvent extraction systems are introduced separately.

4.5.2 *Production of High-Purity La_2O_3*

P350 is mostly used to separate light rare earth elements in HNO_3 solution. The major rare earth solvent extraction process includes high-purity lanthanum process, Pr/Nd separation process, and high-purity scandium process.

In China P350/kerosene-$RE(NO_3)_3$-HNO_3 system is mainly used to produce high-purity lanthanum. Figure 4.34 shows the high-purity lanthanum process using P350-HNO_3 solvent extraction system. The La_2O_3 in the feed accounts for ~50 % of the REO. The rest of REO are Pr_6O_{11} and Nd_2O_3 with minor amount of CeO_2 and other rare earth oxides. The process starts with 38 stages of PrNd extraction followed by 22 stages of scrubbing. The majority of lanthanum remains

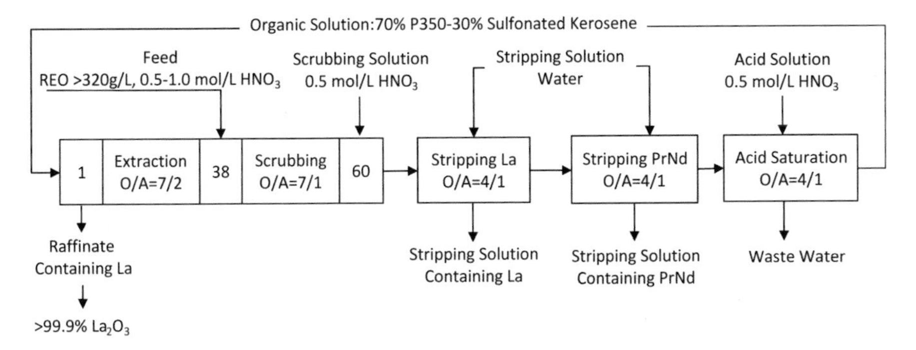

Fig. 4.34 High-purity La process using P350-HNO_3 solvent extraction system

in the raffinate with trace amount of impurity. High-purity La_2O_3 up to 99.99 % can be produced from this raffinate. The organic solution after scrubbing loaded with PrNd and a small amount of La is stripped by ten stages of water stripping to recover La and ten stages of water stripping to recover PrNd. Before recycled back to extraction, the organic solution goes through six stages of water saturation to restore the extraction capability of the organic solution. In comparison with TBP, P350 system does not need high feed acid concentration. Also, La recovery up to 99 % can be achieved using P350 as extractant.

4.5.3 Pr-Nd Separation Process

As indicated by Fig. 4.33, high separation factors can be obtained at high concentration due to the self-salting effects of rare earth nitrate. Figure 4.35 shows the Pr-Nd separation process using P350-HNO_3 solvent extraction system at high rare earth concentration. The feed contains about 357 g/L REO. The REO composition is 50 % Nd_2O_3, 18 % Pr_6O_{11}, 2 % CeO_2, 2 % Sm_2O_3, and 20 % La_2O_3. P350-HNO_3 system is used to extract Nd first. After scrubbing, stripping, and a series of treatment, Nd_2O_3 product can be obtained with about 85 % purity. The Nd extraction raffinate contains LaCePr and is extracted again to separate Pr. Pr_6O_{11} product can be produced with 95–98 % purity. The Pr extraction raffinate contains the majority of remaining La and Ce that can be recovered as a mixed oxide or as feed for continuous solvent extraction separation.

4.5.4 P350 Extracting Scandium

P350 has very good extraction capability and selectivity for rare earth in HNO_3 solution. However, in HCl solution, P350 has almost no extraction for rare earth elements. In solution with less than 6.0 mol/L HCl, P350 barely extracts scandium.

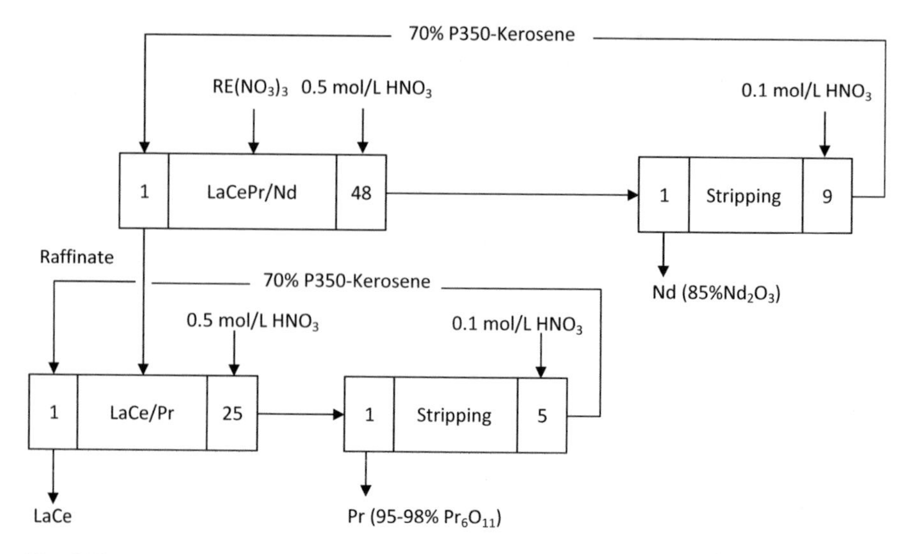

Fig. 4.35 Pr-Nd separation process using P350-HNO_3 solvent extraction system

However, with the increase in HCl concentration over 6 mol/L, the scandium extraction by P350 increases rapidly. The extraction of scandium by P350 from HCl solution is considered as follows by Zhao and Li (1990):

$$Sc^{3+}_{(a)} + 3Cl^-_{(a)} + 3P350_{(o)} = ScCl_3 \cdot P350_{\,(o)} \tag{4.68}$$

In HCl solution, P350 also extracts HCl and H_2O:

$$3H^+_{(a)} + 3Cl^-_{(a)} + 2H_2O + 3P350_{(o)} = 3HCl \cdot 2H_2O \cdot 3P350_{\,(o)} \tag{4.69}$$

Therefore, P350 can be used to remove Th and Zr from scandium HNO_3 solution to produce 99.9 % Sc_2O_3 product (Chen et al. 1981). It can also be used to extract scandium from rare earth HCl leaching solution (Li 1987). Zhang et al. (1991) conducted detailed investigation on high-purity scandium production by P350-HCl extraction system. As shown in Table 4.32, the scandium distribution ratio increases with the increase in P350 concentration, HCl concentration, and O/A ratio. The distribution ratio and percentage of extraction decrease with the increase of scandium concentration in the aqueous feed solution. However, phase disengagement becomes slower at higher P350 concentration in organic phase and lower scandium concentration in aqueous phase.

The extraction isotherm of Sc is shown in Fig. 4.36 at the conditions of 40 % P350-60 % kerosene, O/A = 1/1, and 6.0 mol/L HCl. When the scandium concentration reaches 20 g/L in the aqueous solution, the organic solution is close to saturation. The system has very good selectivity for scandium with very good

Table 4.32 Sc distribution ratio and extraction percentage in P350-HCl system

P350 in kerosene (%)	Sc^{3+} in feed (g/L)	HCl in feed (mol/L)	O/A	*D*	%*E*
20	20.2	5.9	1/1	0.35	26.03
30	20.2	5.9	1/1	0.64	38.95
40	20.2	5.9	1/1	1.04	50.92
50	20.2	5.9	1/1	1.10	58.30
40	1.0	6.0	1/1	4.99	83.31
40	5.0	6.0	1/1	4.15	81.66
40	10.0	6.0	1/1	2.06	67.32
40	20.0	6.0	1/1	0.92	47.92
40	30.0	6.0	1/1	0.56	35.90
40	20.0	4.1	1/1	0.28	22.05
40	20.0	4.9	1/1	0.61	37.70
40	20.0	5.7	1/1	0.88	46.81
40	20.0	6.8	1/1	1.21	54.74
40	20.0	6.0	1/1	0.77	27.80
40	20.0	6.0	2/1	1.43	74.06
40	20.0	6.0	3/1	2.54	88.35

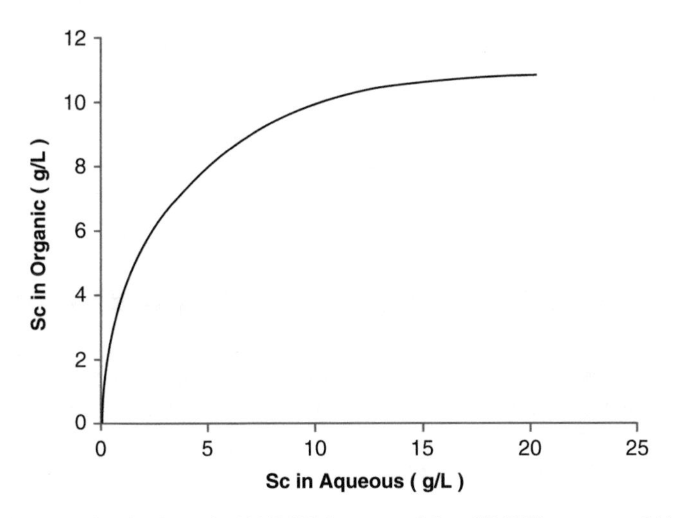

Fig. 4.36 Sc extraction isotherm in 40%P350-kerosene-6.0 mol/L HCl system at O/A = 1/1

separation results for impurities such as Si, Fe, Mn, Mg, Pb, Ni, Al, Ca, Cu, and Ti. As shown in Table 4.33, the scandium extraction percentage is higher than 98 % after three stages of countercurrent extraction under the conditions 40 % P350-60 % kerosene, 19.61 g/L of Sc in feed, 5.8 mol/L of HCl in feed, and O/A = 1.9/1. The purity of scandium product is higher than 99.99 %.

Table 4.33 Three stages of countercurrent extraction by 40 % P350

	Equilibrium aqueous phase		Equilibrium organic phase	
Run	HCl, mol/L	Sc, g/L	Sc, g/L	%*E*
1	5.68	0.310	10.142	99.30
2	5.59	0.396	10.113	97.98
3	5.65	0.396	10.113	97.98
4	5.60	0.327	10.149	98.33
5	5.62	0.293	10.167	98.51
6	5.60	0.268	10.180	98.63
7	5.65	0.261	10.182	98.65
8	5.65	0.264	10.182	98.65

4.6 Cyanex 923 Solvent Extraction System

Cyanex 923 extractant is a colorless mobile liquid. It is a mixture of four trialkylphosphine oxides including trihexylphosphine oxide ($R_3P(O)$), dihexylmonooctylphosphine oxide ($R_2R'P(O)$), dioctylmonohexyl-phosphine oxide ($RR'_2P(O)$), and trioctylphosphine oxide ($R'_3P(O)$) where R is -$(CH_2)_5CH_3$ and R′ is -$(CH_2)_7CH_3$. Cyanex 923 acts as a solvating extractant in metal extraction. The commercial grade Cyanex 923 contains impurity acids that have reduction effects. In cases of metal reduction to be avoided, Cyanex 923 can be purified. The purification of Cyanex 923 can be done by washing with 2 % Na_2CO_3 solution followed by 1.0 mol/L H_2SO_4 solution washing and distilled water washing to neutrality (Lu and Li 1999).

4.6.1 *Cyanex 923 Extracting RE^{3+} in HNO_3 Acid Solution*

Similar toTBP and 350, Cyanex 923 forms neutral complex with rare earth nitrate. The extraction takes place on the interphase between organic solution and aqueous solution. The following equilibrium exists in the purified Cyanex 923-n-octane-HNO_3 system (Chu et al. 1998; Tong et al. 2003):

$$RE^{3+}_{(a)} + 3NO_3^- + 3Cyanex923_{(o)} = RE(NO_3)_3 \cdot 3Cyanex923_{(o)} \quad (4.70)$$

However, there may be different extraction mechanisms depending on the extraction system. Ma et al. (2011a, b) reported that the La^{3+} and Y^{3+} are extracted in terms of $RE(OH)^{2+}$ in unpurified Cyanex 923-*n*-octane-HNO_3 system. The extraction of La^{3+} and Y^{3+} can be represented by Eqs. (4.71) and (4.72):

$$La(OH)^{2+}_{(a)} + 2NO_3^- + 2Cyanex923_{(o)} = RE(OH)(NO_3)_2 \cdot 2Cyanex923_{(o)} \quad (4.71)$$

$$Y(\text{OH})^{2+}_{(a)} + 2\text{NO}_3^- + 2\text{Cyanex923}_{(o)} = \text{RE(OH)(NO}_3)_2 \cdot 2\text{Cyanex923}_{(o)} \quad (4.72)$$

El-Nadi et al. (2007, 2011) reported a different equilibrium reaction through the study of La^{3+}, Pr^{3+}, and Sm^{3+} extraction in unpurified Cyanex 923-kerosene-HNO_3 system. The reaction equation is expressed by Eq. (4.73):

$$\text{RE}^{3+}_{(a)} + 3\text{NO}_3^- + 2\text{Cyanex923}_{(o)} = \text{RE(NO}_3)_3 \cdot 2\text{Cyanex923}_{(o)} \quad (4.73)$$

where RE^{3+} is La^{3+}, Pr^{3+}, or Sm^{3+}.

The study of Pr^{3+} extraction in Cyanex 923-kerosene-HNO_3 by Panda et al. (2014) produced the following extraction reaction:

$$\text{RE}^{3+}_{(a)} + 3\text{NO}_3^- + \text{Cyanex923}_{(o)} = \text{RE(NO}_3)_3 \cdot \text{Cyanex923}_{(o)} \quad (4.74)$$

According to above extraction reactions proposed by different research groups, a general equation can be used to descript the rare earth extraction in Cyanex-HNO_3 system:

$$\text{RE}^{3+}_{(a)} + 3\text{NO}_3^- + n\text{Cyanex923}_{(o)} = \text{RE(NO}_3)_3 \cdot n\text{Cyanex923}_{(o)} \quad (4.75)$$

$$K_{ex} = \frac{[\text{RE(NO}_3)_3 \cdot n\text{Cyanex923}_{(o)}]}{[\text{RE}^{3+}_{(a)}] \times [\text{NO}_3^-]^3 \times [\text{Cyanex923}_{(o)}]^n} \quad (4.76)$$

$$D = K_{ex} \cdot [\text{NO}_3^-]^3 \cdot [\text{Cyanex923}_{(o)}]^n, \quad n = 1 - 3 \quad (4.77)$$

According to Eq. (4.75), hydrogen ion does not participate the extraction. The concentration of H^+ has no effects on the rare earth extraction by Cyanex 923. This was confirmed in El-Nadi and co-worker investigation on La, Pr, and Sm extraction by Cyanex 923 where the concentration of H^+ has negligible effects on rare earth extraction (El-Nadi et al. 2007; El-Nadi 2010).

At 25 °C and pH 2.1–2.3, Chu et al. (1998) used 0.023 mol/L purified Cyanex 923 in n-octane to extract 15 rare earth elements from HNO_3 solution containing 0.5 mol/L $NaNO_3$ and 2.1×10^{-4} mol/L RE^{3+}. The logarithm of distribution ratios of the 15 rare earths is shown in Fig. 4.37. The extraction of the rare earths by Cyanex 923 shows tetrad effects with Y between Ce and Pr but below middle and heavy rare earths. La is at the bottom of the curve. The distribution ratio (lg*D*) curve shows that La can be well separated from the others. Y can be well separated from the rest of other heavy rare earths. The rare earth separation factors in Cyanex 923-HNO_3 are shown in Table 4.34.

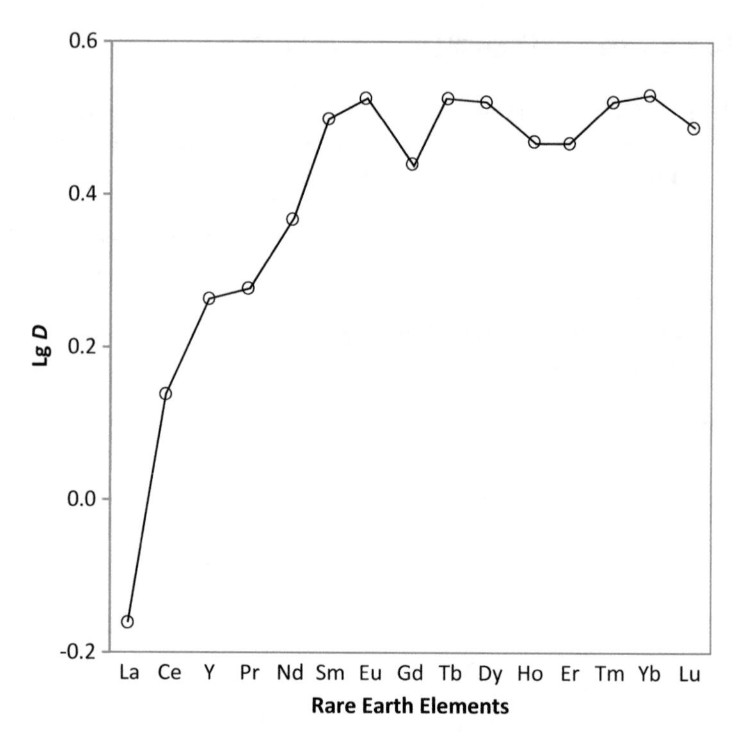

Fig. 4.37 Distribution ratios of the rare earth elements in Cyanex 923-HNO_3 system

4.6.2 Extraction and Stripping of Yb from H_2SO_4 Solution (Wang et al. 2006)

From low-acidity H_2SO_4 solution, the extraction of Yb^{3+} by Cyanex 923 can be represented by the following equation:

$$\begin{aligned} & Yb^{3+}_{(a)} + SO^{2-}_{4(a)} + HSO^{-}_{4(a)} + 2Cyanex923_{(o)} \\ & = Yb(SO_4)(HSO_4)\cdot 2Cyanex923_{(o)} \end{aligned} \quad (4.78)$$

Ionization of H_2SO_4 occurs in the aqueous phase as shown below:

$$H_2SO_4 \overset{k_{a1}}{\rightarrow} H^+ + HSO_4^- \quad (4.79)$$

$$HSO_4^- \overset{k_{a2}}{\rightarrow} H^+ + SO_4^{2-} \quad (4.80)$$

where $k_{a1} = 10^3$ and $k_{a2} = 10^{-1.99}$.

Table 4.34 Separation factors of rare earths in Cyanex 923-HNO_3 system

β	Ce	Pr	Nd	Sm	Eu	Gd	Tb	Dy	Ho	Er	Tm	Yb	Lu	Y
La	1.982	2.716	3.381	4.519	4.842	3.954	4.909	4.764	4.236	4.217	4.732	4.898	4.355	2.780
Ce		1.371	1.706	2.280	2.443	1.995	2.477	2.404	2.138	2.128	2.388	2.472	2.198	1.403
Pr			1.245	1.663	1.782	1.455	1.807	1.754	1.560	1.552	1.742	1.803	1.603	0.977
Nd				1.337	1.432	1.169	1.452	1.409	1.253	1.247	1.400	1.449	1.288	0.822
Sm					1.072	0.875	1.086	1.054	0.938	0.933	1.047	1.084	0.964	0.615
Eu						0.817	1.014	0.984	0.875	0.871	0.977	1.012	0.899	0.574
Gd							1.242	1.205	1.072	1.067	1.197	1.239	1.102	0.703
Tb								0.971	0.863	0.859	0.964	0.998	0.877	0.566
Dy									0.889	0.885	0.993	1.028	0.914	0.583
Ho										1.042	1.169	1.211	1.076	0.687
Y											1.122	1.161	1.033	0.659
Er												1.035	0.920	0.587
Tm													0.889	0.568
Yb														0.638

In H_2SO_4 solution, Yb^{3+} forms complexes with SO_4^{2-}:

$$Yb^{3+}_{(a)} + SO^{2-}_{4(a)} \overset{\beta 1}{\Leftrightarrow} Yb(SO_4)^{+}_{(a)} \tag{4.81}$$

Therefore, the total concentration of Yb^{3+} in aqueous phase is

$$\left[Yb^{3+}_{total}\right] = \left[Yb^{3+}_{(a)}\right] + \left[YbSO^{+}_{4(a)}\right] = \left[Yb^{3+}_{(a)}\right]\left(1 + \beta_1\left[SO^{2-}_{4(a)}\right]\right) \tag{4.82}$$

And, the extraction equilibrium K_{ex} is

$$K_{ex} = \frac{[Yb(SO_4)(HSO_4)\cdot 2Cyanex923_{(o)}] \times \left(1 + \beta_1[SO^{2-}_{4(a)}]\right)}{[Yb^{3+}_{total}] \times [SO^{2-}_{4(a)}] \times [HSO^{-}_{4(a)}] \times [Cyanex923_{(o)}]^2} \tag{4.83}$$

$$\begin{aligned} \lg K_{ex} = {} & \lg D\left(1 + \beta_1\left[SO^{2-}_{4(a)}\right]\right) - 2\lg\left[Cyanex923_{(o)}\right] - \lg\left[SO^{2-}_{4(a)}\right] \\ & - \lg\left[HSO^{-}_{4(a)}\right] \end{aligned} \tag{4.84}$$

Figure 4.38 shows the comparison of stripping of Cyanex 923 loaded with Yb by different acids at various concentrations. The loaded organic solution contains

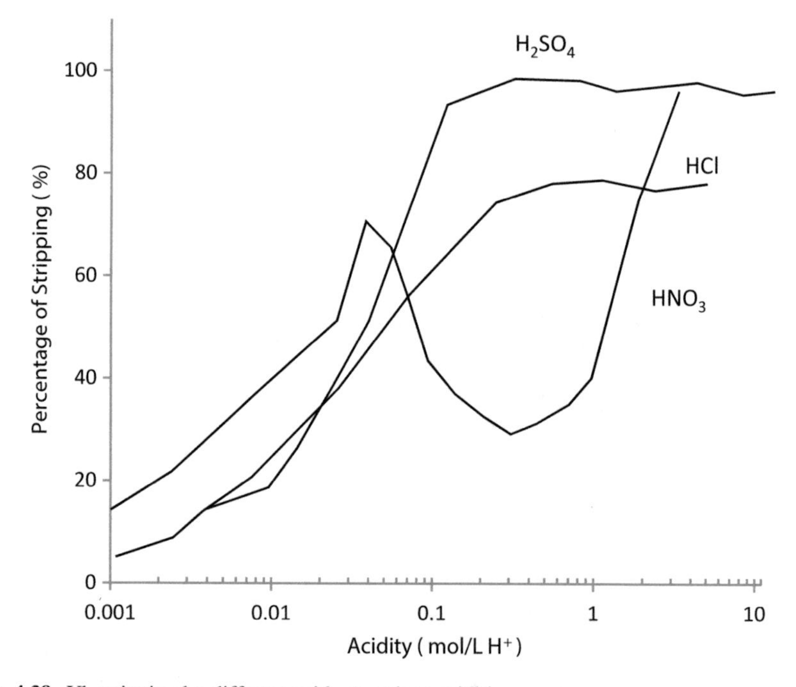

Fig. 4.38 Yb stripping by different acids at various acidities

0.15 mol/L Cyanex 923 and 9.78×10^{-4} mol/L Yb. The O/A ratio of stripping is 1/1. The stripping efficiency of different acids follows the order of $H_2SO_4 > HCl > HNO_3$. At 0.16 mol/L H_2SO_4, 97.4 % Yb can be stripped.

4.6.3 *Ce^{4+} and Th^{4+} Separation*

In HNO_3 acid solution, the extraction of Ce^{4+} and Th^{4+} is similar with the extraction of RE^{3+} by purified Cyanex 923 (Lu and Li 1999):

$$Ce^{4+}_{(a)} + 4NO^{-}_{3(a)} + 2Cyanex923_{(o)} = Ce(NO_3)_4 \cdot 2Cyanex923_{(o)} \tag{4.85}$$

$$Th^{4+}_{(a)} + 4NO^{-}_{3(a)} + 2Cyanex923_{(o)} = Th(NO_3)_4 \cdot 2Cyanex923_{(o)} \tag{4.86}$$

Cyanex 923 also extracts HNO_3:

$$H^{+}_{(a)} + NO^{-}_{3(a)} + Cyanex923_{(o)} = HNO_3 \cdot Cyanex923_{(o)} \tag{4.87}$$

In HNO_3 solution, the extraction of Ce^{4+}, Th^{4+}, and RE^{3+} follows the order of:

$$Ce4+ >> Th4+ >> Gd^{3+} \sim La^{3+}$$

In purified Cyanex 923-n-hexane-H_2SO_4 system, the extraction of Ce^{4+} and Th^{4+} is similar to the extraction of RE^{3+} (Lu et al. 1998a, b):

$$\begin{aligned} &Ce^{4+}_{(a)} + SO^{2-}_{4(a)} + 2HSO^{-}_{4(a)} + 2Cyanex923_{(o)} \\ &\quad = Ce(SO_4)(HSO_4)_2 \cdot 2Cyanex923_{(o)} \end{aligned} \tag{4.88}$$

$$\begin{aligned} &Th^{4+}_{(a)} + SO^{2-}_{4(a)} + 2HSO^{-}_{4(a)} + 2Cyanex923_{(o)} \\ &\quad = Th(SO_4)(HSO_4)_2 \cdot 2Cyanex923_{(o)} \end{aligned} \tag{4.89}$$

$$K_{ex} = D \frac{1 + \sum \beta_i \left[SO^{2-}_{4(a)}\right]^i}{\left[SO^{2-}_{4(a)}\right] \times \left[HSO^{-}_{4(a)}\right]^2 \times \left[Cyanex923_{(o)}\right]^2} \tag{4.90}$$

where β_i is the stability constants of metal ions with ${SO_4}^{2-}$. For Th^{4+}, $\beta_1 = 10^{3.32}$ and $\beta_2 = 10^{5.70}$.

In the sulfuric acid leaching solution of bastnasite, F^- is always present and participates the extraction (Liao et al. 2001a, b; 2002). The equilibrium reaction can be represented by the following equation:

$$
\begin{aligned}
& Ce^{4+}_{(a)} + SO^{2-}_{4(a)} + 2HSO^{-}_{4(a)} + HF_{(a)} + 2Cyanex923_{(o)} \\
& \quad = Ce(HF)(SO_4)(HSO_4)_2 \cdot 2Cyanex923_{(o)}
\end{aligned}
\tag{4.91}
$$

The H_2SO_4 and HF not only participate in the Ce^{4+} extraction but also compete for extraction. The extraction equations are as follows (Liao et al. 2001a, b):

$$H_2SO_{4(a)} + Cyanex923_{(o)} = H_2SO_4 \cdot Cyanex923_{(o)} \tag{4.92}$$

$$HF_{(a)} + Cyanex923_{(o)} = HF \cdot Cyanex923_{(o)} \tag{4.93}$$

Figure 4.39 shows the effects of H_2SO_4 concentration on the extraction of Ce^{4+}, Th^{4+}, La^{3+}, and Gd^{3+}. The extraction of Ce^{4+} decreases with the increase in H_2SO_4 concentration in the range of 0.5–4.0 mol/L. With the increase in H_2SO_4 concentrate over 4.0 mol/L, there is formation of a third phase. The extraction of Th^{4+}, Gd^{3+}, and La^{3+} increases with the increase in H_2SO_4 concentration. In low-acidity H_2SO_4 medium the extraction follows the order of:

$$Ce^{4+} > Th^{4+} > Gd^{3+} > La^{3+}$$

Utilizing the difference of extraction in the purified Cyanex 923-n-hexane-H_2SO_4 system, a process to separate Ce^{4+} with Th^{4+} and the other RE^{3+} from leaching solution of mixed bastnasite and monazite was developed by Li et al. (1998). The schematic process flow is shown in Fig. 4.40. The bastnasite leaching solution contains 14–60 g/L CeO_2, 0.01–1.0 g/L ThO_2, 0–12 g/L F, 0.5–20 g/L H_3BO_3, and 0.5–3.0 mol/L H_2SO_4. The CeO_2 accounts for over 20 % of the total rare earth oxide

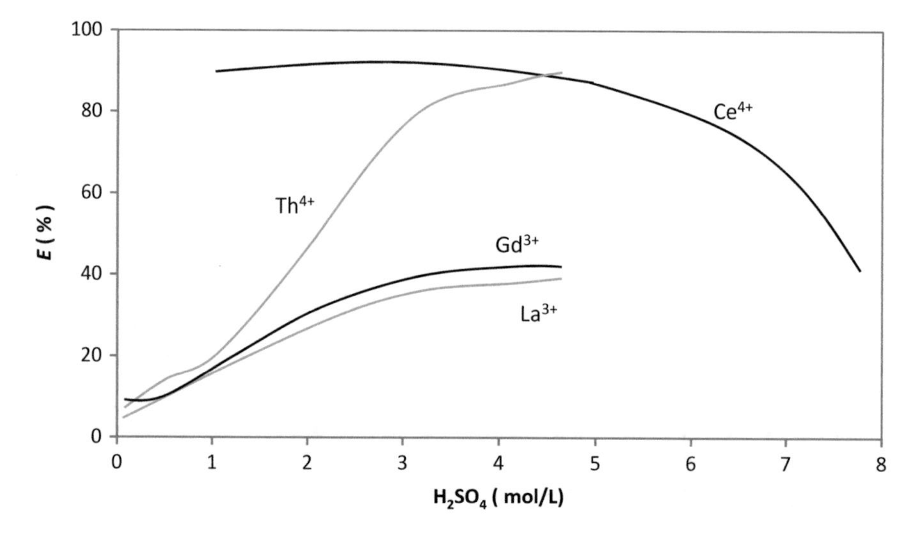

Fig. 4.39 Effects of H_2SO_4 on extraction. Purified Cyanex 923 = 0.11 mol/L in n-hexane, $[Ce^{4+}] = 1.25 \times 10^{-2}$ mol/L, $[Th^{4+}] = 1.38 \times 10^{-4}$ mol/L, $[Gd^{3+}] = 2.20 \times 10^{-4}$ mol/L, $[La^{3+}] = 2.34 \times 10^{-4}$ mol/L (Lu et al. 1998a, b)

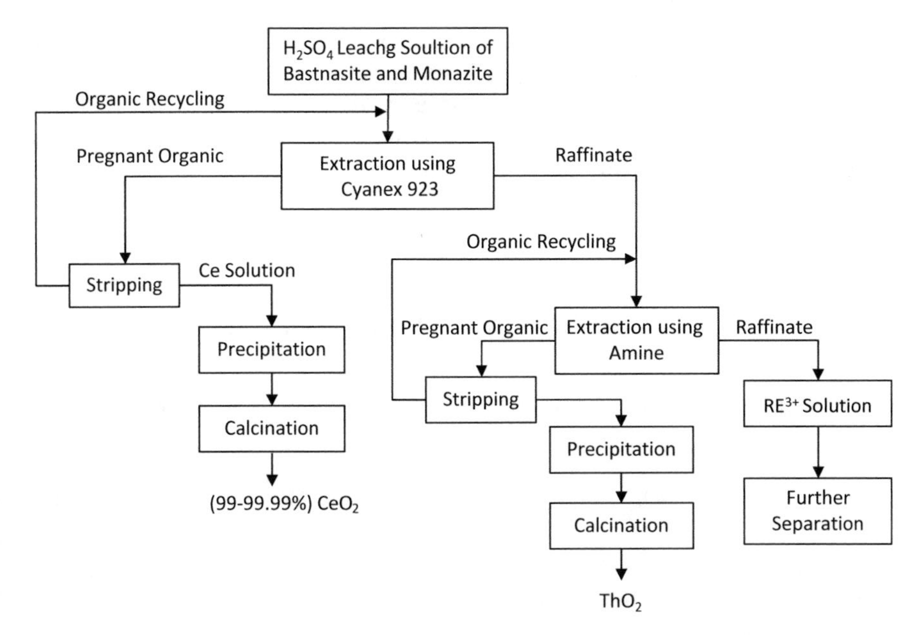

Fig. 4.40 Separation process of Ce^{4+}, Th^{4+}, and RE^{3+} from H_2SO_4 leaching solution of mixed bastnasite and monazite

in the solution. The first step is to use Cyanex 923 to separate the Ce^{4+} from Th^{4+} and other RE^{3+}. The Ce^{4+} is loaded in the organic solution and the Th^{4+} and other RE^{3+} are left in the raffinate. H_2SO_4 solution is used to strip the Ce^{4+} out of the organic solution before the organic solution recycled. The stripping solution loaded with Ce^{4+} is precipitated and calcinated to produce 99–99.99 % CeO_2 product which contains less than 0.01 % ThO_2. The Th^{4+} in the raffinate is extracted by an amine extractant to separate with other RE^{3+}. Less than 0.00001 % ThO_2 will end up in the RE^{3+} product. Over 95 % Ce^{4+} and 99 % Th^{4+} can be recovered.

4.7 Naphthenic Acid Extraction System

Naphthenic acid is a by-product of petroleum refining. The major components of the purified naphthenic acid used for rare earth extraction are cyclopentyl carboxylic acids with the following structure:

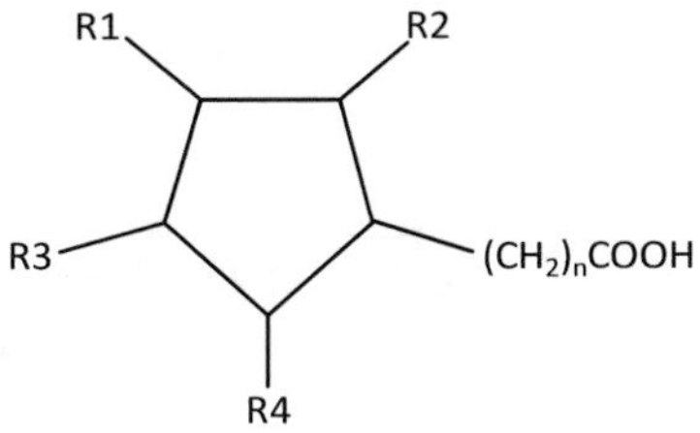

R1 R2 R3 $(CH_2)_nCOOH$ R4

where *n* is normally less than 4. R1, R2, R3, and R4 are alkyl groups or hydrogen.

It belongs to carboxylic acid. In comparison with HEH/EPH, HDEHP, and HBTMPP, the major advantages of naphthenic acid in rare earth solvent extraction are low cost, high loading capacity, and easy stripping. It has been used for rare earth extraction since 1960s. It can be used for rare earth separation and purification, impurity removal, rare earth recovery, and concentration from low-concentration solutions.

4.7.1 Naphthenic Acid Extracting Rare Earth

Naphthenic acid as a weak organic acid can be denoted by HL. The extraction of RE^{3+} by naphthenic acid is cation-exchange mechanism:

$$RE^{3+}_{(a)} + 3HL_{(o)} = RE(HL_2)_{3(o)} + 3H^{+}_{(a)} \tag{4.94}$$

Bauer and Lindstrom (1964) investigated the extraction of 15 rare earth elements from H_2SO_4 solution using naphthenic acid diluted with diethyl ether (DEE) and n-hexanol. Figure 4.41 shows the effects of pH on percentage of rare earth extraction. The extraction increases with the increase in pH. At pH 7.6, rare earth extraction reaches its maximum. However, as shown in Fig. 4.42 the increase of pH does not affect the relative distribution coefficients between individual rare earth significantly. Naphthenic acid is an effective extractant for rare earth

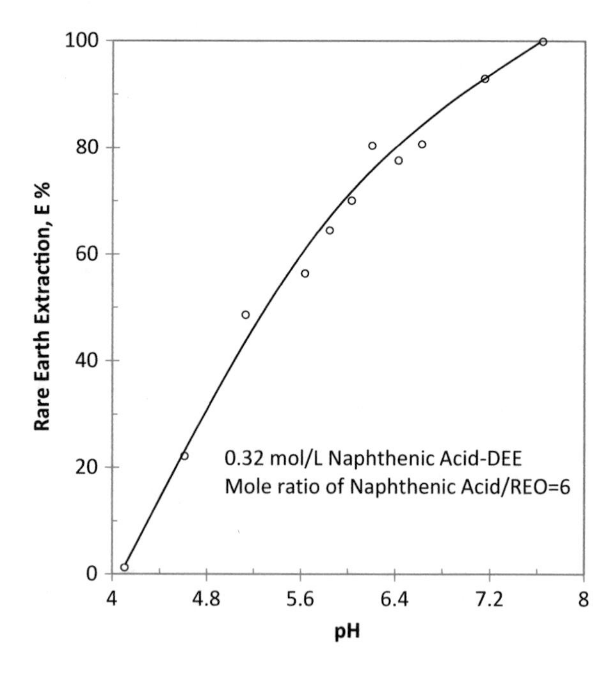

Fig. 4.41 Effects of pH on percentage of rare earth extraction

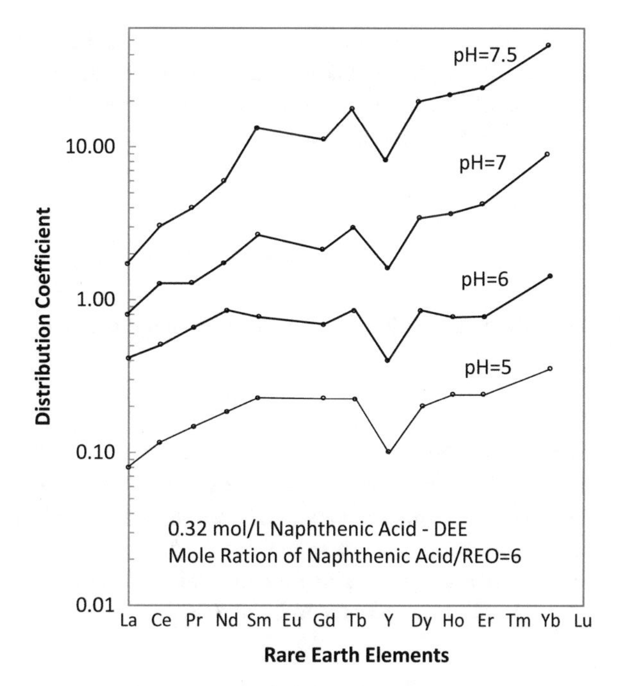

Fig. 4.42 Effects of pH on rare earth distribution coefficient in naphthenic acid system

extraction when diluted with DEE or n-hexanol. Serious emulsion will result in difficult phase disengagement when benzene, xylene, kerosene, or hexane is used as diluent. The addition of chelating agent of EDTA can increase the average separation factor of heavy rare earths to 2.2 while the addition of DTPA can increase the average separation factor of light rare earths to 3.5. With the addition of EDTA and pH below 8, Y is between Tb and Dy in the extraction sequence. When pH is increased to 10, Y shifts to the position close to Nd. This provides the option for pure Y separation by a two-step countercurrent extraction process using naphthenic acid as extractant. The addition of DTPA makes Y extraction with Nd possible at all pH values. Bauer and Lindstrom's work also indicates that the anion of rare earth salts has negligible effects on extraction efficiency of the naphthenic acid system.

In addition to the rare earth extraction, naphthenic acid can extract other metal ions as well. A general extraction sequence is shown below (Fu 1991):

$$Fe^{3+} > Th^{4+} > Z^{4+} > U^{4+} > In^{3+} > Ti^{3+} > Ga^{3+} > UO_2^{2+} > Sn^{2+} > Al^{3+} > Hg^{2+} > Cu^{2+} > Zn^{2+} > Pb^{2+} > Ag^{+} > Cd^{2+} > RE^{3+} > Ni^{2+} > Sr^{2+} > Co^{2+} > Fe^{2+} > Cr^{2+} > Mn^{2+} > Ca^{2+} > Mg^{2+} > Cs^{+}$$

The rare earth extraction sequence in naphthenic acid-mixed alcohol-kerosene-HCl system at pH 4.8–5.1 is as follows (Fu 1991; Tan 1997):

$$Sm^{3+} > Nd^{3+} > Pr^{3+} > Dy^{3+} > Yb^{3+} > Lu^{3+} > Tb^{3+} > Ho^{3+} > Tm^{3+} > Er^{3+} > Gd^{3+} > La^{3+} > Y^{3+}$$

The direct rare earth extraction using naphthenic acid will generate H^+ to increase the aqueous-phase acidity and consequently reduce the rare earth distribution coefficient and extraction-stage efficiency. To overcome the acidity changes and associated issues, saponification is often adopted (Li et al. 2010a, b). The naphthenic acid saponification using ammonium hydroxide can be represented by the following equation.

$$HL_{(o)} + NH_4OH = NH_4L_{(o)} + H_2O \tag{4.95}$$

The rare earth extraction by saponified naphthenic acid is as follows:

$$RE^{3+} + 3NH_4L_{(o)} = REL_{3(o)} + 3NH_4^+ \tag{4.96}$$

Figure 4.43 shows the effects of pH on the rare earth distribution coefficient in a naphthenic acid-mixed alcohol-kerosene-$RECl_3$ system with 80 % saponification (Fu 1991). In comparison with the system in Fig. 4.42, the Y distribution coefficient is the lowest in the pH range of 4.0–5.2. It is worth noting that the rare earth extraction sequence in naphthenic acid extraction system is very complicated. The extraction sequence can be altered by the changes of system parameters such

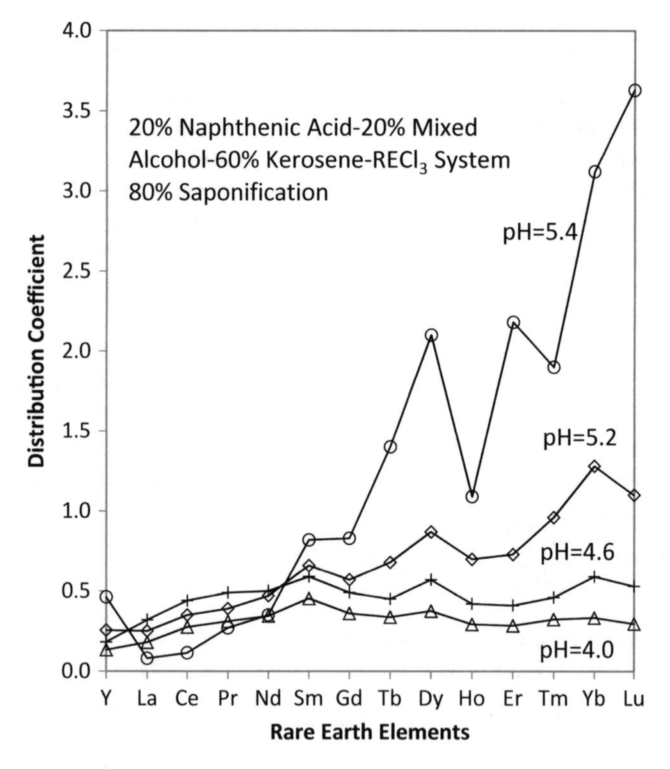

Fig. 4.43 Effects of pH on rare earth distribution coefficient in saponified naphthenic acid system

as degree of saponification, aqueous acidity at equilibrium, concentration of naphthenic acid, O/A ratio, and rare earth concentration. Therefore, strict process control is very important in extraction processes using naphthenic acid.

4.7.2 Separation and Purification of High-Purity Y

There are many Y separation and purification systems. Some of them are summarized in Table 4.35. Regardless of the extraction system, they all utilize the position shifting of the Y distribution coefficient to achieve Y separation from other rare earth elements. As a low-cost extractant with good chemical stability, naphthenic acid is commonly used in high-purity Y separation and purification. There are two-step extraction process and one-step extraction process to produce high-purity Y using naphthenic acid as extractant.

Figure 4.44 shows the two-step extraction process (Dai et al. 1985a, b). The feed contains about 1 mol/L mixed heavy rare earth chloride with >50 % Y at pH 4. The organic solution contains 0.68 mol/L ammonium-saponified naphthenic acid and 0.17 mol/L naphthenic acid diluted in kerosene with 15 % mixed alcohol. In the first step extraction, the organic is fed into the 1st stage and the rare earth is fed in the 19th stage. Scrubbing solution is introduced to the extraction from the 30th stage. The volumetric ratio is organic solution/feed/scrubbing solution = 3.5/0.46/1.0. A crude Y product with >97 % purity and a non-Y product with less than 0.3 % Y_2O_3 are produced. In the first step extraction, the Y is left in the raffinate and the non-Y rare earths are extracted in the organic solution. The organic solution loaded with non-Y rare earths is stripped and treated before recycling. In the second step, feed preparation is performed to remove the excessive NH_4Cl in the first step raffinate and adjust the RE^{3+} concentration to about 1.0 mol/L. The same organic solution is used to separate the crude Y product to a high-purity Y product with >99.99 % purity and a secondary Y product with 65–70 % purity. In the second step, the organic solution is fed in the 1st stage and the feed is introduced in the 32nd stage. Scrubbing solution is fed from the 40th stage. The volumetric ratio of organic solution/feed solution/scrubbing solution is 28.4/3.0/15.

The organic solution treatment after stripping includes the washing using 3.0 mol/L HCl solution at O/A ratio of 4/1 and the washing using deionized water at the O/A ratio of 2/1 until pH >4. Ammonium hydroxide (NH_4OH) solution is used to neutralize the naphthenic acid before recycling. The separation of Y utilizes its low distribution ratio in the system where it is the lowest among all of the rare earths. Table 4.36 lists the distribution ratios and the separation factors of the rare earth in the naphthenic acid extraction system. It can be seen that the distribution ratio of Y is the lowest in both systems with various amounts of naphthenic acid and saponified naphthenic acid. The average separation factor of RE/Y is about 3.0.

The one-step Y separation process also utilizes its low distribution (Dai et al. 1985a, b). As shown in Fig. 4.45, the mixed rare earth is separated in one step to a Y product with >99.99 % purity and a non-Y product containing

Table 4.35 Summary of Y separation systems (Fu 1991)

Y separation system				Separation factor $\beta_{RE/Y}$													
Organic phase	Aqueous phase	Feed	Y Position	La	Ce	Pr	Nd	Sm	Eu	Gd	Tb	Dy	Ho	Er	Tm	Yb	Lu
50 % HDEHP-Shellsol	$RECl_3$-HCl	Mixed RE	Ho-Y-Er				0.13	0.21	0.28	0.29	0.59	0.72	0.86	1.20	1.69	2.25	3.09
250 g/L N263-mixed alcohol	2.5 M $RE(NO_3)_3$ 0.1 M NH_4NO_3	Mixed RE	Tm-Y-Er	>32		>30.6	>10	5.04	4.60	2.84	2.58	2.77	2.65	1.98	1.46	0.95	0.54
300 g/L N262-solvent naphtha	1.5 M $RE(NO_3)_3$ 0.1 M HNO_3	Mixed RE	Er-Y-Tm	50.3	44.6	24.7	17.0	10.4		4.36	2.58	3.30	3.00	1.30	0.61		
300 g/L N262-solvent naphtha	0.5 M $RE(NO_3)_3$ 2.5 M NH_4NO_3-HNO_3	Mixed RE	Er-Y-Tm	62.0		15.2	13.5	3.78		1.74		1.96	1.54	1.21	0.87	0.61	
300 g/L N262-solvent naphtha	0.5 M $RE(NO_3)_3$ 2.5 M $LiNO_3$-HNO_3	Mixed RE	Yb-Y-Lu	65.0		24.0	20.5	13.5		5.05		5.58	4.40	3.17	2.00	1.05	0.80
300 g/L N262-solvent naphtha	0.5 M $RE(NO_3)_3$ 2.5 M $LiNO_3$-HNO_3	Single Y					1.6			1.53				1.34			

300 g/L N262-solvent naphtha	0.32 M RE $(NO_3)_3$ 4 M NH_4SCN-HCl	Mixed RE								1.03	4.50	3.90	7.40	9.50	36.2	85.0	155.0
40 % Kapilon-aromatic diluent	23 g/L RE $(NO_3)_3$ 4 M NH_4NO_3-HNO_3	Mixed RE	Tb-Y-Dy	<0.11	<0.11	<0.11	0.11	0.16	<0.5	0.50	0.83	1.93	2.28	3.91	5.11	6.72	4.50
30%R3N +20 % Kapilon-aromatic diluent	80 g/L RE $(NO_3)_3$ 2.5 M Mg $(NO_3)_2$-HNO_3	Mixed RE	Y-La	>10	>10.6	>10.8	10.6	5.35	8.33	2.32	3.13	2.30	2.17	2.57	2.50	3.06	2.50
15 % R3N +4 % iso-α-acid-aromatic diluent	80 g/L RE $(NO_3)_3$ 3 M $NaNO_3$-HNO_3	Mixed RE	Y-La	6.40	9.12	5.39	4.13	4.09	2.84	2.19	2.66	2.07	1.71	1.86	2.22	1.70	2.06
60 % P350-kerosene	0.3 M RE $(NO_3)_3$ 3.5 M NH_4NO_3 0.7 M EDTA	Y-rich concentrate											3.60	9.58	41.7	54.5	
Salicylic acid-butanol	$RE(NO_3)_3$-NH_4NO_3-HNO_3, pH = 5	Single Y	Nd-Y-Sm														

(continued)

Table 4.35 (continued)

Y separation system				Separation factor $\beta_{RE/Y}$													
Organic phase	Aqueous phase	Feed	Y Position	La	Ce	Pr	Nd	Sm	Eu	Gd	Tb	Dy	Ho	Er	Tm	Yb	Lu
20 % Naphthenic acid-alcohol-kerosene	0.62 M $RECl_3$-HCl, pH = 6	Mixed RE					7.06			4.82		4.03					
0.04 M PAH-butanol	$RECl_3$-HCl, pH = 5.5–7	Y-rich concentrate	Nd-Y-Sm														
20 % Naphthenic acid-alcohol-kerosene	0.67 M RE $(NO_3)_3$ -HNO_3, pH = 5	Mixed RE					3.20	2.54		1.38			1.67	2.47		3.14	

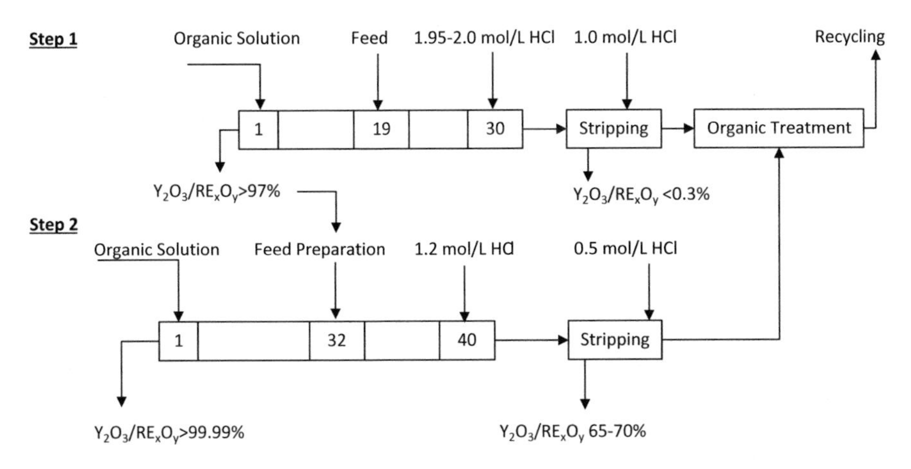

Fig. 4.44 Two-step extraction process of high-purity Y production

Table 4.36 Distribution ratio and separation factor in naphthenic acid extraction system

Organic phase	Naphthenic acid	0.45 mol/L		0.13 mol/L	
	Rare earth	0.142 mol/L		0.240 mol/L	
Aqueous pH		4.19		4.96	
O/A		2/1		2/1	
Temperature		25 °C		25 °C	
		D_{RE}	$\beta_{RE/Y}$	D_{RE}	$\beta_{RE/Y}$
La		0.143	1.22	0.546	2.65
Ce		0.246	2.09	0.711	3.45
Pr		0.340	2.89	0.760	3.69
Nd		0.422	2.35	0.810	3.93
Sm		0.619	5.27	1.041	5.05
Eu		0.529	4.50	0.908	4.41
Gd		0.414	3.52	0.623	3.03
Tb		0.413	3.51	0.681	3.00
Dy		0.370	3.15	0.556	2.70
Ho		0.303	2.58	0.470	2.28
Er		0.261	2.22	0.441	2.14
Tm		0.251	2.13	0.492	2.39
Yb		0.253	2.15	0.532	2.52
Lu		0.232	2.19	0.500	2.42
Y		0.117	1	0.206	1

1.40–1.42 % Y_2O_3. The organic solution loaded with the non-Y rare earth can be stripped by 3.0 mol/L HCl solution in one stage. After HCl acid and water washing, the organic solution contacts with NH_4OH to get saponified before it is recycled back to extraction. Volumetric ratio in extraction is organic solution/feed/scrubbing solution = 25/3.9/7.0. The O/A in stripping is 20–30/1.

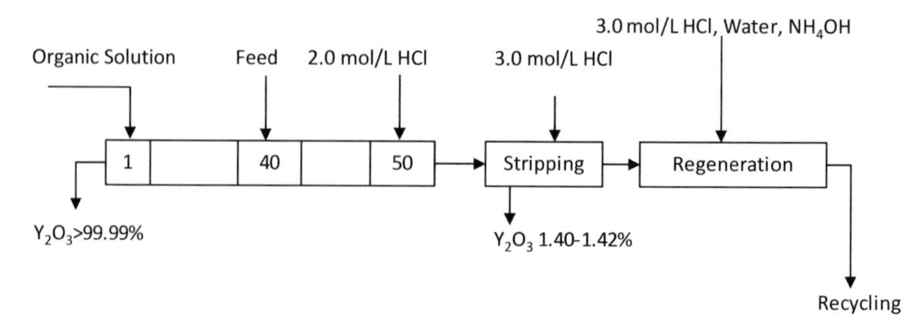

Fig. 4.45 One-step extraction process of high-purity Y production

Three-outlet technology can be adopted in the one-step Y separation process to produce a middle product rich in Er. Industry-scale testing was done by Lu and Ye (1994). In a process with total 73 stages, pure Y ($Y_2O_3 > 99.99$ %) is produced from the first-stage aqueous outlet and non-Y heavy rare earth ($Y_2O_3 < 0.3$ %) is produced from the stripping outlet. A third outlet is opened around 53rd stage where Er is accumulated. The Er product contains about 60 % Er_2O_3 that can be used for further high-purity product separation. In the heavy rare earth product, the Er_2O_3 content is less than 3 %. Extending the number of extraction stages to 110 will produce fluorescent-grade Y product.

Normally, the Y content in the rare earth feed needs to be higher than 60 % Y_2O_3/RE_xO_y to enable effective separation. Xu and Yu (1989) reported that the Y in the mixed rare earth medium Y content (about 30 %) and high La content (about 22 %) cannot be separated by the naphthenic acid-HCl-$RECl_3$ system. Due to the low separation factor between La and Y, La/Y separation using naphthenic acid system is not practical (Tian and Yin 1995). However, naphthenic acid can be used to realize group separation of LaY from the others to produce a high-purity LaY product that can be further separated to high-purity La product and high-purity Y product using other extraction systems (Zhang et al. 1995).

High-purity Y has wide applications in fluorescence, laser, and magneto optic, where higher purity of 99.999 % (fluorescent grade) is required. It can be produced in the naphthenic acid system using extended number of extraction stages (Yang and Liu 2004). As shown in Fig. 4.46, 117 stages are used to produce fluorescent-grade Y. It is common that concentrated ammonium hydroxide is often used for naphthenic acid saponification. However, where ammonium in effluent is strictly regulated, naphthenic acid-sulfoxide-kerosene system can be used. Sodium hydroxide is normally used for the saponification of the naphthenic acid in this system. The addition of either mixed alcohol or sulfoxide is to improve the fluidity of organic solution, to reduce the solubility of naphthenic acid in aqueous solution, to improve the solubility of naphthenic salt in organic solution, and to avoid the polymerization between naphthenic acid and salt.

The naphthenic acid-sulfoxide-kerosene system may be preferable due to the strong odor of mixed alcohol. Tian and coworkers (Tian 1989; Tian et al. 2001) compared these two systems. The comparison is shown in Table 4.37. Except for

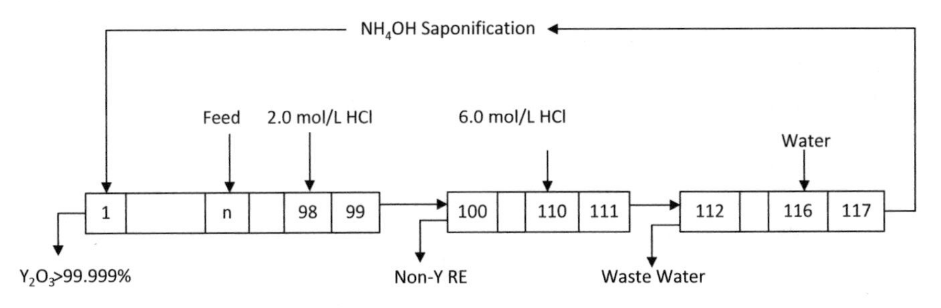

Fig. 4.46 Schematic fluorescent-grade Y_2O_3 production process

Table 4.37 Comparison of naphthenic acid-$RECl_3$ systems with mixed alcohol and petroleum sulfoxide

Extraction system	Loading capacity (mol/L)	Separation factor $\beta_{RE/Y}$	Phase separation (min)	Relative viscosity (mm^2/S)
With petroleum sulfoxide	0.224	2.36	7.0	0.69
With mixed alcohol	0.224	2.07	8.0	1.00

odor, the system with petroleum sulfoxide has slightly lower viscosity and better fluidity. Reliable production of 99.999 % Y_2O_3 in naphthenic acid-sulfoxide-kerosene system has been reported (Zhang 1998; Yang and Liu 2004). To ensure product quality, the raffinate of naphthenic acid extraction containing high-purity Y is always polished using HEH/EHP (Zhang 1998; Liu 2009).

4.7.3 *Impurity Removal*

Naphthenic acid is a weak organic acid. It has extraction ability for rare earth only when the system pH >4.0 (see Fig. 4.41). However, in the feeding stage with low acidity ($[H^+] < 0.01$ mol/L), impurities like Fe^{3+} and Al^{3+}will be extracted preferentially and flow with organic solution. When the organic solution reaches scrubbing stages where the acidity is high ($[H^+] > 0.1$ mol/L), they will be stripped and flow with the aqueous back to the feeding stage where they are extracted again. This will result in the accumulation of impurities like Fe^{3+} and Al^{3+}. At high concentration, they will be hydrolyzed to form gel-like hydroxides. The gel-like precipitates are the major causes of O/W-type emulsions in operation. Another important impurity is Si which exists as SiO_3^{2-} or $SiO_2{\cdot}xH_2O$ and $(H_2SiO_3)_2$. They are not extracted by naphthenic acid but they can react with Fe^{3+} and Al^{3+} to form gel-like precipitates in terms of $Fe(OH)_3{\cdot}SiO_2{\cdot}xH_2O$ and $Al(OH)_3{\cdot}SiO_2{\cdot}xH_2O$. Mixing with organic solution and aqueous solution, these gel-like precipitates will form a negatively charged gel film to prevent the contact between liquid drops resulting in O/W-type emulsion. Therefore, impurities like Al^{3+}, Fe^{3+}, and Si^{4+} must be removed.

The major impurity removal method includes hydrolysis method and solvent extraction method.

Hydrolysis method utilizes the lower precipitation pH of impurity to realize the separation of RE^{3+} from Fe^{3+}, Al^{3+}, etc. Ammonium hydroxide solution is used to raise the solution pH to 4.5–5.0 to precipitate the majority of Fe^{3+} and Al^{3+} and then NH_4HCO_4 solution is used to raise the pH to 5.2–5.4 to complete the impurity precipitation.

Solvent extraction using naphthenic acid is one effective Fe^{3+} and Al^{3+} removal method. At O/A = 1.5, Al^{3+} distribution ratio is 4.035 in 0.25 mol/L saponified naphthenic acid-iso-octyl alcohol-sulfonated kerosene-$RECl_3$-HCl system. The separation factor $\beta_{Al/RE}$ is 36.351. Single-stage Al removal is 85.8 % and two-stage Al removal is 95.0 % (Zeng et al. 2012). However, the separation factor $\beta_{Al/RE}$ is affected by the degree of saponification, feed Al^{3+} concentration, feed pH, O/A ratio, etc. Normally good Al/RE separation can be expected for feed solution with low Al^{3+} concentration and low pH. For feed solution with high Al^{3+} concentration, for example 4.0 g/L at low pH, good Al/RE separation cannot be realized (Li et al. 2010a, b). Emulsion issues occur when the Al^{3+} concentration is high for example at 6.34 g/L in the feed solution. But Al^{3+} can be effectively removed by slowly adding NH_4OH to the extraction system using naphthenic acid without saponification. When the ratio of NH_4OH/feed is >0.14, Al^{3+} can be reduced from 6.34 g/L to 0.02 g/L (Liu et al. 2012). Alternatively, chloronaphthenic acid can be used to avoid emulsion issues for the separation of Al/RE in high Al feed solution (Li et al. 2010a, b; Zhang 2011).

4.7.4 *Rare Earth Recovery from Low-Concentration Solution*

One of the important applications of naphthenic acid extraction system is to recover and concentrate rare earth from low-concentration solutions. The rare earth concentration in the leaching solution of low-grade rare earth ore or recycled secondary rare earth materials is relatively low. For example, the HCl leaching solution of rare earth mine mud contains 5.9 g/L RE^{3+}. It can be upgraded to 56.0 g/L using four stages of countercurrent extraction and two stages of stripping with the extraction system of 25 % naphthenic acid (22 % saponification)-15 % iso-octyl alcohol-60 % sulfonated kerosene. The rare earth recovery is 96 % at O/A = 1 (Zhang et al. 1999a, b). The major impurities in the mine mud leaching solution are Mn^{2+}, Ca^{2+}, Mg^{2+}, K^{+}, Na^{+}, etc. According to the metal extraction sequence by naphthenic acid (see Sect. 4.7.1), the extraction of Mn^{2+} is ahead of the others. Therefore, the separation of RE^{3+}/Mn^{2+} is the major consideration for this solution. Extraction equilibrium indicates that an optimum aqueous pH is 4.2 where D_{RE} is 7.96. At a higher aqueous pH, the extraction of RE^{3+} and Mn^{2+} will all increase and will not favor their separation. The saponification also has important impact on the

Table 4.38 Typical chemical composition of rare earth phosphors (Tu et al., 2012)

Y_2O_3	Eu_2O_3	Tb_4O_7	CeO_2	La_2O_3	Al_2O_3	SiO_2	CaO	MgO	SO_3	Fe_2O_3
16.95	1.50	0.86	1.08	0.07	69.04	0.67	0.18	0.17	0.15	0.13

RE^{3+}/Mn^{2+} separation. The rare earth recovery increases with the increase in saponification. However, precipitation of rare earth and manganese will occur and hinder the extraction. Good RE^{3+}/Mn^{2+} separation can be expected at 22 % saponification. The rare earth stripping in naphthenic acid extraction system is easy. At 25 °C and O/A = 1, two stages of stripping using 3.0 mol/L HCl solution can strip 99.8 % of the extracted rare earth.

The recovery of rare earth elements from recycled materials has drawn increasing attention due to their relatively high rare earth content. For example, rare earth phosphor as shown in Table 4.38 contains 16.95 % Y_2O_3 and 1.50 % Eu_2O_3.

Solvent extraction system with 25 % naphthenic acid-20 % iso-octyl alcohol-55 % sulfonated kerosene is used to separate the HCl leaching solution of the rare earth phosphors at pH 4.5 and O/A = 1. A product with 98.57 % Y_2O_3 and 1.21 % Eu_2O_3 is produced through six stages of cross-flow extraction.

4.7.5 *Effects of Additives on Naphthenic Acid Extraction System*

Additives such as alcohols, TBP, petroleum sulfoxide, and ethers are always used in the naphthenic acid-diluent extraction system. Without additives, emulsification can cause serious phase disengagement problems. The additives do not extract rare earth when used alone. However, apparent synergism or antisynergism can be observed when the additives are added to the naphthenic acid extraction system. For example, rare earth extraction first increases and then decreases with the increase in alcohol concentration. Naphthenic acid as an organic weak acid exists as polymers in nonpolar diluents. The addition of alcohol leads to the dissociation of polymers due to the formation of hydrogen bond between naphthenic acid and alcohol as shown in the following equation:

$$(\mathrm{LH}:\mathrm{HL})_n + 2n\mathrm{ROH} = 2n(\mathrm{HL}\cdots\mathrm{HOR}) \tag{4.97}$$

where *HL* represents the naphthenic acid and *ROH* represents the alcohol.

The dissociation of naphthenic acid polymer leads to the increase of its effective concentration and the corresponding rare earth extraction. However, the addition of excessive alcohol will form more hydrogen bonds and reduce the extraction power of naphthenic acid significantly. Correspondingly, the rare earth extraction decreases at high alcohol concentration:

$$\mathrm{HL}\cdots\mathrm{HOR} + \mathrm{HOR} = \mathrm{ROH}\cdots\mathrm{HL}\cdots\mathrm{HOR} \tag{4.98}$$

Different additives react differently with naphthenic acid to form either synergism or antisynergism. Currently, mixed alcohol is the major additive used in the naphthenic acid extraction system.

4.8 Amine Extraction System

Amines are the organic compounds with a functional group that contains a basic nitrogen atom with a lone pair. Amines are derivatives of ammonia normally containing over 12 carbons with high molecular weight. According to the number of hydrogen atoms replaced by substitute groups, amines are classified into four subcategories: primary amines, secondary amines, tertiary amines, and quaternary ammonium compounds. Their formula structures are shown below:

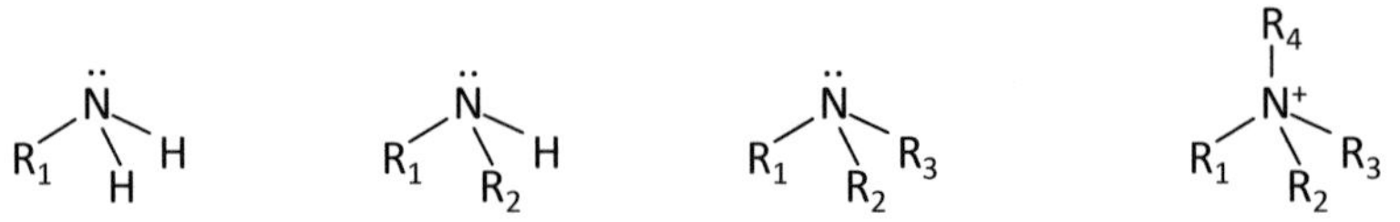

Primary Amine Secondary Amine Tertiary Amine Quaternary Ammonium

where R_1, R_2, R_3, and R_4 are alkyl or aryl groups.

The extraction ability of amines is closely related to their alkalinity. In aqueous solution, the alkalinity of amines is measured by alkalinity constant:

$$\mathrm{R3N} + \mathrm{H_2O} = \mathrm{R3NH^+} + \mathrm{OH^-} \tag{4.99}$$

$$K_\mathrm{B} = \frac{[\mathrm{R3NH^+}] \cdot [\mathrm{OH^-}]}{[\mathrm{R3N}]} \tag{4.100}$$

where R3N is used to represent the amine.

In inert organic solvent, the alkalinity constant is expressed by the reaction constant with acid:

$$\mathrm{R3N} + \mathrm{HA} = \mathrm{R3NH^+} + \mathrm{A^-} \tag{4.101}$$

$$\mathrm{K_B} = \frac{[\mathrm{R3NH^+}] \cdot [\mathrm{A^-}]}{[\mathrm{R3N}] \cdot [\mathrm{HA}]} \tag{4.102}$$

where *HA* represents an inorganic acid.

Metal extraction by amines in acidic media follows anion-exchange mechanism (also referred as to ion-pair mechanism). As shown by the following reaction equation, extraction takes place between the negatively charged metal complex and amine-acid complex:

$$MA_{n(a)}^{(z-n)} + (n-z)R3NHA_{(o)} = \left[(R3NH)_{n-z}MA_n\right]_{(o)} + (n-z)A_{(a)}^{-} \tag{4.103}$$

where z is the charge of the metal cation and n is the number of inorganic acid ligand in the metal complex.

Based on the extractability of metal ions by amines, they can be classified into four groups (Liu 1979).

- Group 1 metal ions such as Mo^{6+}, U^{6+}, V^{5+}, and Zr^{4+} can be extracted by primary, secondary, and tertiary amines.
- Group 2 metal ions like U^{4+}, Th^{4+}, and Ti^{4+} can be extracted by primary and secondary amines. Extraction by tertiary amines is difficult.
- Group 3 metal ions like V^{3+}, Fe^{3+}, and RE^{3+} can be extracted by primary amines. Extraction by secondary amines is difficult. They are not extracted by tertiary amines.
- Group 4 metal ions like Al^{3+}, V^{4+}, and Cr^{3+} and most of the divalent cations like Fe^{2+}, Co^{2+}, and Ni^{2+} are not extractable by amines.

In the 1960s, Rice and Stone (1962) investigated the rare earth extraction with a broad range of 41 amines. Favorable extraction is with some primary amines which include heptadecylamine, dodecylamine, Primene 81-R, and Primene JM-T. The only secondary amine with extraction for rare earth is n-benzylheptadecylamine (BHDA). Primene 81-R is a mixture of primary aliphatic amines with highly branched chains (t-$C_{12}H_{25}NH_2$ to t-$C_{14}H_{29}NH_2$). Primene JM-T is similar to Primene 81-R with greater molecular weight (t-$C_{18}H_{37}NH_2$ to t-$C_{21}H_{43}NH_2$). There is no rare earth extraction with tertiary amines. The favorable extraction is from rare earth sulfate solution. Extraction from chloride, nitrate, and perchlorate solutions is negligible. As shown in Table 4.39, the rare earth separation factors in

Table 4.39 Rare earth separation factors in amine-sulfate solution system

	Amine			Kerosene diluent		Isopropyl ether diluent	
Separation	Type	Concentration (mol/L)	pH	RE^{3+} (mol/L)	Separation factor	RE^{3+} (mol/L)	Separation factor
Pr/Nd	Primene 81-R	0.2	1.6	0.015	1.42	0.015	1.38
	Primene JM-T	0.2	1.6	0.015	1.32	0.015	1.29
	BHDA	0.2	1.6	0.015	1.24	0.015	1.18
Sm/Nd	Primene 81-R	0.2	1.6	0.015	2.09	0.015	2.18
	Primene JM-T	0.2	1.6	0.015	1.62	0.015	1.51
	BHDA	0.2	1.6	0.015	1.28	0.015	1.34
Ce(III)/La	Primene 81-R	0.1	1.6	0.005	1.39	0.015	1.37
	Primene JM-T	0.1	1.6	0.005	1.41	0.015	1.45
	BHDA	0.1	1.6	0.005	0.67	0.015	0.76

the amine-sulfate solution system are mostly in the range of 1–2 which is not sufficiently high to render good separation.

In order to increase the separation factors, a chelating agent such as diethylenetriaminepentaacetic acid (DTPA) can be added to the amine extraction system. For example, the addition of DTPA to Primene JM-T-kerosene extraction system can increase the separation factors to as high as 5.4 for Ce/La, 8.0 for Pr/Ce, 3.7 for Nd/Pr, and 7.0 for Sm/Nd (Bauer et al. 1968).

The acidity of the primary amine extraction system has important impact on the molecular species in both organic and aqueous phases (Coleman 1963). At high acidity, metal extraction is inhibited due to the formation of more stable amine bisulfate:

$$(RNH_3)_2SO_4 + H_2SO_4 = 2RNH_3HSO_4 \tag{4.104}$$

At low acidity, the rare earth ions precipitate due to hydrolysis and amine salt reverts to free base:

$$(RNH_3)_2SO_4 + 2OH^- = 2RNH_2 + SO_4^{2-} + 2H_2O \tag{4.105}$$

Therefore, an appropriate acidity range needs to be determined for effective rare earth extraction in the primary amine extraction system.

N1923 is an effective commercial extractant for separating rare earths, thorium, and iron from sulfuric acid solution. It contains >99 % primary amine with the following structure. Its molecular weight is 280.9 and nitrogen content is 4.99 % (Li and Liu 1988).

$$R-\underset{\displaystyle R'}{\underset{|}{CH}}-\ddot{N}H_2$$

where R and R′ are all C_9–C_{11} alkyl groups.

At appropriate acidity ($0.0001 < H_2SO_4 < 0.5$ mol/L), N1923 reacts with sulfuric acid to form amine sulfate which extracts RE^{3+} into organic phase:

$$RNH_{2(o)} + 2H^+ + SO_4^{2-} = (RNH_3)_2SO_{4(o)} \tag{4.106}$$

$$1.5(RNH_3)_2SO_{4(o)} + RE^{3+} + 1.5SO_4^{2-} = (RNH_3)_3RE(SO_4)_{3(o)} \tag{4.107}$$

where RNH_2 is used to represent N1923.

At high acidity ($H_2SO_4 > 1$ mol/L), as shown by Eq. (4.104) the amine sulfate is converted to amine bisulfate. The extraction of RE^{3+} with amine bisulfate is

$$1.5(RNH_3HSO_4)_{2(o)} + RE^{3+} + 1.5SO_4^{2-} = (RNH_3)_3RE(SO_4)_{3(o)} + 1.5H_2SO_4 \tag{4.108}$$

Li et al. (1987a, b) investigated the extraction of 15 rare earth elements with N1923. The extraction of rare earth decreases with the increase in H_2SO_4 concentration in aqueous phase. At constant acidity, the extraction of rare earth decreases with the increase in atomic number showing “reverse extraction sequence.”

The extraction of trivalent Fe^{3+} by N1923 from H_2SO_4 solution follows the same extraction mechanism as the extraction of trivalent RE^{3+}. The extraction reaction of tetravalent Th^{4+} and Ce^{4+} in sulfuric acid solution by N1923 is shown below:

$$2(RNH_3)_2SO_{4(o)} + M^{4+} + 2SO_4^{2-} = (RNH_3)_4M(SO_4)_{4(o)} \quad (4.109)$$

where M^{4+} represents Th^{4+} or Ce^{4+}.

References

Bauer, D. J., & Lindstrom, R. E. (1964). Naphthenic acid solvent extraction of rare earth sulfates. United States of Department of the Interior. Bureau of Mines, Report of Investigations 6396.

Bauer, D. J., Lindstrom, R. E., & Higbie, K. B. (1968), Extraction behavior of cerium-group lanthanides in a primary amine-chelating agent system. United States Department of the Interior, Bureau of Mines, Report of Investigation 7100.

Chai, J. (1998). Effects of solvent on the extraction of rare earth elements with HDEHP. *Chinese Journal of Rare Metals, 22*(2), 85–89.

Chen, H. (1993). Study of Nd/Sm group separation for XunWu rare earth using P204-HCl system. *Gansu Nonferrous Metal, 8*, 32–36.

Chen, Y., Li, G., & Chen, G. (1981). Solvent extraction separation of Sc and Th. *Rare metals, 6*, 82–83.

Chu, D., Ma, G., & Li, D. (1998). Extraction of rare earth ions from nitrate medium with cyanex 923. *Chinese Journal of Analytical Chemistry, 26*(11), 1346–1349.

Coleman, C. F. (1963). Amines as extractants. *Nuclear Science and Engineering, 17*(2), 274–286.

Dai, Z., Song, W., & Wang, C. (1985). Solvent extraction separation of yttrium and mixed rare earth low in yttrium. Chinese Patent, CN 82102270.

Dai, Z., Wang, C., & Wang, Z. (1985). Solvent extraction separation of high purity yttrium. Chinese Patent, CN 85102220.

Deng, Z., Xu, T., Zheng, C., Li, G., & Zhang, W. (1990). Group separation of middle and heavy rare earths by stripping P204 organic solution. *JiangXi YouSe JinShu, 4*(2), 5–8.

Deng, Z., Xu, Y., & Yang, F. (2003). Optimization study on the extraction separation process of mixed light rare earth. *Jiangxi Nonferrous Metals, 17*(1), 29–31.

Ding, H., Xu, Y., Yang, W., Yuan, S., & Lu, X. (2007). Extraction studies of Th(IV) from HCl solution by DHEHP. *Science Technology and Engineering, 7*(22), 5752–5754.

Ding, Y., & Chen, L. (2001). Study of RE extraction separation flowsheet for bastnasite. *JiangXi Metallurgy, 21*(3–4), 75–78.

El-Hefny, N. E., El-Nadi, Y. A., & Daoud, J. A. (2010). Equilibrium and mechanism of samarium extraction form chloride medium using sodium salt of cyanex 272. *Separation and Purification Technology, 75*, 310–315.

El-Nadi, Y. A. (2010). Effect of diluents on the extraction of praseodymium and samarium by cyanex 923 from acidic nitrate medium. *Journal of Rare Earths, 28*(2), 215–220.

El-Nadi, Y. A., El-Hefny, N. E., & Daoud, J. A. (2007). Extraction of lanthanum and samarium from nitrate medium by some commercial organophosphorus extractants. *Solvent Extraction and Ion Exchange, 25*, 225–240.

Fan, M., Meng, X., Wang, W., Zhang, X., & Zhou, Y. (2002). Industrial experiments of extracting neodymium and removing impurities in P204-HCl system. *Chinese Rare Earths, 23*(3), 63–64.

Fu, Z. (1991). Extraction separation and purification of yttrium oxide with naphthenic acid from hydrochloric acid solution. *Uranium Mining and Metallurgy, 10*(2), 42–47.

Gao, S., Shen, X. H., Chen, Q. D., & Gao, H. (2012). Solvent extraction of thorium (IV) using W/O microemulsion. *Science China-Chemistry, 55*(9), 1712–1718.

Hao, X. (1995). P204-external reflux extraction for middle rare earth separation. *Science &Technology of Baotou Steel (Group) Corporation, 21*(1, supplement), 94–98.

Hao, X., Zhang, L., & Wang, Q. (1995). Middle rare earth separation process in P204-HCl system-external reflux technology. *Chinese Rare Earths, 16*(1), 11–16.

He, J., Tan, H., Cheng, X., Zhang, P., & Cai, R. (1991). Study of Sc extraction mechanism by P204. *Chinese Rare Earths, 12*(2), 22–26.

Hou, S. (2005). Applications of P507 AND p204 in rare earth separation. *Science &Technology of Baotou Steel (Group) Corporation, 31*(supplement), 26–29.

Isogawa, C., Murayama, N., & Shibata, J. (2015). Solvent extraction of scandium with mixed extractant of versatic acid 10 and TBP. *Journal of Engineering Science and Technology*, Special Issue on SOMCHE 2014 & RSCE 2014 Conference, January (2015), 78–85.

Le, S., & Li, D. (1990). Kientics and mechanism of extraction of Er(III) from H_2SO_4 solution with HEH(EHP). *Chinese Journal of Applied Chemistry, 7*(3), 1–5.

Li, D., Ji, E., Xu, X., Yu, D., Zeng, G., & Ni, J. (1987a). The extraction mechanism of RE(III), Fe(III), and Th(IV) in sulfuric acid solution by N1923. *Chinese Journal of Applied Chemistry, 4*(2), 36–41.

Li, D., & Liu, D. (1988). Extraction mechanism of Al (III) in sulfuric acid solution with N1923. *Nonferrous Metals (Extractive Metallurgy), 3*, 29–31.

Li, D., Lu, J., Wei, Z., Wang, H., Meng, S., & Ma, G. (1998). An extraction separation process of Ce and Th from bastnasite leaching solution. Chinese Patent, CN 1254024A.

Li, D., Wang, H., & Chen, Z. (1988). Studies of extraction mechanism of rare earth elements with HEH(EHP). *Acta Chimica Sinica, 46*, 492–495.

Li, D., Wang, H., Zeng, G., & Xue, Z. (1984). Extraction mechanism of Ce(IV) from H_2SO_4 solution by HEH (EHP). *Journal of the Chinese Rare Earth Society, 2*(2), 9–19.

Li, D., Wu, Z., Guo, X., & Ji, E. (1982). Light/heavy rare earth separation using P507 from HNO_3 solution. *Journal of Rare Earth, 1*, 23–27.

Li, D., Zhang, J., & Xu, M. (1985). Studies of extraction mechanism of rare earth compounds with MONO (2-ethyl hexyl) -2 ethyl hexyl phosphonate (HEH/EHP). *Chinese Journal of Applied Chemistry, 2*(2), 17–23.

Li, H., Chen, Z., & Meng, S. (1987b). Extraction equilibria of individual rare earth in HEH(EHP)-kerosene-HCl-$RECl_3$ system. *Journal of The Chinese Rare Earth Society, 5*(3), 71–74.

Li, H., & Cheng, Z.H. (1990). Thermodynamics of rare earth solvent extraction: $Nd(NO_3)_3$-P350 system. *Rare Metals, 5*, 332–337.

Li, H.G. (2005). *Hydrometallurgy*. ZhongNan University Press, ChangSha, China.

Li, J., Zhang, X., Chang, H., & Wu, W. (2010a). Research on extraction separation of rare-earths and aluminum by chloronaphthenic acid system. *Mulipurpose Utilization of Mineral Resources, 5*, 43–45.

Li, J., Zhang, X., Xu, Y., & Wu, W. (2010b). Research on distribution ratio and separation coefficient of single rare earth and aluminum in naphthenic acid system. *Nonferrous Metals (Extractive Metallurgy), 4*, 33–36.

Li, J. (1987). A process to extraction scandium from rare earth Ore. Chinese Patent, CN85106255A.

Li, L. (2011). *Rare earth extraction and separation* (pp. 194–321). Inner Mongolia: Inner Mongolia Science and Technology Press.

Li, Q., & Li, D. (1995). Extraction mechanism of Th(IV) with HBTMPP from HCL acid medium. *Chinese Journal of Applied Chemistry, 12*(4), 58–61.

Liao, F., Zhu, S., & Tao, H. (2007). Study on the extraction and separation of Nd (III) and Y(III) with Cyanex272. *Journal of Instrumental Analysis, 26*(4), 563–565.
Liao, W., Yu, G., & Li, D. (2001a). Extraction mechanism of Ce(IV) and F(I) in the separation process of Bastnasite leach solution by cyanex 923. *Acta Metallurgica Sinica (English Letters), 14*(1), 21–26.
Liao, W., Yu, G., & Li, D. (2002). Kinetics of Ce(IV) extraction from H_2SO_4-HF medium with Cyanex 923. *Talanta, 56*, 613–618.
Liao, W., Yu, G., & Li, D. (2001b). Solvent extraction of Ce(IV) and F(I) from sulfuric acid leaching of Bastnasite by Cyanex 923. *Solvent Extraction and Ion Exchange, 19*(2), 243–259.
Liu, C. (1979). Amine extractants. *Organic Chemistry, 1*, 70–89.
Liu, J., & Wang, Y. (2014). Extraction of praseodymium (III) and neodymium (III) in saponified P705-HCl-Kerosene system. *CIESC Journal, 65*(1), 264–270.
Liu, M. (2009). Discussion of high purity Y_2O_3 production process. *Chemical Engineering & Equipment, 10*, 31–32.
Liu, Z., Zhu, W., & Guo, Q. (2012). Study on process of extracting aluminum from high-alumina rare earth solution. *Materials Research and Application, 6*(4), 256–258.
Long, Z., Huang, X., Huang, W., & Zhang, G. (2000). Ce4+ extraction mechanism from rare earth sulfate solution containing fluorine with DEHPA. *Journal of the Chinese Rare Earths Society, 18*(1), 18–20.
Lu, J., & Li, D. (1999). Separation of Ce(IV) and Th (IV) from RE(III) in HNO_3 solution by Cyanex 923 extractant. *Acta Metallurgica Sinica (English Letters), 12*(2), 191–197.
Lu, J., Ma, G., Li, D., & Ni, M. (1998a). Extraction kinetics of Er with HBTMPP. *Chinese Journal of Applied Chemistry, 15*(3), 43–46.
Lu, J., Wei, Z., Li, D., Ma, G., & Jiang, Z. (1998b). Recovery of Ce(IV) and Th (IV) from rare earths (III) WITH Cyanex 923. *Hydrometallurgy, 50*, 77–78.
Lu, Y., & Ye, Z. (1994). Industrial scale testing of New rare earth extraction process using naphthenic acid to produce Y. *E, and Heavy Rare Earth Products in One-step, Chinese Rare Earths, 15*(1), 14–17.
Ma, E. (1989). The development of rare earth solvent extraction research. *Rare Metals and Alloys, 99*, 30–35.
Ma, G., & Li, D. (1992). Extraction separation of scandium (III) and Iron (III) from HCl solution with HBTMPP. *Chinese Journal of Analytical Chemistry, 20*(10), 1113–1116.
Ma, H., Han, Q., Li, W., & Ma, W. (2011a). Study on Y(III) extraction and stripping performance of Cyanex923 in nitric acid medium. *Chemistry & Bioengineering, 28*(7), 58–60.
Ma, H., Li, W., Li, J., & Ma, W. (2011b). Study of extraction performance of Cyanex923 for La(III) in nitrate medium. *Chemistry & Bioengineering, 28*(11), 42–44.
Panda, N., Devi, N. B., & Mishra, S. (2014). Solvent extraction of Pr(III) from acidic nitrate medium using Cyanex 921 and Cyanex 923 as extractants in kerosene. *Turkish Journal of Chemistry, 38*, 504–511.
Peppard, D. F., Ferraro, J. R., & Mason, G. W. (1958a). Hydrogen bonding in organophosphoric. *Journal of Inorganic and Nuclear Chemistry, 7*(3), 231–244.
Peppard, D. F., Mason, G. W., Driscoll, W. J., & Sironen, R. J. (1958b). Acidic esters of orthophosphoric acid as selective extractants for metallic cations—tracer studies. *Journal of Inorganic and Nuclear Chemistry, 7*(3), 276–285.
Peppard, D. F., Mason, G. W., Maier, J. L., & Driscoll, W. J. (1957). Fractional extraction of the lanthanides as their Di-alkyl orthophosphates. *Journal of Inorganic and Nuclear Chemistry, 4*(5–6), 334–343.
Qiao, J., Liu, Z., & Hao, X. (2002). Studies on extraction of light rare earths with HDEHP in sulphuric acid medium. *Chinese Rare Earths, 23*(4), 29–32.
Rice, A. C., & Stone, C. A. (1962), Amines in Liquid-Liquid Extraction of Rare Earth Elements, United States Department of the Interior, Bureau of Mines, Report of Investigation 5923.
Shen, C., Xie, T., & Li, D. (1985). Process of separating mixed rare earth separation using Saponified HEH(EHP), Chinese Patent, CN 85 102210 B.

Song, N., Liao, W., Tong, S., Jia, Q., Liu, W., & Shi, Y. (2009). Solvent extraction of rare earths with mixtures of HDEHP and sec-nonylphenoxy acetic acid. *Chinese Journal of Analytical Chemistry, 37*(11), 1633–1637.

Sun, D. (1994). The rule of transition of extraction kinetics behavior of rare earth elements with HEH(EHP). *Acta Chimica Sinica, 52*, 1095–1099.

Tan, A.F. (1997). Production of high purity Y by naphtenicacid solvent extraction method. *Rare Metals and Hard Alloys, 19*(2), 48–51.

Tian, J. (1989). High purity Y_2O_3 separation from heavy rare earth using naphthenic acid-petroleum sulphoxide-$RECl_3$ system. *Chinese Rare Earths, 10*(6), 22–28.

Tian, J., & Yin, J. (1995). La/Y separation in naphthenic acid-mixed alcohol-$RECl_3$ system. *Jiangxi Metallurgy, 15*(4), 28–29.

Tian, J., Zhao, Q., Yin, J., Cai, H., Chen, S., & Cheng, R. (2001). Yttrium separation for mixed RECl3 solution using naphthenic aicd-petroleum sulphoxide system. *Hydrometallurgy of China, 20*(1), 37–40.

Tian, Y., Zhai, Y., Xiuying, Z., Zhang, X., & Xiao, F. (1998). Extraction of Sc^{3+} by P204 in HCl system. *Journal of Northeastern University (Natural Science), 19*(2), 162–165.

Tong, H., Lei, J., & Li, D. (2003). Mass transfer kinetics of extraction of RE(III) with Cyanex 923. *Journal of WuHan University of Technology, 25*(6), 7–9.

Tu, Y., Wang, X., Mei, G., Lu, K., & Weng, X. (2012). Extraction and separation of Y and Eu from waste rare earth phospher. *Moden Mining, 520*(8), 29–31.

Wang, B., Guo, X., Fan, H., Wu, Y., Gao, J., & Kang, J. (1989). Solid-liquid extraction separation and its application in thorium and rare earth-TBP-NH_4SCN-paraffin Wax system. *Chinese Journal of Rare Metals, 22*(4), 241–245.

Wang, J., Wu, G., Liu, L., Liu, J., & Chen, J. (2002). Study of D2EHPA in vitriol medium of separating Non-RE impurities from RE-chloride by extraction. *Chinese Rare Earths, 23*(6), 69–70.

Wang, L., Long, Z., Huang, X., Peng, X., Han, Y., & Cui, D. (2009a). Extraction of trace rare earth from phosphoric acid. *Journal of the Chinese Rare Earth Society, 27*(2), 228–233.

Wang, L., Long, Z., Huang, X., Peng, X., Han, Y., & Cui, D. (2009b). Extraction kinetics of trace rare earth from phosphoric acid. *Journal of the Chinese Rare Earth Society, 27*(6), 812–815.

Wang, W., Wang, X., Meng, S., Li, H., & Li, D. (2006). Extraction and stripping of Yb(III) from H_2SO_4 medium by Cyanex 923. *Journal of Rare Earths, 24*, 685–689.

Wang, X.T., Sun, D.C., Zhang, Z.F., & Wang, J.J. (1986). P350 properties and mechanism on rare earth solvent extraction. *Rare Earth, 3*, 25–28.

Wang, X. (1998). P204 separating and concentrating scanadium. *Liaoning Chemical Industry, 27*(6), 320–322.

Wang, Z., Meng, S., Song, W., Guo, C., Qi, J., & Li, D. (1995). Extraction separation of rare earth elements (III) with Bis(2,4,4-trimethylpentyl) phosphinic acid. *Chinese Journal of Analytical Chemistry, 23*(4), 391–394.

Wu, J. (1988). Solvent extraction separation of rare earth elements. *Resource Processing Technologies, 35*(2), 108–114.

Xiong, Y., Liu, S., & Li, D. (2006a). Kinetics Y(III) extraction with C272 using a constant interfacial cell with laminar flow. *Journal of Alloys and Compounds, 408–412*(2006), 1056–1060.

Xiong, Y., Wang, Y. G., & Li, D. (2004). Kinetics of extraction and stripping of Y(III) by C272 as an acidic extractant using a constant interfacial cell with laminar flow. *Solvent Extraction and Ion Exchange, 22*(5), 833–851.

Xiong, Y., Wu, D., & Li, D. (2006b). Mass transfer kinetics of Y(III) using a constant interfacial cell with laminar flow. Part II extraction with C272. *Hydrometallurgy, 82*, 184–189.

Xu, BX. (2003). Handbook of rare earth separation. Processing, Designing and Optimization, and Rare Earth Material Application. Jinlin Music Press, p. 400–550.

Xu, G., & Yuan, C. (1987). *Solvent extraction of rare earth* (pp. P195–P204). Beijing: China Science Press.

Xu, X., & Yu, Z. (1989). Naphthenic acid-HCl-$RECl_3$ system separating mixed rare earth with high La but low Y. *Chinese Rare Earths, 10*(3), 54–56.

Yan, C. H., Liao, C., Jia, J., Wu, S., & Li, B. (1999). Comparison of the ecumenical and technical indices on rare earth separation processes of Ion-adsorptive deposit by solvent extraction. *Journal of the Chinese Rare Earth Society, 17*(3), 256–262.

Yang, G., Li, G., Luo, Y., Jiang, G., Zhao, Z., Ma, H., Guo, H., & Heaping, Z. (1998). Study of technology on removing impurity Sm2O3 from Nd2O3. *Journal of the Chinese Rare Earth Society, 19*(6), 15–19.

Yang, J., Ling, C., Han, Q., & Han, X. (2013). A light rare earth separation method, Chinese Patent, CN 102912157 A.

Yang, Q., & Liu, Z. (2004). Extraction of high purity Y_2O_3 with naphthenic acid. *Chinese Rare Earths, 25*(3), 35–38.

Zang, L., & Wang, Q. (1995). The new application of RE's saponification technique in the separation of LRE. *Journal of the Chinese Rare Earth Society, 16*(3), 28–31.

Zeng, Q., Zeng, Q., & Chang, Q. (2012). Study on the extraction and separation of rare earth and aluminum in naphthenic acid system. *Nonferrous Metals Science and Engineering, 3*(2), 17–19.

Zhang, F., Ma, G., & Li, D. (1999a). Transfer and extraction performances of Er and Y with HBTMPP in hollow fibers. *Chinese Journal of Applied Chemistry, 16*(2), 84–86.

Zhang, G., Huang, X., Gu, B., Hu, K., & Luo, Y. (1988). Rare Earth separation from H_2SO_4 Solution, Chinese Patent, CN 86105043 A.

Zhang, L. (1998). Study of producing fluorescent grade yttrium oxide in naphthenic acid-petroleum sulphoxide-hyrochloric acid system. *Chinese Rare Earths, 19*(6), 19–22.

Zhang, L., Hao, X., & Wang, Q. (1993). Study on extraction of rare earths with HDEHP in HCl acid medium. *Chinese Rare Earths, 14*(3), 22–29.

Zhang, P., Jiang, F., & Lu, S. (1999b). To separate rare earth from the aqueous chloinate of mine Mud by naphthenic acid extraction—extraction technology. *Chemistry World, 6*, 326–330.

Zhang, W., Zhang, L., You, S., Feng, S., & Hou, S. (1991). P350-HCl system extracting high purity scandium process. *Journal of Rare Earth, 12*(4), 18–21.

Zhang, X., & Li, D. (1993). Extraction of rare earth ions (III) with Bis(2,4,4-trimethylpentyl) phosphinic acid. *Chinese Journal of Applied Chemistry, 10*(4), 72–74.

Zhang, X. (2011). Experimental research on the distribution ratio and separation coefficient of rare earth and aluminum in saponified chloronaphthenic acid system. *Chinese Rare Earths, 32*(2), 93–97.

Zhang, Z., He, P., Zhang, C., Bao, F., & Fang, J. (1995). Industrial separation testing of LaY/RE separation by naphthenic acid extraction system. *Chinese Rare Earths, 16*(2), 28–32.

Zhao, Y., & Li, D. (1990). Mechanism of extraction of scandium by Di (1-methylheptyl)methyl phosphonate. *Chinese Journal of Applied Chemistry, 7*(2), 1–5.

Zhong, S., Tao, M., Jiang, R., & Li, J. (2001). New extractional process for separation of the rare earth mineral with middle Y and rich Eu. *Chinese Rare Earths, 22*(2), 26–29.

Zhou, J., Yan, C., & Liao, C. (1998). Process of eliminating fluorine and extracting cerium (IV) from Mianning Bastnaesite. *Chinese Rare Earths, 19*(3), 9–17.

Chapter 5
Cascade Solvent Extracting Principles and Process Design

5.1 Introduction

The common rare earth solvent extraction systems are introduced in Chap. 4. After the solvent extraction system is determined based on the separation factors and distribution ratios of the rare earths, separation processes are to be designed according to the feed characteristics, product specifications, and separation targets. Cascade solvent extraction processes are commonly used for industrial rare earth separation. This chapter introduces the rare earth cascade extraction principles and equations developed by Professor Guangxian Xu (Xu and Yuan 1987; Xu 1995) at the Peking University, China.

A cascade solvent extraction process involves multistage contacts of aqueous phase and organic phase in order to achieve targeted separation of one or one group of solutes from the other. According to the flow patterns of organic phase and aqueous phase, the cascade extraction is normally classified into counter-current extraction, cross flow extraction, fractional extraction, and circulating extraction etc.

The research and development of rare earth cascade extraction process focuses on the distribution of rare earths between organic phase and aqueous phase in each stage at different process configurations and different process parameters so that the relationship between separation targets and process parameters can be determined.

The nomenclature used in this chapter is summarized in Table 5.1.

J. Zhang et al., *Separation Hydrometallurgy of Rare Earth Elements*,
DOI 10.1007/978-3-319-28235-0_5

Table 5.1 Nomenclature

Sign	Name	Unit
A	Solute(s) easily extractable	
$A_{(a)}$	Mass flow rate of A in aqueous phase	mmol/min or g/min
$A_{(o)}$	Mass flow rate of A in organic phase	mmol/min or g/min
$A_{F(a)}$	Mass flow rate of A in aqueous feed	mmol/min or g/min
$[A_{F(a)}]$	Concentration of A in aqueous feed	mol/L or g/L
$[A_{i(a)}]$	Concentration of A in aqueous phase in stage i	mol/L or g/L
$[A_{i(o)}]$	Concentration of A in organic phase in stage i	mol/L or g/L
$[A_{n+m(o)}]$	Concentration of A in organic outlet	mol/L or g/L
a	Concentrating factor of A	
B	Solute(s) difficult to extract	
$B_{(a)}$	Mass flow rate of B in aqueous phase	mmol/min or g/min
$B_{(o)}$	Mass flow rate of B in organic phase	mmol/min or g/min
$B_{F(a)}$	Mass flow rate of B in aqueous feed	mmol/min or g/min
$[B_{F(a)}]$	Concentration of B in aqueous feed	mol/L or g/L
$[B_{i(a)}]$	Concentration of B in aqueous phase in stage i	mol/L or g/L
$[B_{i(o)}]$	Concentration of B in organic phase in stage i	mol/L or g/L
$[B_{1(a)}]$	Concentration of B in aqueous outlet	mol/L or g/L
b	Concentrating factor of B	
D_A	Average distribution ratio of A in extraction	
D_B	Average distribution ratio of B in extraction	
D'_A	Average distribution ratio of A in scrubbing	
D'_B	Average distribution ratio of B in scrubbing	
E_A	Average extraction factor of A in extraction: $E_A = A_{(o)}/A_{(a)}$	
E_B	Average extraction factor of B in extraction: $E_B = B_{(o)}/B_{(a)}$	
E'_A	Average extraction factor of A in scrubbing: $E'_A = A'_{(o)}/A'_{(a)}$	
E'_B	Average extraction factor of B in scrubbing: $E'_B = B'_{(o)}/B'_{(a)}$	
E_M	Average extraction factor in extraction: $E_M = (A_{(o)} + B_{(o)})/(A_{(a)} + B_{(a)})$	
E'_M	Average extraction factor in scrubbing: $E'_M = (A'_{(o)} + B'_{(o)})/(A'_{(a)} + B'_{(a)})$	
f_A	Mole fraction of A in feed	
f'_A	Mole fraction of solutes (A + B) in organic outlet	
f_B	Mole fraction of B in feed: $f_B = 1 - f_A$	
f'_B	Mole fraction of solutes (A + B) in aqueous outlet: $f'_B = 1 - f'_A$	
i	Stage i in extraction	
j	Stage j in scrubbing	
J_S	Extraction reflux ratio: $J_S = S_{(o)}/M_{1(a)}$	
J_w	Washing reflux ratio: $J_S = W_{(a)}/M_{n+m(o)}$	

(continued)

Table 5.1 (continued)

Sign	Name	Unit
M	Mixture of A and B	
$M_{F(o)}$	Organic feed mass flow rate	mmol/min or g/min
$M_{F(a)}$	Aqueous feed mass flow rate	mmol/min or g/min
$[M_{1(a)}]$	Concentration of A and B in aqueous outlet	mol/L or g/L
$[M_{n+m(o)}]$	Concentration of A and B in organic outlet	mol/L or g/L
$M_{1(a)}$	Mass flow rate of A and B in aqueous outlet	mmol/min or g/min
$M_{n+m(o)}$	Mass flow rate of A and B in organic outlet	mmol/min or g/min
m	Number of scrubbing stages excluding feeding stage	
n	Number of extraction stages including feeding stage	
$P_{A(o)}$	Product purity of A in organic outlet	
$P_{B(o)}$	Mole fraction of B in organic outlet: $P_{B(o)} = 1 - P_{A(o)}$	
$P_{B(a)}$	Product purity of B in aqueous outlet	
$P_{A(a)}$	Mole fraction of A in aqueous outlet: $P_{A(a)} = 1 - P_{B(a)}$	
$Q_{A(o)}$	Daily Production of A in organic outlet	kg/day
$Q_{B(a)}$	Daily production of B in aqueous outlet	kg/day
$Q_{F(a)}$	Daily capacity of feed in terms of REO	kg/day
R	Organic phase/aqueous phase ratio in extraction: $R = V_o/(V_{aF} + V_w)$	
R'	Organic phase/aqueous phase ratio in scrubbing: $R' = V_o/V_w$	
r	Volume ratio of mixing/settling	
$S_{(o)}$	Maximum extraction of solutes (A and B)	mol/min or g/min
$S_{B(o)}$	Maximum extraction of B	mol/min or g/min
t	Mixing time of extraction	min
t'	Mixing time of scrubbing	min
V	Effective volume of mixing in each stage	L
V_t	Total volume of mixing and settling in extraction	L
V'_t	Total volume of mixing and settling in scrubbing	L
V_a	Aqueous volume or flow rate	mL or mL/min
V_{aF}	Aqueous feed volume or flow rate	mL or mL/min
V_{oF}	Organic feed volume or flow rate	mL or mL/min
V_o	Organic phase volume or flow rate	mL or mL/min
V_w	Scrubbing solution volume or flow rate	mmol/min or g/min
$W_{(a)}$	Maximum scrubbing of solutes	mL or mL/min
Y_A	A Recovery	
Y_B	B Recovery	
β	Average separation factor in extraction: $\beta = D_A/D_B = E_A/E_B$	
β'	Average separation factor in scrubbing: $\beta' = D'_A/D'_B = E'_A/E'_B$	

(continued)

Table 5.1 (continued)

Sign	Name	Unit
ϕ_A	Extraction residual fraction of A in fractional extraction: $\phi_A = A_1/A_F$	
ϕ_B	Extraction residual fraction of B in fractional extraction: $\phi_B = B_1/B_F$	
$1-\phi_A$	Extraction fraction of A in fractional extraction: $1-\phi_A = A_{n+m(o)}/A_F$	
$1-\phi_B$	Extraction fraction of B in fractional extraction: $1-\phi_B = B_{n+m(o)}/B_F$	
ψ_A	Extraction residual fraction of A in countercurrent extraction	
ψ_B	Extraction residual fraction of B in countercurrent extraction	

5.2 Assumptions of Xu's Cascade Extraction Principles

Xu's cascade extraction principles are based on five basic assumptions for rare earth separation:

1. *Extractability of Rare Earth Elements*
   According to the order of extractability, rare earth elements are represented by A, B, C . . ., and so on. The two-outlet separation process only considers the separation of A and B. A is the one or one group of rare earth elements with relatively high extractability and B is the one or one group of rare earth elements with relatively low extractability. For example, in the separation of LaCePr/Nd, A will be used to represent Nd and B will be used to represent LaCePr.
   In the three-outlet process, a middle component is also considered. For example for the separation of La/CePr/Nd, Nd will be represented by A which is easily extractable, CePr will be represented by B as the middle component, and La will be represented by C which is relatively difficult to extract.
2. *Average Separation Factor* (β)
   Separation factor is defined as the ratio of distribution ratios of A and B in the separation system.

$$\beta_{A/B} = \frac{D_A}{D_B} \tag{5.1}$$

   Due to the variation of distribution ratios in different stages with slightly different conditions, the A/B separation factors (β_i) vary slightly in different stages in a rare earth fractional extraction process. In cascade process design, the average separation factor $\beta_{A/B}$ is used. When the separation factor in extraction is different with the separation factor in scrubbing, two average separation factors $\beta_{A/B}$ and $\beta'_{A/B}$ are used respectively.
3. Constant Extraction Factor (E)
   The extraction factor E in a continuous operation is defined as the ratio of the mass flow rate of a solute in the organic phase to the mass flow rate of the solute in the aqueous phase.

$$E = \frac{\text{Mass flow rate of the solute in organic phase}}{\text{Mass flow rate of the solute in aqueous phase}} \tag{5.2}$$

In rare earth separation, it is a common practice to control the rare earth concentration in the organic phase in each stage to be as close as possible. Therefore, the extraction factor E_{RE} in extraction and E'_{RE} in scrubbing are approximately constant.

4. *Rare Earth Composition in Feeding Stage*
   In the design of the two-outlet process, it is assumed that the aqueous feed and aqueous phase in the feeding stage have the same rare earth composition. If it is organic feeding, the organic feed and the organic phase in the feeding stage are assumed to have the same rare earth composition.
5. *Constant Flow Ratio* (R)
   It is assumed that the extraction stages and the scrubbing stages all have constant organic flow to aqueous flow ratios.

5.3 Countercurrent Solvent Extraction Process

5.3.1 *Extraction Residual Fraction* ϕ *and Recovery* Y

For large volume operation and more efficient use of solvent, countercurrent extraction is employed. As shown in Figure 5.1, the organic phase flows from stage 1 to stage n while the aqueous phase flows oppositely from stage n to stage 1. The countercurrent operation conserves the mass transfer driving force and renders optimal performance.

In rare earth solvent extraction of A and B, the objective of a countercurrent process is to extract A to the organic phase and leave B in the aqueous phase. However, in reality partial A is always left in the aqueous phase and partial B is always extracted into the organic phase. The extraction residual fraction ϕ is

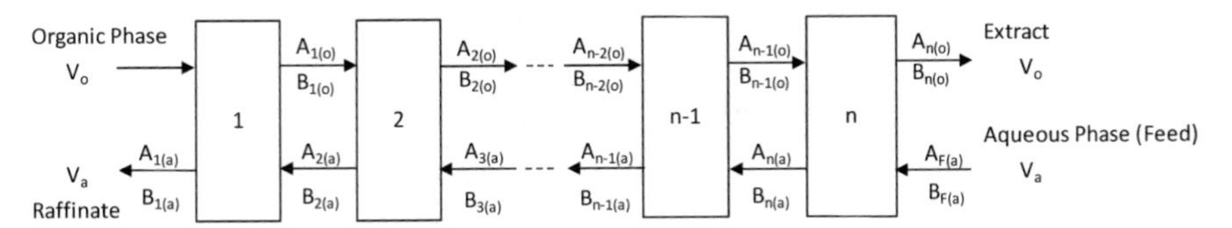

Fig. 5.1 Countercurrent extraction process and solute distribution. V_o: organic flow rate; V_a: aqueous flow rate; $A_{(o)}$: mass flow rate of A in organic phase; $A_{(a)}$: mass flow rate of A in aqueous phase; $B_{(o)}$: mass flow rate of B in organic phase; $B_{(a)}$: mass flow rate of B in aqueous phase

defined as the ratio of the mass flow rate of A in the aqueous outlet to the mass flow rate of A in the aqueous feed.

$$\phi_A = \frac{\text{Mass flow rate of A in aqueous outlet}}{\text{Mass flow rate of A in aqueous feed}} = \frac{[A_{1(a)}]V_a}{[A_{F(a)}]V_a} = \frac{[A_{1(a)}]}{[A_{F(a)}]} \quad (5.3)$$

where $[A_{(a)}]$ is the aqueous concentration of A and V_a is the aqueous flow rate.

In stage 1 in Fig. 5.1, the mass balance of A is:

$$[A_{2(a)}]V_a = [A_{1(a)}]V_a + [A_{1(o)}]V_o \quad (5.4)$$

The flow ratio R is:

$$R = \frac{V_o}{V_a} \quad (5.5)$$

The extraction factor of A in any stage i is:

$$E_A = \frac{[A_{i(o)}]V_o}{[A_{i(a)}]V_a} = \frac{[A_{i(o)}]}{[A_{i(a)}]}R = D_A R \quad (5.6)$$

Therefore, Eq. (5.4) can be expressed as:

$$[A_{2(a)}] = [A_{1(a)}](1 + E_A) \quad (5.7)$$

In stages 1 and 2, the mass balance of A is:

$$\begin{aligned} [A_{3(a)}] &= [A_{1(a)}] + [A_{2(o)}]R \\ &= [A_{1(a)}] + [A_{2(a)}]E_A \\ &= [A_{1(a)}](1 + E_A + E_A^2) \end{aligned} \quad (5.8)$$

In all stages, the mass balance of A is:

$$\begin{aligned} [A_{F(a)}] &= [A_{1(a)}] + [A_{n(o)}]R \\ &= [A_{1(a)}] + [A_{n(a)}]E_A \\ &= [A_{1(a)}](1 + E_A + E_A^2 + \cdots + E_A^n) \end{aligned} \quad (5.9)$$

Multiply E_A, Eq. (5.9) can be converted to:

$$E_A[A_{F(a)}] = [A_{1(a)}](E_A + E_A^2 + E_A^3 \cdots + E_A^{n+1}) \quad (5.10)$$

Therefore, Eq. (5.10) – Eq. (5.9) = :

$$(E_A - 1)[A_{F(a)}] = [A_{1(a)}](E_A^{n+1} - 1) \tag{5.11}$$

when $E_A \neq 1$, the extraction residual fraction is:

$$\phi_A = \frac{[A_{1(a)}]}{[A_{F(a)}]} = \frac{E_A - 1}{E_A^{n+1} - 1} \tag{5.12}$$

when $E_A = 1$, the extraction residual fraction is:

$$\phi_A = \frac{[A_{1(a)}]}{[A_{F(a)}]} = \frac{1}{n + 1} \tag{5.13}$$

The same relationships exist for solute B.

$$\phi_B = \frac{E_B - 1}{E_B^{n+1} - 1} \tag{5.14}$$

The recovery of A is the ratio of its mass flow rate in organic outlet to its mass flow rate in aqueous feed. Therefore,

$$Y_A = \frac{[A_{n(o)}]V_o}{[A_{F(a)}]V_a} = \frac{[A_{F(a)}]V_a - [A_{1(a)}]V_a}{[A_{F(a)}]V_a} = 1 - \phi_A \tag{5.15}$$

The recovery of B (Y_B) is the ratio of its mass flow rate in aqueous outlet to its mass flow in the feed. Therefore,

$$Y_B = \frac{[B_{1(a)}]V_a}{[B_{F(a)}]V_a} = \phi_B \tag{5.16}$$

Normally, $E_B < 1$ in two-outlet processes and $E_B{}^{n+1} \ll 1$, therefore:

$$Y_B = \phi_B = \frac{E_B - 1}{E_B^{n+1} - 1} \approx 1 - E_B \tag{5.17}$$

5.3.2 *Concentrating Factor and Purity*

The concentrating factor of a solute is used to measure its purity change after n stages of countercurrent extraction. For the easily extractable solute A, it is defined as the ratio of the A/B ratio in organic outlet to the A/B ratio in aqueous feed.

$$a=\frac{[A_{n(o)}]/[B_{n(o)}]}{[A_{F(a)}]/[B_{F(a)}]}=\frac{([A_{F(a)}]-[A_{1(a)}])}{R[A_{F(a)}]}\cdot\frac{R[B_{F(a)}]}{([B_{F(a)}]-[B_{1(a)}])}=\frac{(1-\phi_A)}{(1-\phi_B)} \tag{5.18}$$

The purity of A in the organic phase outlet is:

$$P_A=\frac{[A_{n(o)}]}{[A_{n(o)}]+[B_{n(o)}]}=\frac{a[A_{F(a)}]}{a[A_{F(a)}]+[B_{F(a)}]} \tag{5.19}$$

The concentrating factor of B is defined as the ratio of B/A concentration ratio in aqueous outlet to the B/A concentration ratio in feed.

$$b=\frac{[B_{1(a)}]/[A_{1(a)}]}{[B_{F(a)}]/[A_{F(a)}]}=\frac{\phi_B}{\phi_A} \tag{5.20}$$

The purity of B in aqueous outlet is:

$$P_B=\frac{[B_{1(a)}]}{[B_{1(a)}]+[A_{1(a)}]}=\frac{b[B_{F(a)}]}{b[B_{F(a)}]+[A_{F(a)}]} \tag{5.21}$$

5.3.3 *Example of Application*

Using P350 extraction system to separate La/Pr with the presence of salting out agent, the separation factor $\beta_{Pr/La}=5.0$. By manipulating the organic/aqueous flow ratio R, the extraction factors of La and Pr can be controlled to $E_A=E_{Pr}=2.5$ and $E_B=E_{La}=0.5$. If the feed composition is 50 % La and 50 % Pr, what is the expected La purity and recovery after ten stages of countercurrent extraction?

Assuming the stage efficiency is 90 %, the effective number of stages is:

$$n=10\times 90\%=9$$

Then the extraction residual fractions of Pr and La are:

$$\phi_{Pr}=\phi_A=\frac{E_A-1}{E_A^{n+1}-1}=\frac{2.5-1}{2.5^{10}-1}=1.57\times 10^{-4}$$

$$\phi_{La} = \phi_B = 1 - E_B = 1 - 0.5 = 0.5$$

The concentrating factor of La is:

$$b = \frac{\phi_B}{\phi_A} = \frac{0.5}{1.57 \times 10^{-4}} = 3.18 \times 10^3$$

The purity of La is:

$$P_{La} = \frac{b[B_{F(a)}]}{b[B_{F(a)}] + [A_{F(a)}]} = \frac{3.18 \times 10^{-3} \times 50\%}{3.18 \times 10^{-3} \times 50\% + 50\%} = 99.97\%$$

The recovery of La is:

$$Y_{La} = Y_B = \phi_B = 1 - E_B = 50\%$$

It can be seen that the La product has good purity but with low recovery. If 70 % La recovery is desired for the same feed composition in the same extraction stages, what is the expected purity?

To reach 70 % recovery, it will need to reduce the La extraction factor E_B to 0.3. This can be easily done by changing the organic/aqueous flow ratio R. However, the change of R will also affect the Pr extraction factor E_A. According to Eq. (5.6), E_A and E_B are:

$$E_A = D_A R, \quad E_B = D_B R$$

Therefore:

$$\frac{E_A}{E_B} = \frac{D_A R}{D_B R} = \beta_{A/B}$$

Assuming separation factor is constant, the Pr extraction factor will become:

$$E_A = E_B \beta_{A/B} = 0.3 \times 5 = 1.5$$

Therefore,

$$\phi_{Pr} = \phi_A = \frac{E_A - 1}{E_A^{n+1} - 1} = \frac{1.5 - 1}{1.5^{10} - 1} = 8.82 \times 10^{-3}$$

$$\phi_{La} = \phi_B = 1 - E_B = 1 - 0.3 = 0.7$$

$$b = \frac{\phi_{B}}{\phi_{A}} = \frac{0.7}{8.82 \times 10^{-3}} = 79.3$$

$$P_{La} = P_{B} = \frac{b[B_{F(a)}]}{b[B_{F(a)}] + [A_{F(a)}]} = \frac{79.3 \times 50\%}{79.3 \times 50\% + 50\%} = 98.75\%$$

It can be seen that the increase of La recovery from 50 to 70 % will reduce the La purity from 99.97 to 98.75 % by the same number of extraction stages. If both 70 % recovery and 99.97 % purity are desired for the same feed, can this be achieved by increasing the number of extraction stages?

To reach 70 % La recovery with the same product purity, it will need $E_{B} = 0.3$ and the same concentrating factor $b = 3.18 \times 10^{3}$.

$$Y_{La} = Y_{B} = \phi_{B} = 1 - E_{B} = 70\%, \ E_{B} = 0.3$$

$$b = \frac{\phi_{B}}{\phi_{A}} = 3.18 \times 10^{3}$$

Therefore,

$$\phi_{B} = 0.7, \ \phi_{A} = 2.2 \times 10^{-4}, \text{ and } E_{A} = 1.5$$

$$\phi_{A} = \frac{E_{A} - 1}{E_{A}^{n+1} - 1} = \frac{1.5 - 1}{1.5^{n+1} - 1} = 2.2 \times 10^{-4}, n = 18.06$$

The required theoretical number of stages is 18.06.

Assuming the same stage efficiency of 90 %, the actual number of stages is 21. It indicates that increasing the number of extraction stages to 21, 70 % La recovery can be achieved with the purity of 99.97 %. Another question is what is the maximum La recovery with the same product purity? And how many extraction stages are needed to reach the maximum La recovery?

Based on above analysis and calculations, to increase the La recovery will need to reduce the La extraction factor E_{B}. Accordingly, the Pr extraction factor E_{A} will be reduced as well when the separation factor $\beta_{A/B}$ is constant. However, E_{A} cannot be less than 1. Therefore, E_{B} cannot be less than $E_{A}/\beta_{A/B} = 1/5 = 0.2$. The maximum La recovery will be $1 - 0.2 = 80$ %. When $E_{A} = 1$, according to Eqs. (5.13) and (5.20), the extraction residual fraction φ_{A} and the theoretical number of extraction stages are:

$$\phi_{A} = \frac{\phi_{B}}{b} = \frac{0.8}{3.18 \times 10^{3}} = \frac{1}{n + 1}, n = 3974$$

If 90 % stage efficiency is used, it will need 4416 stages of extraction (3974 theoretical stages).

The example indicates that it is not practical to seek high product purity and high recovery at the same time in rare earth separation by using only countercurrent solvent extraction process.

5.4 Cross Flow Solvent Extraction Process

The representation of cross flow solvent extraction processes is shown in Fig. 5.2. It includes cross flow extraction, cross flow washing (scrubbing), and cross flow stripping.

In the cross flow extraction process, fresh solvent contacts with the feed in all stages as the feed flows from the first stage to the last stage. The purpose of cross flow extraction is to extract the component A as complete as possible from the aqueous phase to produce high purity B in the aqueous outlet. When the separation factor $\beta_{A/B}$ is large, a pure B product can be produced. However, the recovery of B is low and the consumption of organic solvent is large. Therefore, it is not commonly used in industrial production.

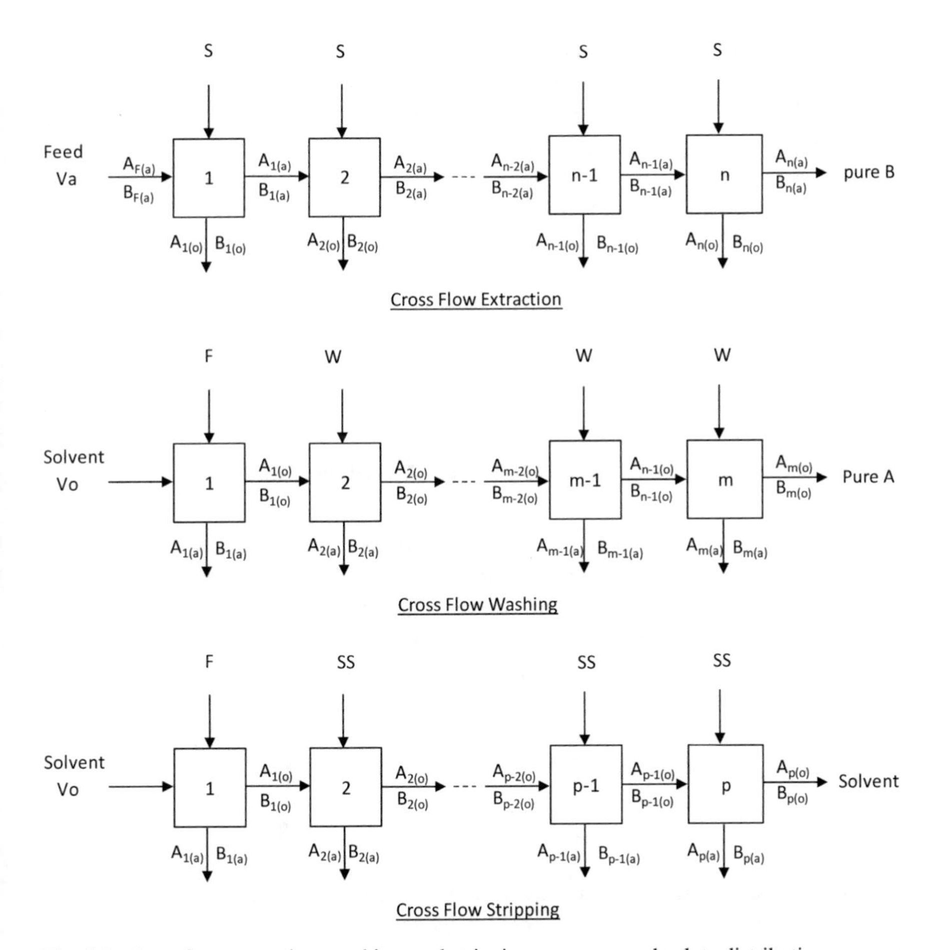

Fig. 5.2 Cross flow extraction, washing, and stripping processes and solute distribution

In the cross flow washing (scrubbing), fresh washing solution contacts with the loaded organic phase in every washing stage. When the separation factor $\beta_{A/B}$ is large, a pure product of A can be produced. However, the recovery of A is low and the consumption of washing solution is large. It can be used for the separation of rare earth elements and non-rare earth impurities.

The cross flow stripping is similar to the cross flow washing. Due to the high consumption of stripping solution, it is only used in the production of those with high purity or difficult to strip.

In industrial rare earth separation, cross flow processes are rarely used. Therefore, cross flow process design and calculations are not discussed in this chapter.

5.5 Fractional Solvent Extraction Process

As discussed in Sect. 5.3, countercurrent solvent extraction process can produce high purity product but cannot reach high recovery at the same time. In the 1980s, Xu and coworkers developed rare earth cascade solvent extraction principles and the methods for designing and calculating the optimum process parameters. These principles and methods have been successfully used in industrial rare earth separation processes over 30 years. With the adoption of advanced computing technology, the application of these principles and methods have been progressed and extended to new rare earth separation process design and simulation. Xu's rare earth cascade solvent extraction principles and methods are discussed in details in the following sections.

5.5.1 *Material Balance of Aqueous Feeding System*

Figure 5.3 shows the fractional extraction process with two solutes, A and B. Organic solution is fed from the first stage and aqueous solution is discharged from the first stage. The feed is introduced in the nth stage, which is referred as to the feeding stage. The organic solution is discharged from the $(n+m)$th stage and the scrubbing solution is introduced in the $(n+m)$th stage. The stages 1 to n are extraction stages and the stages $(n+1)$ to $(n+m)$ are scrubbing stages.

Assuming M is the mass flow rate of the total solutes (A + B), the organic phase and aqueous phase have the following relations in any stage i.

$$M_{i(a)} = A_{i(a)} + B_{i(a)} \quad M_{i(o)} = A_{i(o)} + B_{i(o)} \tag{5.22}$$

Using M_F to represent the mass flow rate of the feed and assuming $M_F = 1$ mmol/min or 1 g/min, the mole fractions or mass fractions (f_A and f_B) of A and B have the following relationships:

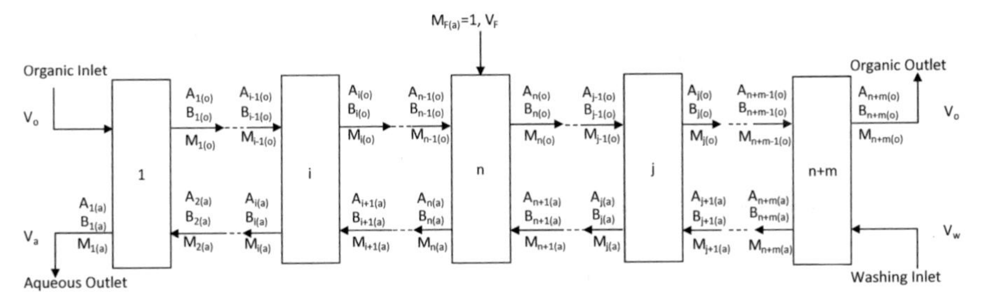

Fig. 5.3 Fractional solvent extraction process with two solutes A and B

$$f_{\mathrm{A}} + f_{\mathrm{B}} = 1,\ M_{\mathrm{F}} = M_{1(\mathrm{a})} + M_{n+m(\mathrm{o})} \tag{5.23}$$

$$f_{\mathrm{A}} = \frac{A_{1(\mathrm{a})} + A_{n+m(\mathrm{o})}}{M_{\mathrm{F}}},\ f_{\mathrm{B}} = \frac{B_{1(\mathrm{a})} + B_{n+m(\mathrm{o})}}{M_{\mathrm{F}}} \tag{5.24}$$

The recovery of A in organic outlet (Y_{A}) and the recovery of B in the aqueous outlet (Y_{B}) are:

$$Y_{\mathrm{A}} = \frac{A_{n+m(\mathrm{o})}}{f_{\mathrm{A}} M_{\mathrm{F}}},\ Y_{\mathrm{B}} = \frac{B_{1(\mathrm{a})}}{f_{\mathrm{B}} M_{\mathrm{F}}} \tag{5.25}$$

The purity of A in organic outlet and the purity of B in aqueous outlet are:

$$P_{\mathrm{A}_{n+m(o)}} = \frac{A_{n+m(\mathrm{o})}}{M_{n+m(\mathrm{o})}} = \frac{A_{n+m(\mathrm{o})}}{A_{n+m(\mathrm{o})} + B_{n+m(\mathrm{o})}},\ P_{B_{1(\mathrm{a})}} = \frac{B_{1(\mathrm{a})}}{M_{1(\mathrm{a})}} = \frac{B_{1(\mathrm{a})}}{A_{1(\mathrm{a})} + B_{1(\mathrm{a})}} \tag{5.26}$$

Therefore, the mass flow rates of organic outlet $M_{n+m(\mathrm{o})}$ and aqueous outlet $M_{1(\mathrm{a})}$ are:

$$M_{n+m(\mathrm{o})} = \frac{A_{n+m(\mathrm{o})}}{P_{\mathrm{A}_{n+m(\mathrm{o})}}} = \frac{f_{\mathrm{A}} M_{\mathrm{F}} Y_{\mathrm{A}}}{P_{\mathrm{A}_{n+m(\mathrm{o})}}},\ M_{1(\mathrm{a})} = \frac{B_{1(\mathrm{a})}}{P_{B_{1(\mathrm{a})}}} = \frac{f_{\mathrm{B}} M_{\mathrm{F}} Y_{\mathrm{B}}}{P_{B_{1(\mathrm{a})}}} \tag{5.27}$$

Have f'_{A} as the solute fraction in organic outlet and f'_{B} as the solute fraction in aqueous outlet, then:

$$f'_{\mathrm{A}} = \frac{M_{n+m(\mathrm{o})}}{M_{\mathrm{F}}} = \frac{f_{\mathrm{A}} Y_{\mathrm{A}}}{P_{\mathrm{A}_{n+m(\mathrm{o})}}},\ f'_{\mathrm{B}} = \frac{M_{1(\mathrm{a})}}{M_{\mathrm{F}}} = \frac{f_{\mathrm{B}} Y_{\mathrm{B}}}{P_{\mathrm{B}_{1(\mathrm{a})}}} \tag{5.28}$$

According to Eqs. (5.23) and (5.28),

$$f'_{\mathrm{A}} + f'_{\mathrm{B}} = \frac{M_{n+m(\mathrm{o})} + M_{1(\mathrm{a})}}{M_{\mathrm{F}}} = 1 \tag{5.29}$$

When the organic product A and aqueous product B are all high purity, the following approximate relations exist:

$$P_{A_{n+m(o)}} \approx 1,\ P_{B_{1(a)}} \approx 1,\ Y_A \approx 1,\ Y_B \approx 1,\ f_A^{'} \approx f_A,\ f_B^{'} \approx f_B \tag{5.30}$$

In extraction, the mass balance is:

$$M_{i+1(a)} = M_{i(o)} + M_{1(a)},\ i = 1,\ 2,\ 3 \cdots n-1 \tag{5.31}$$

In scrubbing, the mass balance is:

$$M_{j+1(a)} = M_{j(o)} - M_{n+m(o)},\ j = n,\ n+1,\ \cdots,\ n+m-1 \tag{5.32}$$

Equations (5.31) and (5.32) are operating line equations. Similar operating line equations also exist for solute A and B:

$$A_{i+1(a)} = A_{i(o)} + A_{1(a)},\ B_{i+1(a)} = B_{i(o)} + B_{1(a)} \tag{5.33}$$

$$A_{j+1(a)} = A_{j(o)} - A_{n+m(o)},\ B_{j+1(a)} = B_{j(o)} + B_{n+m(o)} \tag{5.34}$$

The extraction factors of A, B, and their total are:

$$E_{Ai} = \frac{A_{i(o)}}{A_{i(a)}},\ E_{Bi} = \frac{B_{i(o)}}{B_{i(a)}},\ E_{Mi} = \frac{M_{i(o)}}{M_{i(a)}} = \frac{A_{i(o)} + B_{i(o)}}{A_{i(a)} + B_{i(a)}} \tag{5.35}$$

The maximum extraction of M to organic solution is detonated by S_o. The maximum scrubbing of M to washing solution is detonated by W_a. According to the mass balance in scrubbing, S_o and W_a have the following relation:

$$W_a = S_o - M_{n+m(o)} \tag{5.36}$$

Since M_F is assumed as 1,

$$f_A^{'} = \frac{M_{n+m(o)}}{M_F} = M_{n+m(o)},\ W_a = S_o - f_A^{'} \tag{5.37}$$

The ratio of $W_a/M_{n+m(o)}$ is defined as washing reflux ratio and detonated by J_W.

$$J_W = \frac{W_a}{M_{n+m(o)}} = \frac{W_a}{f_A^{'}} \tag{5.38}$$

According to Xu's assumption of constant extraction factor, extraction factor in scrubbing is:

$$E_{\mathrm{M}}' = \frac{M_{j(\mathrm{o})}}{M_{j(\mathrm{a})}} = \frac{S_{\mathrm{o}}}{W_{\mathrm{a}}} = \frac{W_{\mathrm{a}} + M_{n+m(\mathrm{o})}}{W_{\mathrm{a}}} = 1 + \frac{1}{J_{\mathrm{W}}} \text{ or } J_{\mathrm{W}} = \frac{1}{E_{\mathrm{M}}' - 1} \tag{5.39}$$

$M_{i(\mathrm{a})}$ ($i \neq 1$) in each stage of extraction is very close. Therefore, L_{a} is used to detonate $M_{i(\mathrm{a})}$ ($i \neq 1$) in all stages with the same total solutes. Then, the total solute balance in extraction can be expressed by:

$$L_{\mathrm{a}} = S_{\mathrm{o}} + M_{1(\mathrm{a})} = W_{\mathrm{a}} + M_{n+m(\mathrm{o})} + M_{1(\mathrm{a})} = W_{\mathrm{a}} + 1 \tag{5.40}$$

The ratio of $S_{\mathrm{o}}/M_{1(\mathrm{a})}$ is defined as the extraction reflux ratio and detonated by J_{S}.

$$J_{\mathrm{S}} = \frac{S_{\mathrm{o}}}{M_{1(\mathrm{a})}} \tag{5.41}$$

The extraction factor in extraction is:

$$E_{\mathrm{M}} = \frac{M_{i(\mathrm{o})}}{M_{i(\mathrm{a})}} = \frac{S_{\mathrm{o}}}{L_{\mathrm{a}}} = \frac{S_{\mathrm{o}}}{S_{\mathrm{o}} + M_{1(\mathrm{a})}} = \frac{J_{\mathrm{S}}}{1 + J_{\mathrm{S}}} \text{ or } J_{\mathrm{S}} = \frac{E_{\mathrm{M}}}{1 - E_{\mathrm{M}}} \tag{5.42}$$

According to above equations, the following relationships between E_{M} and E'_{M} can be obtained:

$$E_{\mathrm{M}} = \frac{E_{\mathrm{M}}' f_{\mathrm{A}}'}{E_{\mathrm{M}}' - f_{\mathrm{B}}'} \tag{5.43}$$

$$E_{\mathrm{M}}' = \frac{E_{\mathrm{M}} f_{\mathrm{B}}'}{E_{\mathrm{M}} - f_{\mathrm{A}}'} \tag{5.44}$$

The following relationships between J_{S} and J_{W} can be obtained:

$$J_{\mathrm{S}} = \frac{f_{\mathrm{A}}'}{f_{\mathrm{B}}'}(J_{\mathrm{W}} + 1) \tag{5.45}$$

$$J_{\mathrm{W}} = J_{\mathrm{S}} \frac{f_{\mathrm{B}}'}{f_{\mathrm{A}}'} - 1 \tag{5.46}$$

5.5.2 *Material Balance of Organic Feeding System*

The difference between aqueous feeding and organic feeding is illustrated in Fig. 5.4. Figure 5.4a shows the distribution and material balance of solutes

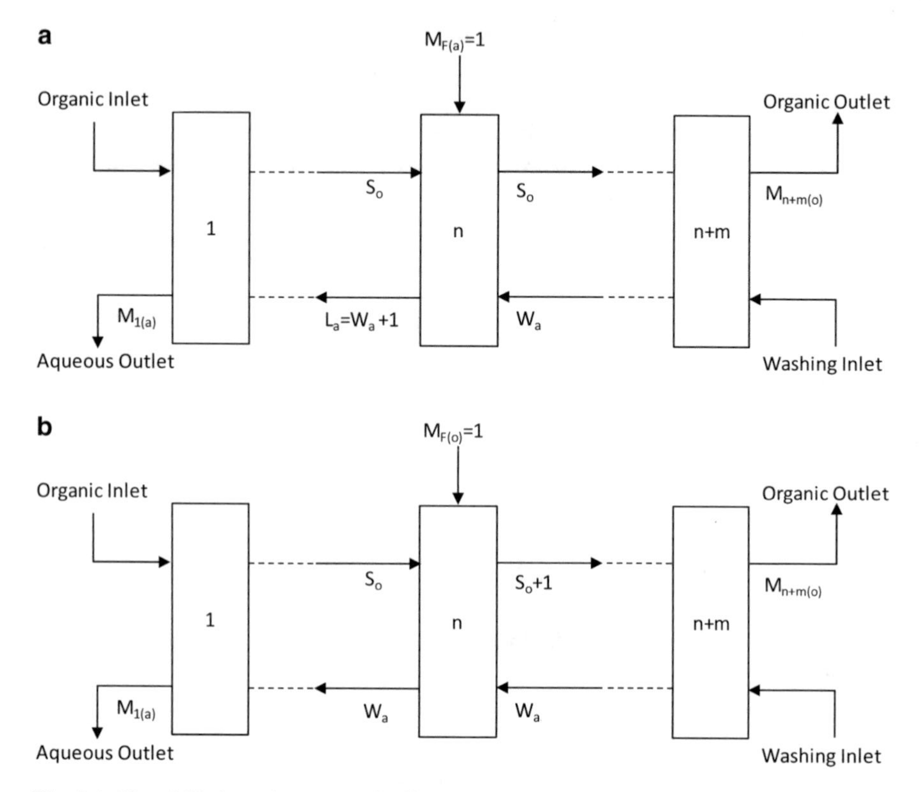

Fig. 5.4 Material balance in aqueous feeding system and organic feeding system

(A and B) in the aqueous feeding system. Figure 5.4b shows the distribution and material balance of solutes in the organic feeding system.

Based on the material balance in the organic feeding system in Fig. 5.4b, the following relationships between E_M and E'_M and between J_S and J_W can be obtained:

$$E_M = \frac{1 - E'_M f'_B}{f'_A} \tag{5.47}$$

$$E'_M = \frac{1 - E_M f'_A}{f'_B} \tag{5.48}$$

$$J_S = J_W \frac{f'_A}{f'_B} - 1 \tag{5.49}$$

$$J_W = \frac{f'_B}{f'_A}(J_S + 1) \tag{5.50}$$

5.5.3 Concentrating Factor

The concept of concentrating factor is introduced in Sect. 5.3 for the countercurrent solvent extraction process. In fractional solvent extraction process, the same concept applies. It is used to measure the separation of A and B.

The concentrating factor of A in fractional solvent extraction is the ratio of $[A]/[B]$ ratio in organic outlet to $[A]/[B]$ ratio in feed:

$$a = \frac{\left[A_{n+m(\mathrm{o})}\right] / \left[B_{n+m(\mathrm{o})}\right]}{\left[A_{\mathrm{F(a)}}\right] / \left[B_{\mathrm{F(a)}}\right]} = \frac{P_{\mathrm{A}_{n+m(\mathrm{o})}} / \left(1 - P_{\mathrm{A}_{n+m(\mathrm{o})}}\right)}{f_{\mathrm{A}} / f_{\mathrm{B}}} \tag{5.51}$$

Therefore, the purity of A in organic outlet can be expressed by the A concentrating factor and the mass fractions of A and B in feed.

$$P_{\mathrm{A}_{n+m(\mathrm{o})}} = \frac{af_{\mathrm{A}}}{af_{\mathrm{A}} + f_{\mathrm{B}}} \tag{5.52}$$

The concentrating factor of B in fractional solvent extraction is the ratio of $[B]/[A]$ ratio in aqueous outlet to $[B]/[A]$ ratio in feed:

$$b = \frac{\left[B_{1(\mathrm{a})}\right] / \left[A_{1(\mathrm{a})}\right]}{\left[B_{\mathrm{F(a)}}\right] / \left[A_{\mathrm{F(a)}}\right]} = \frac{P_{\mathrm{B}_{1(\mathrm{a})}} / \left(1 - P_{\mathrm{B}_{1(\mathrm{a})}}\right)}{f_{\mathrm{B}} / f_{\mathrm{A}}} \tag{5.53}$$

Therefore, the purity of B in aqueous outlet can be expressed by the B concentrating factor and the mass fractions of A and B in feed.

$$P_{\mathrm{B}_{1(\mathrm{a})}} = \frac{bf_{\mathrm{B}}}{bf_{\mathrm{B}} + f_{\mathrm{A}}} \tag{5.54}$$

Total concentrating factor of A and B is defined as the product of a and b:

$$ab = \frac{P_{\mathrm{A}_{n+m(\mathrm{o})}} P_{\mathrm{B}_{1(\mathrm{a})}}}{\left(1 - P_{\mathrm{A}_{n+m(\mathrm{o})}}\right)\left(1 - P_{\mathrm{B}_{1(\mathrm{a})}}\right)} \tag{5.55}$$

5.5.4 Extraction Residual Fraction

The extraction residual fractions of A and B are the ratios of mass flow rates of A and B in aqueous outlet to those in feed.

$$\phi_A = \frac{A_{1(a)}}{A_{F(a)}}, \ \phi_B = \frac{B_{1(a)}}{B_{F(a)}}, \tag{5.56}$$

The extraction fractions of A and B are:

$$1 - \phi_A = \frac{A_{n+m(o)}}{A_{F(a)}}, \ 1 - \phi_B = \frac{B_{n+m(o)}}{B_{F(a)}} \tag{5.57}$$

Since the ratio of A/B equals to the ratio of $[A]/[B]$ in each phase, the following relations between concentrating factors and extraction residual factors exist:

$$a = \frac{1 - \phi_A}{1 - \phi_B}, \ b = \frac{\phi_B}{\phi_A} \tag{5.58}$$

$$\phi_A = \frac{a - 1}{ab - 1}, \ \phi_B = 1 - \frac{b - 1}{ab - 1} \tag{5.59}$$

In a fractional extraction process with high separation efficiency, it is common that $a \gg 1$ and $b \gg 1$. Therefore:

$$\phi_A \approx \frac{1}{b}, \ \phi_B \approx 1 - \frac{1}{a} \tag{5.60}$$

$$a \approx \frac{1}{1 - \phi_B}, \ b \approx \frac{1}{\phi_A} \tag{5.61}$$

5.5.5 Example of Application

A rare earth solution contains 25 % Pr_6O_{11} and 75 % Nd_2O_3 is separated in N263-DTPA system by ten stages of extraction and ten stages of scrubbing. Two products of 99.9 % Nd_2O_3 and 99.8%Pr_6O_{11} are obtained. What are their concentrating factors, extraction residual fractions, total concentrating factor, and their recoveries?

In the N263-DTPA extraction system, Pr is A that is relatively easy to extract and Nd is B that is relatively difficult to extract. Therefore:

$$f_A = 0.25, \quad f_B = 0.75, \quad P_{An+m(o)} = 0.998, \quad P_{B1(a)} = 0.999$$

According to Eqs. (5.51) and (5.53), the concentrating factors are:

$$a = \frac{P_{\mathrm{A}_{n+m(\mathrm{o})}} / \left(1 - P_{\mathrm{A}_{n+m(\mathrm{o})}}\right)}{f_{\mathrm{A}}/f_{\mathrm{B}}} = \frac{0.998/(1-0.998)}{0.25/0.75} = 1347$$

$$b = \frac{P_{\mathrm{B}_{1(\mathrm{a})}} / \left(1 - P_{\mathrm{B}_{1(\mathrm{a})}}\right)}{f_{\mathrm{B}}/f_{\mathrm{A}}} = \frac{0.999/(1-0.999)}{0.75/0.25} = 333$$

$$ab = 1347 \times 333 = 4.48 \times 10^5$$

According to Eq. (5.60), the extraction residual factors are:

$$\phi_{\mathrm{A}} \approx \frac{1}{b} = \frac{1}{333} = 3.0 \times 10^{-3}$$

$$\phi_{\mathrm{B}} \approx 1 - \frac{1}{a} = 1 - \frac{1}{1347} = 0.999$$

The recoveries of A and B are:

$$Y_{\mathrm{A}} = 1 - \phi_{\mathrm{A}} = 1 - 0.003 = 99.7\%$$

$$Y_{\mathrm{B}} = \phi_{\mathrm{B}} = 0.999 = 99.9\%$$

5.6 Extrema Equations

Based on the Xu's assumptions for cascade solvent extraction and the principle of material balance, Xu and coworkers developed the extreme value equations for the fractional solvent extraction system in order to achieve effective separation of A and B.

5.6.1 Aqueous Feeding System

The minimum extraction factor in extraction of an aqueous feeding fractional extraction system is:

$$(E_{\mathrm{M}})_{\min} = \frac{(\beta f_{\mathrm{A}} + f_{\mathrm{B}})(P_{\mathrm{B}1} - f_{\mathrm{B}})}{(\beta f_{\mathrm{A}} + f_{\mathrm{B}})P_{\mathrm{B}1} - f_{\mathrm{B}}} \tag{5.62}$$

E_{M} must meet the following conditions in order to achieve effective separation.

$$(E_{\mathrm{M}})_{\min} < E_{\mathrm{M}} < 1 \tag{5.63}$$

When the aqueous outlet product is high purity B, $P_{B1(a)} \approx 1$. Equation (5.62) can be simplified to:

$$(E_M)_{min} \approx \frac{(\beta f_A + f_B)(1 - f_B)}{(\beta f_A + f_B) - f_B} = f_A + \frac{f_B}{\beta} \tag{5.64}$$

The separation factor E'_M in scrubbing must meet the following conditions to achieve effective washing:

$$1 < E'_M < \left(E'_M\right)_{max} \tag{5.65}$$

$$\left(E'_M\right)_{max} = \frac{f_B - P_{B_{n+m(o)}}}{\frac{f_B}{\beta f_A + f_B} - P_{B_{n+m(o)}}} \tag{5.66}$$

When the organic outlet product is high purity A, $P_{B_{n+m(o)}} \approx 0$. Equation (5.66) can be simplified to:

$$\left(E'_M\right)_{max} \approx \beta f_A + f_B \tag{5.67}$$

When the product in aqueous outlet and the product in organic outlet are all high purity products, there are five extreme value equations for the aqueous feeding system.

$$1 > E_M > (E_M)_{min} \approx f_A + \frac{f_B}{\beta} \tag{5.68}$$

$$1 < E'_M < \left(E'_M\right)_{max} \approx \beta f_A + f_B \tag{5.69}$$

$$J_S > (J_S)_{min} = \frac{(E_M)_{min}}{1 - (E_M)_{min}} \approx \frac{\beta f_A + f_B}{f_B(\beta - 1)} \tag{5.70}$$

$$S_o > (S_o)_{min} = M_{1(a)}(J_S)_{min} \approx \frac{1}{(\beta - 1)} + f_A \tag{5.71}$$

$$W_a > (W_a)_{min} = (S_o)_{min} - M_{n+m(o)} \approx \frac{1}{\beta - 1} \tag{5.72}$$

where $(S_o)_{min}$ is minimum extraction and $(W_a)_{min}$ is minimum scrubbing in comparison with the maximum extraction S_o and maximum scrubbing W_a.

5.6.2 *Organic Feeding System*

Xu and coworkers also developed five extreme value equations for the organic feeding fractional solvent extraction system that produces high purity products in both aqueous outlet and organic outlet.

$$1 > E_{\mathrm{M}} > (E_{\mathrm{M}})_{\min} \approx \frac{1}{\beta f_{\mathrm{A}} + f_{\mathrm{B}}} \tag{5.73}$$

$$1 < E'_{\mathrm{M}} < \left(E'_{\mathrm{M}}\right)_{\max} \approx \frac{1}{\frac{f_{\mathrm{A}}}{\beta} + f_{\mathrm{B}}} \tag{5.74}$$

$$J_{\mathrm{S}} > (J_{\mathrm{S}})_{\min} \approx \frac{1}{f_{\mathrm{B}}(\beta - 1)} \tag{5.75}$$

$$S_{\mathrm{o}} > (S_{\mathrm{o}})_{\min} = f_{\mathrm{B}}(J_{\mathrm{S}})_{\min} \approx \frac{1}{(\beta - 1)} \tag{5.76}$$

$$W_{\mathrm{a}} > (W_{\mathrm{a}})_{\min} = (S_{\mathrm{o}})_{\min} + f_{\mathrm{B}} \approx \frac{1}{(\beta - 1)} + f_{\mathrm{B}} \tag{5.77}$$

5.7 Number of Stages of Fractional Solvent Extraction

In the design of a cascade solvent extraction process, the most important parameters are the number of stages and the flow rate of each phase. This section introduces the calculation of the number of stages for different systems.

5.7.1 *System with Constant Extraction Factor of Individual Solute*

As shown in Fig. 5.5, a fractional solvent extraction includes n stages of extraction and m stages of scrubbing. The feeding stage is included in extraction for aqueous feeding and it is included in scrubbing for organic phase.

According to Xu's assumption about the feeding stage phase composition, it can be assumed that the aqueous phase from the $n+1$ stage and the organic phase from the $n-1$ stage have the same rare earth composition as the feed. For the two component system, this assumption can be expressed by the following equation:

$$\frac{\left[A_{n+1(\mathrm{a})}\right]}{\left[B_{n+1(\mathrm{a})}\right]} = \frac{\left[A_{\mathrm{F(a)}}\right]}{\left[B_{\mathrm{F(a)}}\right]} = \frac{\left[A_{n-1(\mathrm{o})}\right]}{\left[B_{n-1(\mathrm{o})}\right]} \tag{5.78}$$

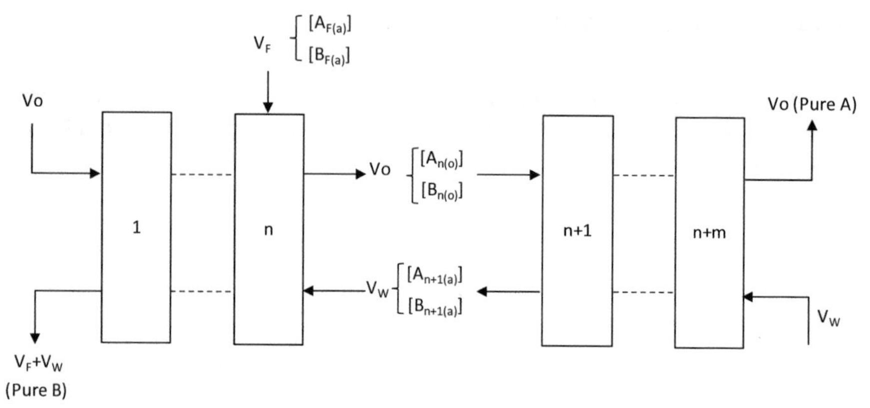

Fig. 5.5 Fractional solvent extraction process

The extraction with stage from 1 to n can be considered as an independent countercurrent extraction system. Therefore, the equations of extraction residual fractions of A and B in the countercurrent extraction process will be applicable here:

$$\phi_{\mathrm{A}} = \frac{E_{\mathrm{A}} - 1}{E_{\mathrm{A}}^{n+1} - 1} \tag{5.79}$$

$$\phi_{\mathrm{B}} = \frac{E_{\mathrm{B}} - 1}{E_{\mathrm{B}}^{n+1} - 1} \tag{5.80}$$

According to Eq. (5. 20), the concentrating factor of B is:

$$b = \frac{\phi_{\mathrm{B}}}{\phi_{\mathrm{A}}} = \frac{\frac{E_{\mathrm{B}}-1}{E_{\mathrm{B}}^{n+1}-1}}{\frac{E_{\mathrm{A}}-1}{E_{\mathrm{A}}^{n+1}-1}} = \frac{(E_{\mathrm{A}}^{n+1} - 1)(1 - E_{\mathrm{B}})}{(E_{\mathrm{B}}^{n+1} - 1)(1 - E_{\mathrm{A}})} \tag{5.81}$$

Since $E_{\mathrm{A}} > 1$ and $E_{\mathrm{B}} < 1$, $E_{\mathrm{A}}^{n+1} \gg 1$ and $E_{\mathrm{B}}^{n+1} \ll 1$. Therefore, the concentrating factor b can be simplified to the following equation:

$$b = \frac{E_{\mathrm{A}}^{n+1}(1 - E_{\mathrm{B}})}{(E_{\mathrm{A}} - 1)} = \frac{E_{\mathrm{A}}^{n}(E_{\mathrm{A}} - E_{\mathrm{B}}E_{\mathrm{A}})}{(E_{\mathrm{A}} - 1)} \approx E_{\mathrm{A}}^{n} \tag{5.82}$$

Therefore, the number of extraction stages can be obtained:

$$n = \log b / \log E_{\mathrm{A}} = \log b / \log \beta E_{\mathrm{B}} \tag{5.83}$$

The scrubbing with stage from $n + 1$ to m can also be considered as an independent countercurrent reverse extraction system. The concentrating factor of A can be expressed by the following equation.

$$a = \frac{1}{\left(E_{\mathrm{B}}'\right)^{m+1}} \tag{5.84}$$

Therefore, the number of the scrubbing stages can be calculated by the following equation.

$$m = \log a / \log \frac{1}{E_{\mathrm{B}}'} - 1 = \log a / \log \frac{\beta'}{E_{\mathrm{A}}'} - 1 \tag{5.85}$$

Equations (5.83) and (5.85) are the two basic equations for the calculation of the number of stages in fractional extraction. They apply to the systems with constant extraction factors of individual solutes in both extraction and scrubbing. However, the extraction factors of individual solutes are not constants in real production. Only the extraction factors of total solutes are close to constants. The following are relationships between the extraction factor of total solutes and individual solutes:

$$E_{\mathrm{B}} = \frac{E_{\mathrm{M}}}{\beta - (\beta - 1)P_{\mathrm{B(a)}}} \tag{5.86}$$

$$E_{\mathrm{A}}' = \frac{E_{\mathrm{M}}'}{\beta - (\beta - 1)P_{\mathrm{A(o)}}} \tag{5.87}$$

If B is the major component in the feed, the majority of the extraction stages has $P_{\mathrm{B(a)}} > 0.90$. Therefore, $E_{\mathrm{B}} \approx E_{\mathrm{M}}$ is relatively constant. The number of extraction stages calculated from Eq. (5.83) is reliable but the number of scrubbing stages calculated from Eq. (5.85) is not sufficient. If A is the major component of the feed, most of the scrubbing stages have $P_{\mathrm{A(o)}} > 0.90$ and $E_{\mathrm{A}}' \approx E_{\mathrm{M}}'$. The number of scrubbing stages calculated from Eq. (5.85) is reliable but the number of extraction stages obtained from Eq. (5.83) will be low.

5.7.2 *System with Constant Extraction Factor of Mixed Solutes*

5.7.2.1 Purity Equilibrium Line Equation

Based on Xu's assumption of average separator factor, the separation factor can be expressed by the following equation:

$$\beta = \frac{E_{\mathrm{A}}}{E_{\mathrm{B}}} = \frac{A_{\mathrm{(o)}}}{A_{\mathrm{(a)}}} / \frac{B_{\mathrm{(o)}}}{B_{\mathrm{(a)}}} \tag{5.88}$$

According to the definition, the purity of A in organic outlet and aqueous outlet can be expressed by the following equations:

$$P_{A_{(o)}} = \frac{A_{(o)}}{A_{(o)} + B_{(o)}}, \quad P_{A_{(a)}} = \frac{A_{(a)}}{A_{(a)} + B_{(a)}} \tag{5.89}$$

Therefore, the purity equilibrium line equations of A can be obtained from above equations:

$$P_{A_{(o)}} = \frac{\beta P_{A(a)}}{1 + (\beta - 1)P_{A(a)}}, \quad P_{A_{(a)}} = \frac{P_{A_{(o)}}}{\beta - (\beta - 1)P_{A_{(o)}}} \tag{5.90}$$

Since $P_{A(o)} = 1 - P_{B(o)}$ and $P_{A(a)} = 1 - P_{B(a)}$, the purity equilibrium line equations of B can be obtained as follows:

$$P_{B_{(o)}} = \frac{P_{B(a)}}{\beta - (\beta - 1)P_{B(a)}}, \quad P_{B_{(a)}} = \frac{\beta P_{B_{(o)}}}{1 + (\beta - 1)P_{B_{(o)}}} \tag{5.91}$$

5.7.2.2 Purity Operating Line Equation in Extraction

In extraction the operating line equation of A is:

$$A_{i+1(a)} = A_{i(o)} + A_{1(a)} \tag{5.92}$$

Since $A_{i+1(a)} = M_{i+1(a)}P_{A_{i+1(a)}}$, $A_{i(o)} = M_{i(o)}P_{A_{i(o)}}$, $A_{1(a)} = M_{1(a)}P_{A_{1(a)}}$, the operating line equation can be expressed by purity and total mass flow rate of solutes:

$$P_{A_{i+1(a)}} = P_{A_{i(o)}}\frac{M_{i(o)}}{M_{i+1(a)}} + P_{A_{1(a)}}\frac{M_{1(a)}}{M_{i+1(a)}} \tag{5.93}$$

In extraction the operating line equation of total solute M is:

$$M_{i+1(a)} = M_{i(o)} + M_{1(a)} \tag{5.94}$$

Therefore, Eq. (5.93) can be converted to:

$$P_{A_{i+1(a)}} = P_{A_{i(o)}}\frac{M_{i(o)}}{M_{i+1(a)}} + P_{A_{1(a)}}\left(1 - \frac{M_{1(o)}}{M_{i+1(a)}}\right) \tag{5.95}$$

For the system with constant extraction factor of total solutes, the extraction factor can be expressed by the following equation:

$$E_{\mathrm{M}} = \frac{M_{i(\mathrm{o})}}{M_{i+1(\mathrm{a})}} \tag{5.96}$$

Therefore, the purity operating line equation of A in extraction can be obtained from Eqs. (5.95) and (5.96).

$$P_{\mathrm{A}_{i+1(\mathrm{a})}} = E_{\mathrm{M}} P_{\mathrm{A}_{i(\mathrm{o})}} + P_{\mathrm{A}_{1(\mathrm{a})}} (1 - E_{\mathrm{M}}) \tag{5.97}$$

For solute B, the same purity operating line equation in extraction exists:

$$P_{\mathrm{B}_{i+1(\mathrm{a})}} = E_{\mathrm{M}} P_{\mathrm{B}_{i(\mathrm{o})}} + P_{\mathrm{B}_{1(\mathrm{a})}} (1 - E_{\mathrm{M}}) \tag{5.98}$$

When B in the aqueous outlet is high purity, $P_{\mathrm{A}_{1(\mathrm{a})}} \approx 0$. Equation (5.97) can be simplified.

$$P_{\mathrm{A}_{i+1(\mathrm{a})}} = E_{\mathrm{M}} P_{\mathrm{A}_{i(\mathrm{o})}} \tag{5.99}$$

5.7.2.3 Purity Operating Line Equation in Scrubbing

The operating line equation of B in scrubbing is:

$$B_{j+1(\mathrm{a})} = B_{j(\mathrm{o})} + B_{n+m(\mathrm{o})} \tag{5.100}$$

Since $B_{j+1(\mathrm{a})} = M_{j+1(\mathrm{a})} P_{\mathrm{B}_{j+1(\mathrm{a})}}$, $B_{j(\mathrm{o})} = M_{j(\mathrm{o})} P_{\mathrm{B}_{j(\mathrm{o})}}$, $B_{n+m(\mathrm{o})} = M_{n+m(\mathrm{o})} P_{\mathrm{B}_{n+m(\mathrm{o})}}$, the operating line equation can be expressed as:

$$P_{\mathrm{B}_{j+1(\mathrm{a})}} = P_{\mathrm{B}_{j(\mathrm{o})}} \frac{M_{j(\mathrm{o})}}{M_{j+1(\mathrm{a})}} + P_{\mathrm{B}_{n+m(\mathrm{o})}} \frac{M_{n+m(\mathrm{o})}}{M_{j+1(\mathrm{a})}} \tag{5.101}$$

For a system with constant extraction factor of total solutes, the extraction factor in scrubbing can be expressed by the following equation:

$$E'_{\mathrm{M}} = \frac{M_{j(\mathrm{o})}}{M_{j+1(\mathrm{a})}} \tag{5.102}$$

Therefore, the purity operating line equation of B in scrubbing can be obtained:

$$P_{\mathrm{B}_{j+1(\mathrm{a})}} = E'_{\mathrm{M}} P_{\mathrm{B}_{j(\mathrm{o})}} - \left(E'_{\mathrm{M}} - 1\right) P_{\mathrm{B}_{n+m(\mathrm{o})}} \tag{5.103}$$

For A, the same operating line equation in scrubbing also exists:

$$P_{\mathrm{A}_{j+1(\mathrm{a})}} = E'_{\mathrm{M}} P_{\mathrm{A}_{j(\mathrm{o})}} - \left(E'_{\mathrm{M}} - 1\right) P_{\mathrm{A}_{n+m(\mathrm{o})}} \tag{5.104}$$

When A in the organic outlet is high purity, $P_{B_{n+m(o)}} \approx 0$. Equation (5.103) can be simplified:

$$P_{B_{j+1(a)}} = E'_{M} P_{B_{j(o)}} \tag{5.105}$$

5.7.2.4 The Number of Extraction Stages

Combine Eqs. (5.90) and (5.99), the following equation can be obtained:

$$P_{A_{i+1(a)}} = E_{M} P_{A_{i(o)}} = \frac{\beta E_{M} P_{Ai(a)}}{1 + (\beta - 1) P_{Ai(a)}} \tag{5.106}$$

The change rate of A purity with respect to stages in aqueous phase can be calculated by the following equation:

$$\frac{\Delta P_{A_{i(a)}}}{\Delta i} = \frac{P_{A_{i+1(a)}} - P_{A_{i(a)}}}{i + 1 - i} = \frac{\beta E_{M} P_{Ai(a)}}{1 + (\beta - 1) P_{Ai(a)}} - P_{A_{i(a)}} \tag{5.107}$$

or

$$\frac{\Delta i}{\Delta P_{A_{i(a)}}} = \frac{1 + (\beta - 1) P_{Ai(a)}}{(\beta E_{M} - 1) P_{Ai(a)} - (\beta - 1) P^{2}_{A_{i(a)}}} \tag{5.108}$$

When n is a large number, Eq. (5.108) is approximately equal to the differentiation of stage with respect to purity:

$$\frac{\mathrm{d}i}{\mathrm{d}P_{A_{(a)}}} \approx \frac{\Delta i}{\Delta P_{A_{(a)}}} = \frac{1 + (\beta - 1) P_{A(a)}}{(\beta E_{M} - 1) P_{A(a)} - (\beta - 1) P^{2}_{A_{(a)}}} \tag{5.109}$$

Integrate Eq. (5.109); the following equation can be obtained:

$$\begin{aligned} n = \int_{0}^{n} \mathrm{d}i = \int_{P_{A1(a)}}^{P_{An(a)}} \frac{\mathrm{d}P_{A_{(a)}}}{(\beta E_{M} - 1) P_{A(a)} - (\beta - 1) P^{2}_{A_{(a)}}} + \int_{P_{A1(a)}}^{P_{An(a)}} \frac{\mathrm{d}P_{A_{(a)}}}{\frac{(\beta E_{M} - 1)}{\beta - 1} - P_{A(a)}} \\ = n_1 + n_2 \end{aligned} \tag{5.110}$$

$$n_1 = \int_{P_{\mathrm{A1(a)}}}^{P_{\mathrm{A}n\mathrm{(a)}}} \frac{\mathrm{d}P_{\mathrm{A(a)}}}{(\beta E_{\mathrm{M}} - 1)P_{\mathrm{A(a)}} - (\beta - 1)P_{\mathrm{A(a)}}^2}$$

$$= \frac{1}{\beta E_{\mathrm{M}} - 1} \ln \frac{P_{\mathrm{A}_{n\mathrm{(a)}}} / \left(P^*_{\mathrm{A(a)}} - P_{\mathrm{A}_{n\mathrm{(a)}}}\right)}{P_{\mathrm{A}_{1\mathrm{(a)}}} / \left(P^*_{\mathrm{A(a)}} - P_{\mathrm{A}_{1\mathrm{(a)}}}\right)} \tag{5.111}$$

$$n_2 = \int_{P_{\mathrm{A1(a)}}}^{P_{\mathrm{A}n\mathrm{(a)}}} \frac{\mathrm{d}P_{\mathrm{A(a)}}}{\frac{(\beta E_{\mathrm{M}} - 1)}{\beta - 1} - P_{\mathrm{A(a)}}} = \ln \frac{\left(P^*_{\mathrm{A(a)}} - P_{\mathrm{A}_{1\mathrm{(a)}}}\right)}{\left(P^*_{\mathrm{A(a)}} - P_{\mathrm{A}_{n\mathrm{(a)}}}\right)} \tag{5.112}$$

$$P^*_{\mathrm{A(a)}} = \frac{(\beta E_{\mathrm{M}} - 1)}{\beta - 1} + \frac{(1 - E_{\mathrm{M}})\beta E_{\mathrm{M}} P_{\mathrm{A}_{1\mathrm{(a)}}}}{(\beta E_{\mathrm{M}} - 1) + (1 - E_{\mathrm{M}})(\beta - 1)P_{\mathrm{A}_{1\mathrm{(a)}}}} \approx \frac{(\beta E_{\mathrm{M}} - 1)}{\beta - 1} \tag{5.113}$$

where $P^*_{\mathrm{A(a)}}$ is the intersection of purity equilibrium line and purity operating line.

Using Taylor series expansion, the following can be obtained:

$$\beta E_{\mathrm{M}} - 1 \approx \ln \beta E_{\mathrm{M}} \tag{5.114}$$

$$\frac{P_{\mathrm{A}_{n\mathrm{(a)}}} / \left(P^*_{\mathrm{A(a)}} - P_{\mathrm{A}_{n\mathrm{(a)}}}\right)}{P_{\mathrm{A}_{1\mathrm{(a)}}} / \left(P^*_{\mathrm{A(a)}} - P_{\mathrm{A}_{1\mathrm{(a)}}}\right)} = \frac{P_{\mathrm{A}_{n\mathrm{(a)}}} / \left(1 - P_{\mathrm{A}_{n\mathrm{(a)}}} / P^*_{\mathrm{A(a)}}\right)}{P_{\mathrm{A}_{1\mathrm{(a)}}} / \left(1 - P_{\mathrm{A}_{1\mathrm{(a)}}} / P^*_{\mathrm{A(a)}}\right)} \approx b \tag{5.115}$$

Therefore,

$$n = \log b / \log \beta E_{\mathrm{M}} + 2.303 \log \frac{P^*_{\mathrm{A(a)}} - P_{\mathrm{A}_{1\mathrm{(a)}}}}{P^*_{\mathrm{A(a)}} - P_{\mathrm{A}_{n\mathrm{(a)}}}} \tag{5.116}$$

Equation (5.116) is the equation for the calculation of the number of extraction stages in fractional extraction with constant extraction factor of total solutes. The following equation can be used to perform rough estimate of the number of extraction stages.

$$n = \log b / \log \beta E_{\mathrm{M}} \tag{5.117}$$

5.7.2.5 The Number of Scrubbing Stages

Following the similar derivation in Sect. 5.7.2.4, the equation for the calculation of the number of scrubbing stages can be obtained:

$$m+1=\log a/\log\frac{\beta'}{E'_{\mathrm{M}}}+2.303\log\frac{P^*_{\mathrm{B}_{(o)}}-P_{\mathrm{B}_{n+m(o)}}}{P^*_{\mathrm{B}_{(o)}}-P_{\mathrm{B}_{n(o)}}} \quad (5.118)$$

$$P^*_{\mathrm{B}_{(o)}}=\frac{\beta'/E'_{\mathrm{M}}-1}{\beta'-1}+\frac{\beta'(1-1/E'_{\mathrm{M}})P_{\mathrm{B}_{n+m(o)}}}{\beta'-E'_{\mathrm{M}}+(E'_{\mathrm{M}}-1)(\beta'-1)P_{\mathrm{B}_{n+m(o)}}}\approx\frac{\beta'/E'_{\mathrm{M}}-1}{\beta'-1} \quad (5.119)$$

Equation (5.118) is the equation for the calculation of the number of scrubbing stages. The following Eq. (5.120) can be used to perform rough estimate for the number of scrubbing stages.

$$m=\log a/\log\frac{\beta'}{E'_{\mathrm{M}}}-1 \quad (5.120)$$

Equations (5.116) and (5.118) have been approved correct in over ten years of rare earth production and in hundreds of simulating experiments.

5.7.2.6 Extrema of Extraction Factor

For a system with aqueous feeding, it can be assumed that the A purity ($P_{\mathrm{A}_{n(a)}}$) in the aqueous phase of the feeding stage equals to the mass fraction of A in the feed (f_{A}). If $P^*_{\mathrm{A}_{(a)}}=P_{\mathrm{A}_{n(a)}}=f_{\mathrm{A}}$, $n\to\infty$ according to Eq. (5.116). Therefore, $P^*_{\mathrm{A}_{(a)}}$ must be $>f_{\mathrm{A}}$. According to Eq. (5.113), the minimum extraction factor in extraction can be obtained:

$$(E_{\mathrm{M}})_{\min}=f_{\mathrm{A}}+f_{\mathrm{B}}/\beta \quad \text{(Aqueous Feeding)} \quad (5.121)$$

If $E_{\mathrm{M}}\le(E_{\mathrm{M}})_{\min}$ in extraction, the targeted separation cannot be achieved. Following the similar derivation, the maximum extraction factor in scrubbing can be obtained:

$$\left(E'_{\mathrm{M}}\right)_{\max}=\beta' f_{\mathrm{A}}+f_{\mathrm{B}} \quad \text{(Aqueous Feeding)} \quad (5.122)$$

Similarly, if $E'_{\mathrm{M}}\ge(E'_{\mathrm{M}})_{\max}$ in scrubbing, the targeted separation cannot be achieved.

For a system with organic feeding, it can be assumed that the B purity ($P^*_{\mathrm{B}_{(o)}}$) in the organic phase of the feeding stage equals to the mass fraction of B in feed (f_{B}). According to Eq. (5.118), $P^*_{\mathrm{B}_{(o)}}$ must be $>f_{\mathrm{B}}$. According to Eq. (5.119), the maximum extraction factor in scrubbing and the minimum extraction factor in extraction can be obtained:

$$\left(E'_{\mathrm{M}}\right)_{\max} = \frac{1}{f_{\mathrm{B}} + f_{\mathrm{A}}/\beta'} \quad \text{(Organic Feeding)} \tag{5.123}$$

$$(E_{\mathrm{M}})_{\min} = \frac{1}{\beta f_{\mathrm{B}} + f_{\mathrm{A}}} \quad \text{(Organic Feeding)} \tag{5.124}$$

5.8 Optimum Equations

5.8.1 Optimum Extraction Factor

In rare earth solvent extraction, the primary objective is to produce products with high purity at high recovery. The secondary objective is to reduce the production cost by increasing throughput. In extraction, the E_{M} must be larger than $(E_{\mathrm{M}})_{\min}$. The higher the E_{M}, the higher the B purity or the lesser the number of stages required to achieve the same product purity. However, J_{S} will increase with the increase of E_{M} resulting in capacity reduction of the same size facility. Therefore, an optimum E_{M} is required to increase the facility capacity while maintaining the targeted product purity and recovery.

If B is the major component of the feed, then $P_{\mathrm{A}n(\mathrm{a})} \ll P^{*}_{\mathrm{A(a)}}$. Equation (5.116) can be simplified to Eq. (5.117) but Eq. (5.118) cannot be simplified to Eq. (5.120). In this case, the process is referred to as extraction controlled process. For an extraction controlled process, the daily production of B can be expressed by the following equation.

$$Q_{\mathrm{B}} = 1.44\frac{V}{t}\frac{\left[B_{1(\mathrm{a})}\right]}{1+R} \tag{5.125}$$

where Q_{B} is the daily production of B with the unit of kg/day, V is the volume of mixer of each stage (L), t is contact time (min), R is phase ratio, $[B_1]$ is the concentration of B in aqueous outlet (g/L).

Using V_{tE} to denote the total volume of mixer and settler in extraction and r to denote the volume ratio of settler to mixer, the volume of mixer of each stage can be expressed by the total volume of extraction.

$$V = \frac{V_{\mathrm{tE}}}{n(1+r)} \tag{5.126}$$

According to Eq. (5.17), the following equation can be obtained:

$$\frac{\left[B_{1(\mathrm{a})}\right]}{\left[B_{n(\mathrm{a})}\right]} \approx 1 - E_{\mathrm{B}} \tag{5.127}$$

Substituting Eqs. (5.126) and (5.127) into Eq. (5.125), we have:

$$Q_B = 1.44\frac{V_{tE}\left[B_{n(a)}\right](1 - E_B)}{n(1 + r)t(1 + R)} \quad (5.128)$$

Substituting Eq. (5.83) into Eq. (5.128), we have:

$$Q_B = 1.44\frac{V_{tE}\left[B_{n(a)}\right](1 - E_B)\ln\beta E_B}{(1 + r)t(1 + R)\ln b} \quad (5.129)$$

Differentiating Eq. (5.129) and making $\frac{\partial Q_B}{\partial E_B} = 0$, we gave:

$$\ln\beta E_B = \frac{1 - E_B}{E_B} \quad (5.130)$$

Using Taylor series expansion, the following can be obtained:

$$\ln\beta E_B = (\beta E_B - 1) - \frac{1}{2}(\beta E_B - 1)^2 + \cdots \approx \beta E_B - 1 \quad (5.131)$$

Substituting the approximation of Eq. (5.131) into Eq. (5.130), the following can be obtained:

$$E_B = \frac{1}{\sqrt{\beta}} \quad (5.132)$$

To maximum the daily production of B (Q_B), Eqs. (5.130) and (5.132) must be met when the total extraction volume (V_{tE}) and the targeted separation efficiency (b) remain unchanged. Therefore, Eqs. (5.130) and (5.132) are the optimum extraction factor equations for extraction.

If A is the major component of the feed, $P_{Bn(o)} \ll P^*_{B(o)}$. Equation (5.118) can be simplified to Eq. (5.120) but Eq. (5.116) cannot be simplified to Eq. (5.117). In this case, the process is referred to as scrubbing controlled process. Following similar derivation, the optimum extraction factor equation for scrubbing can be obtained:

$$E'_A = \sqrt{\beta'} \quad (5.133)$$

5.8.2 *Optimum Reflux Ratio Equation*

If B is the major component in the feed and B is the high purity product in aqueous outlet, the B purity in most of the extraction stages are higher than 0.90, i.e., $P_B > 0.90$. According to Eq. (5.86), in most of the extraction stages there is $E_B = E_M$. Only in the stages close to the feeding stages, there is $E_B < E_M$.

In extraction, the stages where $P_B < 0.90$ are not many. Therefore, E_M is only slightly higher than the average E_B in extraction. In reality, the E_B obtained from Eq. (5.132) is also slightly higher than its actual value. Therefore, it is reasonable to make the following optimum extraction factor equation in extraction:

$$E_M = E_B = \frac{1}{\sqrt{\beta}} \tag{5.134}$$

According to Eq. (5.42), the optimum extraction reflux ratio equation can be obtained:

$$J_S = \frac{E_M}{1 - E_M} = \frac{1}{\sqrt{\beta} - 1} \tag{5.135}$$

If A is the major component of the feed and A is the high purity product in organic outlet, the optimum extraction factor in scrubbing can be obtained:

$$E'_M = E'_A = \sqrt{\beta'} \tag{5.136}$$

Therefore, according to Eq. (5.39), the optimum washing reflux ratio can be expressed by the following equation.

$$J_W = \frac{1}{E'_M - 1} = \frac{1}{\sqrt{\beta'} - 1} \tag{5.137}$$

5.8.3 Controlling Conditions and Variables

In rare earth cascade solvent extraction process, the products purity increases with the increases of J_S and J_W. However, the increase of the reflux ratio reduces the production capacity. To ensure the high purity of product B, the extraction reflux ratio determined by Eq. (5.135) should be used. To ensure the high purity of product A, the washing reflux ratio obtained from Eq. (5.137) should be used. However, J_S and J_W are not independent variables. To ensure the high purity of both A and B, J_S from Eq. (5.135) should be used when it is smaller than J_W from Eq. (5.137). The process is extraction controlled. Otherwise, J_W from Eq. (5.137) should be used when it is smaller than J_S from Eq. (5.135). The process is scrubbing controlled.

5.8.3.1 Aqueous Feeding System

The mass balance of an aqueous feeding system can be represented by the following Table 5.2.

Table 5.2 Mass balance of aqueous feeding system

	Solvent S ↓			Feed $M_{F(a)}$ ↓			Washing W ↓	
Stage	1	⋯	$n-1$	n	$n+1$	⋯	$n+m$	
Organic phase			S_o	S_o	S_o		$M_{n+m(o)}$	→
Aqueous phase	$M_{1(a)}$		$W_a+M_{F(a)}$	$W_a+M_F=S_o+M_1$	$W_a=S_o-M_{n+m(a)}$			
Extraction factor	$E_M=\frac{S_o}{S_o+M_{1(a)}}$				$E'_M=\frac{S_o}{S_o-M_{n+m(o)}}$			

According to Eq. (5.46), the washing reflux is:

$$J_W = J_S \frac{f'_B}{f'_A} - 1 = J_S \frac{f'_B}{1 - f'_B} - 1 \tag{5.138}$$

When $J_W = J_S$, the mole fraction of solutes (A and B) in aqueous outlet f'_B is:

$$f'_B = \frac{1 + J_S}{1 + 2J_S} = \frac{1}{1 + E_M} = \frac{\sqrt{\beta}}{\sqrt{\beta} + 1} \tag{5.139}$$

When:

$$f'_B > \frac{1 + J_S}{1 + 2J_S} = \frac{1}{1 + E_M} = \frac{\sqrt{\beta}}{\sqrt{\beta} + 1} \tag{5.140}$$

The extraction reflux ratio is smaller than the washing reflux ratio: $J_s < J_w$. The process is extraction controlled.

When:

$$f'_B < \frac{1 + J_S}{1 + 2J_S} = \frac{1}{1 + E_M} = \frac{\sqrt{\beta}}{\sqrt{\beta} + 1} \tag{5.141}$$

The washing reflux ratio is smaller than the extraction reflux ratio: $J_W < J_S$. The process is scrubbing controlled.

5.8.3.2 Organic Feeding System

The mass balance of an organic feeding system can be represented by Table 5.3.

Table 5.3 Mass balance of organic feeding system

	Solvent S ↓				Feed $M_{F(a)}$ ↓		Washing W ↓	
Stage	1	···	$n-1$	n	$n+1$	···	$n+m$	
Organic phase			S_o	$S_o+M_{F(o)}$	$S_o+M_{F(o)}$		$M_{n+m(o)}$	→
Aqueous phase ←	$M_{1(a)}$		W_a	$W_a=S_o+M_{1(a)}$	$W_a=S_o+M_{F(o)}-M_{n+m(a)}$			
Extraction factor	$E_M=\frac{S_o}{W_{(a)}}$			$E'_M=\frac{S_o+M_{F(o)}}{W_{(a)}}$				

According to Eq. (5.50), the washing reflux ratio is:

$$J_W = \frac{f'_B}{1-f'_B}(J_S+1) \tag{5.142}$$

When $J_W = J_S$, the mole fraction of solutes (A and B) in aqueous outlet f'_B is:

$$f'_B = \frac{J_S}{1+2J_S} = \frac{E_M}{1+E_M} = \frac{1}{\sqrt{\beta}+1} \tag{5.143}$$

When:

$$f'_B > \frac{J_S}{1+2J_S} = \frac{E_M}{1+E_M} = \frac{1}{\sqrt{\beta}+1} \tag{5.144}$$

The extraction reflux ratio is smaller than the washing reflux ration: $J_s < J_w$. The process is extraction controlled.

When:

$$f'_B < \frac{J_S}{1+2J_S} = \frac{E_M}{1+E_M} = \frac{1}{\sqrt{\beta}+1} \tag{5.145}$$

The washing reflux ratio is smaller than the extraction reflux ratio: $J_W < J_S$. The process is scrubbing controlled.

5.8.3.3 Determination of Optimum Extraction Factor

When the cascade solvent extraction process is extraction controlled, we have:

$$f_B' > \frac{1 + J_S}{1 + 2J_S} = \frac{1}{1 + E_M} = \frac{\sqrt{\beta}}{\sqrt{\beta} + 1} \quad \text{(Aqueous Feeding)} \tag{5.146}$$

or:

$$f_B' > \frac{J_S}{1 + 2J_S} = \frac{E_M}{1 + E_M} = \frac{1}{\sqrt{\beta} + 1} \quad \text{(Organic Feeding)} \tag{5.147}$$

The extraction factors can be calculated by the following equations:

$$E_M = \frac{1}{\sqrt{\beta}} > (E_M)_{min} \tag{5.148}$$

$$E_M' = \frac{E_M f_B'}{E_M - f_A'} < (E_M)_{max} \quad \text{(Aqueous Feeding)} \tag{5.149}$$

or:

$$E_M' = \frac{1 - E_M f_A'}{f_B'} < \left(E_M'\right)_{max} \quad \text{(Organic Feeding)} \tag{5.150}$$

When the cascade solvent extraction process is scrubbing controlled, we have:

$$f_B' < \frac{1 + J_S}{1 + 2J_S} = \frac{1}{1 + E_M} = \frac{\sqrt{\beta}}{\sqrt{\beta} + 1} \quad \text{(Aqueous Feeding)} \tag{5.151}$$

or:

$$f_B' < \frac{J_S}{1 + 2J_S} = \frac{E_M}{1 + E_M} = \frac{1}{\sqrt{\beta} + 1} \quad \text{(Organic Feeding)} \tag{5.152}$$

The extraction factors can be calculated by the following equations:

$$E_M' = \sqrt{\beta} < \left(E_M'\right)_{max} \tag{5.153}$$

$$E_M = \frac{E_M' f_A'}{E_M' - f_B'} > (E_M)_{min} \quad \text{(Aqueous Feeding)} \tag{5.154}$$

or:

$$E_M = \frac{1 - E_M' f_B'}{f_A'} > (E_M)_{min} \quad \text{(Organic Feeding)} \tag{5.155}$$

When the cascade solvent extraction process is in a middle point, i.e., $J_S = J_W$:

$$f'_B = \frac{1+J_S}{1+2J_S} = \frac{1}{1+E_M} = \frac{\sqrt{\beta}}{\sqrt{\beta}+1} \quad \text{(Aqueous Feeding)} \tag{5.156}$$

or:

$$f'_B = \frac{J_S}{1+2J_S} = \frac{E_M}{1+E_M} = \frac{1}{\sqrt{\beta}+1} \quad \text{(Organic Feeding)} \tag{5.157}$$

Then:

$$E_M = \frac{1}{\sqrt{\beta}} = (E_M)_{min} \tag{5.158}$$

$$E'_M = \sqrt{\beta} = \left(E'_M\right)_{max} \tag{5.159}$$

5.9 Application of the Extrema Equations

In Sect. 5.6 the derivation of extrema equations of the aqueous feeding system, $(W_a)_{min}$ is a constant with the value of $1/(\beta-1)$ when the products in both outlets are high purity. In organic feeding system $(S_o)_{min}$ is a constant with the value of $1/(\beta-1)$ when the products in both outlets are high purity. Xu and coworkers defined the following two equations in the cascade solvent extraction process design.

$$W_a = \frac{1}{\beta^k - 1} \quad 0 < k < 1 \quad \text{(Aqueous Feeding)} \tag{5.160}$$

$$S_o = \frac{1}{\beta^k - 1} \quad 0 < k < 1 \quad \text{(Organic Feeding)} \tag{5.161}$$

Therefore, the following equations can be obtained:

$$E_M = f'_A + \frac{f'_B}{\beta^k} \quad \text{(Aqueous Feeding)} \tag{5.162}$$

$$E'_M = \beta^k f'_A + f'_B \quad \text{(Aqueous Feeding)} \tag{5.163}$$

$$E_M = \frac{1}{\beta^k f'_B + f'_A} \quad \text{(Organic Feeding)} \tag{5.164}$$

$$E'_M = \frac{1}{f'_B + f'_A/\beta^k} \quad \text{(Organic Feeding)} \tag{5.165}$$

The reagent consumption is determined by the amount of extraction S_o and the amount of washing W_a. When the throughput and mixing time are determined, S_o and W_a also dictate the cost of cell filling of each stage. According to Eqs. (5.160) and (5.161), the increase of k will result in the decrease of S_o and W_a. Therefore, the number of extraction stages (n) and the number of scrubbing stages (m) obtained from Eqs. (5.116) and (5.118) will increase. As a results of the decrease of S_o and W_a, the single stage reagent consumption will drop while the number of stages will increase. The optimum k value can be obtained from Eqs. (5.134), (5.136), and (5.160, 5.161, 5.162, 5.163, 5.164, and 5.165). Although the optimum k value is related to the feed composition, feeding model, and separation targets, normally the optimum k value is 0.70 based on the process design of hundreds of systems. Therefore, different k values between 0.5 and 0.9 can be used in actual process design. Equations (5.116), (5.118), and (5.160, 5.161, 5.162, 5.163, 5.164, and 5.165) can be used to determine a series of process parameters for the determination of single stage reagent consumption, cost, and capital investment. A final group of optimum process parameters can be determined based on a full economic assessment.

The above method has been approved reliable and valid in the process design of many extraction systems such as naphthenic acid, D2EHPA, HEH/EHP, and other acidic extraction systems.

5.10 Design Procedure of Optimum Rare Earth Cascade Solvent Extraction Process

Figure 5.6 shows the general design procedures of the optimum cascade solvent extraction process. It consists of (1) extraction system determination and separation factor measurement, (2) separation target specification, (3) optimum process parameter determination, (4) number of stage calculation, (5) flow ratio determination, (6) mass balance table construction, and overall process evaluation. In the following sections, each stage of the design procedures is discussed.

5.10.1 Extraction System Determination and Separation Factor Measurement

It is important to select a suitable extraction system by single stage testing to determine organic phase composition, degree of saponification, feed concentration and acidity, washing solution concentration and acidity, etc. Once a suitable

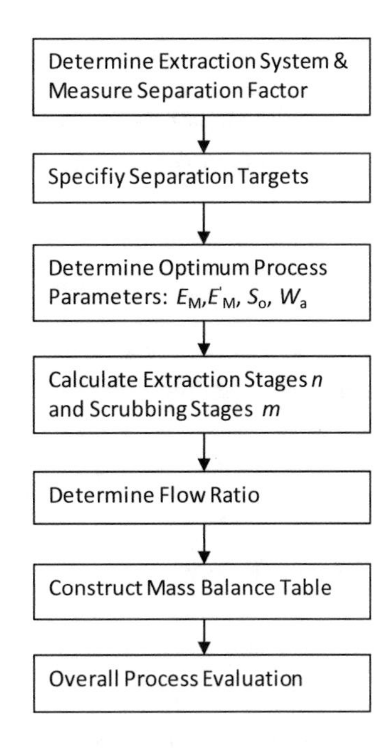

Fig. 5.6 Design procedure of optimum rare earth cascade solvent extraction process

extraction system is determined, the average separation factor in extraction and scrubbing are to be measured.

$$\beta = \frac{E_A}{E_B}, \ \beta' = \frac{E'_A}{E'_B} \tag{5.166}$$

If β and β' are close, normally the smaller one is used in the process design.

5.10.2 *Separation Target Specification*

According to the feed composition, separation strategy is to be determined. The rare earth elements are classified into to A and B based on their extractability in the extraction system. A is the one or one group of rare earths that are easily extracted. B is the one or one group of rare earths that are relative difficult to extract. The mole fraction of A and B can be calculated.

$$f_A = \frac{\text{Total mole of A}}{\text{Total mole of A and B}}, \ f_B = \frac{\text{Total mole of B}}{\text{Total mole of A and B}}, \ f_A + f_B = 1 \tag{5.167}$$

5.10.2.1 A as Major Product

If A is the major product, specify the purity of A in organic outlet to be $P_{A_{n+m(o)}}$ and the recovery of A to be Y_A. Then the concentrating factor of A can be calculated by Eq. (5.51):

$$a = \frac{[A_{n+m(o)}]/[B_{n+m(o)}]}{[A_{F(a)}]/[B_{F(a)}]} = \frac{P_{A_{n+m(o)}}/\left(1 - P_{A_{n+m(o)}}\right)}{f_A/f_B} \tag{5.51}$$

The concentrating factor of B can be calculated by the following equation:

$$b = \frac{[B_{1(a)}]/[A_{1(a)}]}{[B_{F(a)}]/[A_{F(a)}]} = \frac{a - Y_A}{a(1 - Y_A)} \tag{5.168}$$

Then the B purity in aqueous outlet can be calculated by Eq. (5.54):

$$P_{B_{1(a)}} = \frac{bf_B}{bf_B + f_A} \tag{5.54}$$

The A purity in aqueous outlet is $P_{A1(a)} = 1 - P_{B1(a)}$.

The solute fraction in organic outlet and aqueous outlet can be calculated by Eq. (5.28):

$$f'_A = \frac{f_A Y_A}{P_{A_{n+m(o)}}},\ f'_B = \frac{f_B Y_B}{P_{B_{1(a)}}},\ f'_B + f'_A = 1 \tag{5.28}$$

5.10.2.2 B as Major Product

If B is the major product, specify the purity of B in aqueous outlet to be $P_{B1(a)}$ and recovery to be Y_B. Then the concentrating factor of B can be calculated by Eq. (5.53):

$$b = \frac{[B_{1(a)}]/[A_{1(a)}]}{[B_{F(a)}]/[A_{F(a)}]} = \frac{P_{B_{1(a)}}/\left(1 - P_{B_{1(a)}}\right)}{f_B/f_A} \tag{5.53}$$

The concentrating factor of A can be calculated by the following equation:

$$a = \frac{[A_{n+m(o)}]/[B_{n+m(o)}]}{[A_{F(a)}]/[B_{F(a)}]} = \frac{b - Y_B}{b(1 - Y_B)} \tag{5.169}$$

The A purity in organic outlet can be calculated by Eq. (5.52):

$$P_{A_{n+m(o)}} = \frac{af_A}{af_A + f_B} \tag{5.52}$$

The B purity in organic outlet is: $P_{B_{n+m(o)}} = 1 - P_{A_{n+m(o)}}$.

The solute fraction in organic outlet and aqueous outlet can be calculated by Eq. (5.28).

5.10.2.3 A and B both as Major Products

If A and B are both as major products, specify product purity separately that A purity in organic outlet is $P_{A_{n+m(o)}}$ and B purity in aqueous outlet is $P_{B1(a)}$. Then the A concentrating factor and B concentrating factor can be calculated by Eqs. (5.51) and (5.53) separately. The recovery of A and B can be calculated by Eq. (5.25):

$$Y_A = \frac{A_{n+m(o)}}{f_A M_F},\ Y_B = \frac{B_{1(a)}}{f_B M_F} \tag{5.25}$$

The solute fraction in organic outlet and aqueous outlet can be calculated by Eq. (5.28).

5.10.3 Optimum Process Parameter Determination

Firstly, aqueous feeding or organic feeding needs to be determined. As shown in Fig. 5.7, normally the first cascade solvent extraction process uses aqueous feeding which contains multiple rare earth elements. The organic outlet of the first cascade process with more than one rare earth elements can be fed directly to the second cascade solvent extraction process.

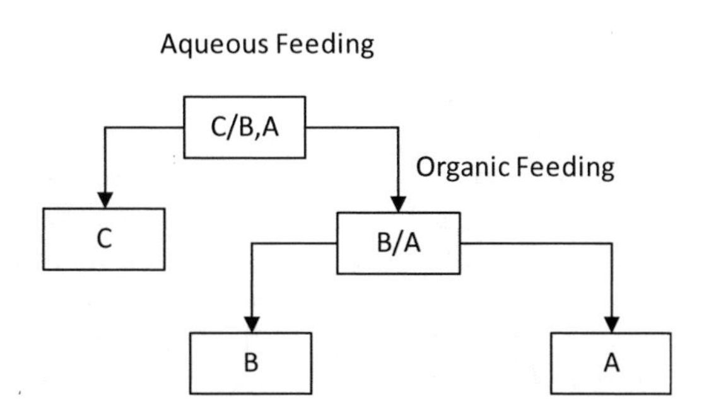

Fig. 5.7 Example of multiple rare earth separation

Table 5.4 Extraction factor determination from optimum equations

Aqueous feeding	If $f'_B > \frac{\sqrt{\beta}}{\sqrt{\beta}+1}$, it is extraction controlled.	If $f'_B < \frac{\sqrt{\beta}}{\sqrt{\beta}+1}$, it is scrubbing controlled.
	$E_M = \frac{1}{\sqrt{\beta}}$ (5.148) $E'_M = \frac{E_M f'_B}{E_M - f'_A}$ (5.149)	$E'_M = \sqrt{\beta}$ (5.153) $E_M = \frac{E'_M f'_A}{E'_M - f'_B}$ (5.154)
	$S_o = \frac{E_M M_1}{1-E_M} = \frac{E_M f'_B}{1-E_M}$, $W_a = S_o - M_{n+m(o)} = S_o - f'_A$	
Organic feeding	If $f'_B > \frac{1}{\sqrt{\beta}+1}$, it is extraction controlled.	If $f'_B < \frac{1}{\sqrt{\beta}+1}$, it is scrubbing controlled.
	$E_M = \frac{1}{\sqrt{\beta}}$ (5.148) $E'_M = \frac{1-E_M f'_A}{f'_B}$ (5.150)	$E'_M = \sqrt{\beta}$ (5.153) $E_M = \frac{1-E'_M f'_B}{f'_A}$ (5.155)
	$S_o = \frac{E_M M_1}{1-E_M} = \frac{E_M f'_B}{1-E_M}$, $W_a = S_o + 1 - M_{n+m(o)} = S_o + 1 - f'_A = S_o + f'_B$	

Table 5.5 Extraction factor determination from extrema equations

	In extraction	In scrubbing
Condition	$(E_M)_{min} < E_M < 1$ (5.63)	$1 < E'_M < (E'_M)_{max}$ (5.65)
Aqueous feeding	$W_a = \frac{1}{\beta^k - 1}$, $0 < k < 1$ (5.160) $M_F = 1.0$ (mmol/min) $S_o = W_a + f'_A$(mmol/min) $E_M = \frac{S_o}{W_a+1}$ $E'_M = \frac{S_o}{W_a}$ (5.170)	
Organic feeding	$S_o = \frac{1}{\beta^k - 1}$, $0 < k < 1$ (5.161) $M_F = 1.0$(mmol/min) $W_a = S_o + f'_B$(mmol/min) $E_M = \frac{S_o}{W_a}$ $E'_M = \frac{S_o+1}{W_a}$ (5.171)	

Secondly, the optimum process parameters are to be determined. The important parameters include extraction factors in extraction and scrubbing, maximum extraction S_o, and maximum washing W_a. They can be determined either by the optimum equations in Sect. 5.8 or the extrema equations in Sect. 5.6. The summary of the equations for the optimum extraction factor calculation is shown in Table 5.4. The summary of the extrema equations for the determination of extraction factor is shown in Table 5.5.

Normally the k value in Table 5.5 can be set at 0.70 for the calculation of other parameters. However, a few different k values can be used to obtain a few sets of other parameters for comparison in order to determine the optimum process parameters.

5.10.4 Determination of the Number of Stages

The derivation of equations for the calculation of the number of stages is discussed in Sect. 5.7.2. Table 5.6 summarizes the equations used to determine the extraction stages n and scrubbing stages m.

5.10.5 Determination of Flow Ratio

In above procedures, S_o and W_a are calculated on the basis of $M_F = 1$ m mol/min. If the total rare earth concentration in the feed, in the organic phase, and the acid concentration in the scrubbing solution are denoted by $[M_F]$, $[M_{(o)}]$, and [HA], the feed flow rate, organic flow rate, and scrubbing solution flow rate will be:

$$V_F = \frac{M_F}{[M_F]} \quad (\text{mL/min})$$

$$V_o = \frac{S_o}{[M_{(o)}]} \quad (\text{mL/min})$$

Assuming 3 mol HA acid can wash 1 mol RE^{3+}, then $V_w = \frac{3W_a}{[\text{HA}]}$ (mL/min)

The ratio of $V_F/V_o/V_w$ is the flow ratio.

5.10.6 Material Distribution and Balance Table Construction

A material distribution and balance table is always constructed to summarize all process parameters, material balance, total rare earth and concentration in the two phases of each stage, flow ratio, etc. Table 5.7 is a typical material distribution and balance table of an aqueous feeding system.

Table 5.8 is a typical material distribution and balance table of an organic feeding system.

Table 5.6 Equations for calculating n and m

B is the major component in feed and B is the high purity product in aqueous outlet	A is the major component in feed and A is the high purity product in organic outlet
$n = \log b/\log \beta E_{\mathrm{M}}$ (5.117)	$n = \log b/\log \beta E_{\mathrm{M}} + 2.303\log \frac{P^{*}_{A_{(a)}} - P_{A_{1(a)}}}{P^{*}_{A_{(a)}} - P_{A_{n(a)}}}$ (5.116) $P^{*}_{A_{(a)}} = \frac{(\beta E_{\mathrm{M}}-1)}{\beta-1} + \frac{(1-E_{\mathrm{M}})\beta E_{\mathrm{M}} P_{A_{1(a)}}}{(\beta E_{\mathrm{M}}-1)+(1-E_{\mathrm{M}})(\beta-1)P_{A_{1(a)}}}$ (5.113) When $P_{A_{1(a)}}$ is small, $P^{*}_{A_{(a)}} \approx \frac{(\beta E_{\mathrm{M}}-1)}{\beta-1}$
$m + 1 = \log a/\log \frac{\beta'}{E'_{\mathrm{M}}} + 2.303\log \frac{P^{*}_{B_{(o)}} - P_{B_{n+m(o)}}}{P^{*}_{B_{(o)}} - P_{B_{n(o)}}}$ (5.118) $P^{*}_{B_{(o)}} = \frac{\beta'/E'_{\mathrm{M}}-1}{\beta'-1} + \frac{\beta'(1-1/E'_{\mathrm{M}})P_{B_{n+m(o)}}}{\beta'-E'_{\mathrm{M}}+(E'_{\mathrm{M}}-1)(\beta'-1)P_{B_{n+m(o)}}}$ (5.119) When $PB_{n+m(o)}$ is small, $P^{*}_{B_{(o)}} \approx \frac{\beta'/E'_{\mathrm{M}}-1}{\beta'-1}$	$m = \log a/\log \frac{\beta'}{E'_{\mathrm{M}}} - 1$ (5.120)

Table 5.7 Material distribution and balance table of an aqueous feeding system

	S_o, V_o, $[M_{(o)}]$ ↓			$M_{F(a)} = V_F[M_{F(a)}]$ ↓			W_a, V_w, $[M_{(a)}]$ ↓
Stage	1	···	$n-1$	n	$n+1$	···	$n+\mathrm{m}$
$M_{(o)}$	$S_o \cdot M_{F(a)}$	···	$S_o \cdot M_{F(a)}$	$S_o \cdot M_{F(a)}$	$S_o \cdot M_{F(a)}$	···	$f'_A \cdot M_{F(a)}$
$M_{(a)}$	$f'_B \cdot M_{F(a)}$	···	$(W_a + 1)M_{F(a)}$	$(W_a + 1)M_{F(a)}$	$W_a \cdot M_{F(a)}$	···	$W_a \cdot M_{F(a)}$
$[M_{(o)}]$	$\frac{S_o \cdot M_{F(a)}}{V_o}$	···	$\frac{S_o \cdot M_{F(a)}}{V_o}$	$\frac{S_o \cdot M_{F(a)}}{V_o}$	$\frac{S_o \cdot M_{F(a)}}{V_o}$	···	$\frac{f'_A \cdot M_{F(a)}}{V_o}$
$[M_{(a)}]$	$\frac{f'_B \cdot M_{F(a)}}{V_f + V_W}$	···	$\frac{(W_a+1)M_{F(a)}}{V_F+V_W}$	$\frac{(W_a+1)M_{F(a)}}{V_F+V_W}$	$\frac{W_a \cdot M_{F(a)}}{V_W}$	···	$\frac{W_a \cdot M_{F(a)}}{V_W}$

Table 5.8 Material distribution and balance table of an organic feeding system

	S_o, V_o, $[M_{(o)}]$ ↓			$M_{F(o)} = V_F[M_{F(o)}]$ ↓			W_a, V_w, $[M_{(a)}]$ ↓
Stage	1	···	$n-1$	n	$n+1$	···	$n+\mathrm{m}$
$M_{(o)}$	$S_o \cdot M_{F(o)}$	···	$S_o \cdot M_{F(o)}$	$(S_o + 1) \cdot M_{F(o)}$	$(S_o + 1) \cdot M_{F(o)}$	···	$f'_A \cdot M_{F(o)}$
$M_{(a)}$	$f'_B \cdot M_{F(o)}$	···	$W_a \cdot M_{F(o)}$	$W_a \cdot M_{F(o)}$	$W_a \cdot M_{F(o)}$	···	$W_a \cdot M_{F(o)}$
$[M_{(o)}]$	$\frac{S_o \cdot M_{F(o)}}{V_o}$	···	$\frac{S_o \cdot M_{F(o)}}{V_o}$	$\frac{(S_o+1) \cdot M_{F(o)}}{V_o+V_F}$	$\frac{(S_o+1) \cdot M_{F(o)}}{V_o+V_F}$	···	$\frac{f'_A \cdot M_{F(o)}}{V_o+V_F}$
$[M_{(a)}]$	$\frac{f'_B \cdot M_{F(o)}}{V_W}$	···	$\frac{W_a \cdot M_{F(o)}}{V_W}$	$\frac{W_a \cdot M_{F(o)}}{V_W}$	$\frac{W_a \cdot M_{F(o)}}{V_W}$	···	$\frac{W_a \cdot M_{F(o)}}{V_W}$

5.10.7 Overall Process Evaluation

The determination of process parameters of a rare earth cascade solvent extraction process are discussed in above Sects. 5.10.1–5.10.6. To achieve individual rare earth element separation from mixed rare earth feed will need multiple rare earth cascade solvent extraction processes. Therefore, multiple sets of process parameters are to be determined. The configuration of the multiple cascade processes, product purity and recovery, process/product flexibility, reagent consumption, capital investment, and waste treatment must be evaluated together to determine the optimum separation process.

5.10.8 Example of Optimum Rare Earth Separation Process Design

The example is to demonstrate the application of the optimum process design of rare earth cascade solvent extraction processes.

Example 1 An acidic extractant system is used to separate Eu_2O_3 and Gd_2O_3. It is required to reach 99.99 % Gd_2O_3 purity. The feed material contains 99 % Gd_2O_3 with major impurities of Eu_2O_3 and Sm_2O_3. The recovery target of Gd_2O_3 is 90 %. What are the process parameters to meet the separation objectives?

Step 1: Separation factors
For the acidic extractant system, the Eu/Gd separation factors are known:

$$\beta_{\mathrm{Eu/Gd}} = \beta'_{\mathrm{Eu/Gd}} = 1.46$$

Step 2: Separation Targets
In the acidic extractant system, the Eu_2O_3 and Sm_2O_3 are easily extractable component A. Gd_2O_3 is relatively difficult to extract and is component B. According to the feed, $f_B = 0.99$, $f_A = 0.01$, $P_{B1(a)} = 0.9999$, $P_{A1(a)} = 1 - P_{B1(a)} = 0.0001$, and $Y_B = 0.90$. The concentrating factors can be calculated by Eqs. (5.53) and (5.169):

$$b = \frac{P_{B_{1(a)}} / \left(1 - P_{B_{1(a)}}\right)}{f_B / f_A} = \frac{0.9999/(1 - 0.9999)}{0.99/0.01} = 101$$

$$a = \frac{b - Y_B}{b(1 - Y_B)} = \frac{101 - 0.90}{101 \times (1 - 0.90)} = 9.9$$

The aqueous fraction and organic fraction can be calculated by Eq. (5.28):

$$f'_B = \frac{f_B Y_B}{P_{B_{1(a)}}} = \frac{0.99 \times 0.90}{0.9999} = 0.89$$

$$f'_A = 1 - f'_B = 1 - 0.89 = 0.11$$

Step 3: Optimum Process Parameters

The optimum process parameters can be determined either by optimum equations or extrema equations. According to Table 5.4, $f'_B = 0.89 > \frac{\sqrt{\beta}}{\sqrt{\beta}+1} = \frac{\sqrt{1.46}}{\sqrt{1.46}-1} = 0.548$ in an aqueous feeding system. The process is extraction controlled. Therefore, the extraction factors can be calculated by Eqs. (5.148) and (5.149).

$$E_M = \frac{1}{\sqrt{\beta}} = \frac{1}{\sqrt{1.46}} = 0.828$$

$$E'_M = \frac{E_M f'_B}{E_M - f'_A} = \frac{0.828 \times 0.89}{0.828 - 0.11} = 1.026$$

Therefore, the extraction and washing can be calculated.

$$S_o = \frac{E_M f'_B}{1 - E_M} = \frac{0.828 \times 0.89}{1 - 0.828} = 4.28, \quad W_a = S_o - f'_A = 4.28 - 0.11 = 4.17$$

Step 4: The number of Stages n and m

Since B is the major component in the feed and B is the high purity product in aqueous outlet, the number of extraction stages can be calculated by Eq. (5.117) according to the equations in Table 5.6.

$$n = \log b / \log \beta E_M = \log 101 / \log(1.46 \times 0.828) = 24.3$$

Therefore, the number of extraction stages is: $n = 25$

In the feeding stage, $P_{Bn(a)} = f_B = 0.99$, $P_{An(a)} = 1 - P_{Bn(a)} = 0.01$. Therefore, the B purity in organic phase in the feeding stage can be calculated by Eq. (5.91):

$$P_{Bn_{(o)}} = \frac{P_{Bn(a)}}{\beta - (\beta - 1) P_{Bn(a)}} = \frac{0.99}{1.46 - (1.46 - 1) \times 0.99} = 0.985$$

The A purity in the organic outlet can be calculated by Eq. (5.52):

$$P_{A_{n+m(o)}} = \frac{a f_A}{a f_A + f_B} = \frac{9.9 \times 0.01}{9.9 \times 0.01 + 0.99} = 0.09$$

The B purity in the organic outlet is:

$$P_{B_{n+m(o)}} = 1 - P_{A_{n+m(o)}} = 1 - 0.09 = 0.91$$

According to Eq. (5.119):

$$\begin{aligned} P^*_{B_{(o)}} &= \frac{\beta'/E'_M - 1}{\beta' - 1} + \frac{\beta'(1 - 1/E'_M)P_{B_{n+m(o)}}}{\beta' - E'_M + (E'_M - 1)(\beta' - 1)P_{B_{n+m(o)}}} \\ &= \frac{1.46/1.026 - 1}{1.46 - 1} + \frac{1.46 \times (1 - 1/1.026) \times 0.91}{1.46 - 1.026 + (1.026 - 1) \times (1.46 - 1) \times 0.91} \\ &= 0.995 \end{aligned}$$

Therefore, the number of scrubbing stages can be calculated by Eq. (5.118):

$$\begin{aligned} m &= \frac{\log a}{\log \frac{\beta'}{E'_M}} + 2.303\log\frac{P^*_{B_{(o)}} - P_{B_{n+m(o)}}}{P^*_{B_{(o)}} - P_{B_{n(o)}}} - 1 \\ &= \log 9.9/\log\frac{1.46}{1.026} + 2.303\log\frac{0.995 - 0.91}{0.995 - 0.985} - 1 = 7.6 \end{aligned}$$

Therefore, the number of scrubbing stages is: $m = 8$

Step 5: Flow Ratio

In the acidic extractant system, the rare earth concentration in organic phase $[M_{(o)}] =$ 0.25 mol/L when the rare earth concentration in the feed $[M_{F(a)}] = 1$ mol/L. The acid concentration in the scrubbing solution is $[HA] = 2.1$ mol/L.

Assuming $M_{F(a)} = 1$ mol/min, the flow rates can be obtained:

$$V_F = \frac{M_{F(a)}}{[M_{F(a)}]} = \frac{1}{1} = 1 \quad (\text{mL/min})$$

$$V_o = \frac{S_o}{[M_{(o)}]} = \frac{4.28}{0.25} = 17.1 \quad (\text{mL/min})$$

$$V_w = \frac{3W_a}{[HA]} = \frac{3 \times 4.17}{2.1} = 5.96 \quad (\text{mL/min})$$

Therefore, the rare earth concentration in each stream can be calculated:

$$[M_{1(a)}] = \frac{f'_B M_{F(a)}}{V_F + V_w} = \frac{0.89 \times 1}{1.00 + 5.96} = 0.128 \quad (\text{mol/L})$$

$$[M_{i(a)}] = \frac{M_{F(a)} + W_a}{V_F + V_w} = \frac{1 + 4.17}{1.00 + 5.96} = 0.743 \quad (\text{mol/L})$$

$i = 1, 2, \cdots, 25$

$$[M_{j(a)}] = \frac{W_a}{V_w} = \frac{4.17}{5.96} = 0.700 \quad (\text{mol/L})$$

$$j = 26,27,\cdots,33$$

$$[M_{33(o)}] = \frac{f'_A M_{F(a)}}{V_o} = \frac{0.11}{17.1} = 0.00643 \quad (\text{mol/L})$$

Step 5 and 6: Material Distribution and Balance Table, and Overall Process Evaluation

The material distribution and balance table can be constructed based on the above results and used for the overall process evaluation.

Example 2 The separation factors of an extraction system are: $\beta = \beta' = 2.00$. Feed composition is $f_A = f_B = 0.50$. A purity ($P_{An+m(o)}$) is required to be 99.9 % while B purity ($P_{B1(a)}$) is required to reach 99.99 %. Design the process parameters of a cascade solvent extraction process to meet the separation target.

Step 1: Concentrating Factor

Since the separator factors are given and the separation targets are specified, the design can start with the calculation of concentrating factors and the aqueous fraction and organic fraction. The concentrating factors of A and B can be calculated according to Eqs. (5.51) and (5.53).

$$a = \frac{P_{A_{n+m(o)}}/\left(1 - P_{A_{n+m(o)}}\right)}{f_A/f_B} = \frac{0.999/(1 - 0.999)}{0.5/0.5} = 999$$

$$b = \frac{P_{B_{1(a)}}/\left(1 - P_{B_{1(a)}}\right)}{f_B/f_A} = \frac{0.9999/(1 - 0.9999)}{0.5/0.5} = 9999$$

Since the product A and B are both high purities, the organic fraction and aqueous fraction can be determined according to Eq. (5.30):

$$f'_A \approx f_A = 0.5, \quad f'_B \approx f_B = 0.5$$

Step 2: Optimum Process Parameter Determination

1. *Using Optimum Equations*

According to Table 5.4,

$$f'_B = 0.5 < \frac{\sqrt{\beta}}{\sqrt{\beta} + 1} = \frac{\sqrt{2}}{\sqrt{2} + 1} = 0.586$$

The process is scrubbing controlled. Therefore, the extraction factors can be calculated using Eqs. (5.14) and (5.15).

$$E'_{M} = \sqrt{\beta} = \sqrt{2} = 1.414$$

$$E_{M} = \frac{E'_{M}f'_{A}}{E'_{M} - f'_{B}} = \frac{1.414 \times 0.5}{1.414 - 0.5} = 0.773$$

Therefore, the extraction and washing can be obtained:

$$S_{o} = \frac{E_{M}f'_{B}}{1 - E_{M}} = \frac{0.773 \times 0.5}{1 - 0.773} = 1.703, \quad W_{a} = S_{o} - f'_{A} = 1.703 - 0.5 = 1.203$$

Since $P_{A1(a)} = 1 - P_{B1(a)} = 1 - 0.9999 = 0.0001$ and $P_{Bn+m(o)} = 1 - P_{An+m(o)} = 1 - 0.999 = 0.001$, the intersection of the purity equilibrium line and purity operating line can be calculated by the approximation equation in Eqs. (5.113) and (5.119).

$$P^{*}_{A_{(a)}} \approx \frac{(\beta E_{M} - 1)}{\beta - 1} = \frac{2.0 \times 0.773 - 1}{2.0 - 1} = 0.546$$

$$P^{*}_{B_{(o)}} \approx \frac{\beta'/E'_{M} - 1}{\beta' - 1} = \frac{2.0/1.414 - 1}{2.0 - 1} = 0.414$$

According to Xu's assumption regarding the feeding stage, the purity of A and B in the aqueous phase is the same as that in the feed.

$$P_{A_{n(a)}} = 0.5, \quad P_{B_{n(a)}} = 0.5$$

The B purity in organic phase in the feeding stage can be calculated by Eq. (5.91):

$$P_{Bn_{(o)}} = \frac{P_{Bn(a)}}{\beta - (\beta - 1)P_{Bn(a)}} = \frac{0.5}{2.0 - (2.0 - 1) \times 0.5} = 0.333$$

The number of extraction stage can be calculated according to Eq. (5.116):

$$n = \log b/\log \beta E_{M} + 2.303\log\frac{P^{*}_{A_{(a)}} - P_{A_{1(a)}}}{P^{*}_{A_{(a)}} - P_{A_{n(a)}}}$$

$$= \log 9999/\log(2.0 \times 0.773) + 2.303\log\frac{0.546 - 0.0001}{0.546 - 0.5} = 23.7$$

The number of scrubbing stage can be calculated according to Eq. (5.118):

$$m+1 = \log a/\log\frac{\beta'}{E'_{\mathrm{M}}} + 2.303\log\frac{P^*_{\mathrm{B_{(o)}}} - P_{\mathrm{B}_{n+m(o)}}}{P^*_{\mathrm{B_{(o)}}} - P_{\mathrm{B}_{n(o)}}}$$
$$= \log 999/\log\frac{2.0}{1.414} + 2.303\log\frac{0.414 - 0.001}{0.414 - 0.333} = 21.6$$

The numbers of extraction stage and scrubbing stage are 24 and 21 based on the optimum equations.

2. *Using Extrema Equations*

According to the Eq. (5.160) and Eq. (5.170) in Table 5.5, the washing and extraction can be calculated:

$$W_{\mathrm{a}} = \frac{1}{\beta^k - 1} = \frac{1}{2.0^{0.7} - 1} = 1.601$$
$$S_{\mathrm{o}} = W_{\mathrm{a}} + f'_{\mathrm{A}} = 1.601 + 0.5 = 2.101$$

Therefore, the extraction factors can be determined according to Eq. (5.170) in Table 5.5:

$$E_{\mathrm{M}} = \frac{S_{\mathrm{o}}}{W_{\mathrm{a}} + 1} = \frac{2.101}{1.601 + 1} = 0.808$$
$$E'_{\mathrm{M}} = \frac{S_{\mathrm{o}}}{W_{\mathrm{a}}} = \frac{2.101}{1.601} = 1.312$$

Therefore, the intersection of the purity equilibrium line and the purity operating line can be calculated according to Eqs. (5.113) and (5.119):

$$P^*_{\mathrm{A_{(a)}}} \approx \frac{(\beta E_{\mathrm{M}} - 1)}{\beta - 1} = \frac{2.0 \times 0.808 - 1}{2.0 - 1} = 0.616$$
$$P^*_{\mathrm{B_{(o)}}} \approx \frac{\beta'/E'_{\mathrm{M}} - 1}{\beta' - 1} = \frac{2.0/1.312 - 1}{2.0 - 1} = 0.524$$

Therefore, n and m can be calculated:

$$n = \log b/\log\beta E_{\mathrm{M}} + 2.303\log\frac{P^*_{\mathrm{A_{(a)}}} - P_{\mathrm{A_{1(a)}}}}{P^*_{\mathrm{A_{(a)}}} - P_{\mathrm{A}_{n(a)}}}$$
$$= \log 9999/\log(2.0 \times 0.808) + 2.303\log\frac{0.616 - 0.0001}{0.616 - 0.5} = 20.9$$

$$m+1=\log a/\log\frac{\beta'}{E'_M}+2.303\log\frac{P^*_{B_{(o)}}-P_{B_{n+m(o)}}}{P^*_{B_{(o)}}-P_{B_{n(o)}}}$$

$$=\log 999/\log\frac{2.0}{1.312}+2.303\log\frac{0.524-0.001}{0.524-0.333}=17.4$$

Therefore, the numbers of extraction stages and scrubbing stages determined through extrema equations are 21 and 17.

Example 3 This example and the question are the same as those in Example 2 except the feeding is organic. Design the optimum process parameters of the organic feeding system.

Step1: a, b, f'_A, and f'_B
The same calculation can be done as above in Example 2.

Step2: Optimum Process Parameter Determination
The optimum process parameter can be determined either by optimum equations or extrema equations.

For organic feeding system, according to Table 5.4,

$$f'_B=0.5>\frac{1}{\sqrt{\beta}+1}=\frac{1}{\sqrt{2}+1}=0.414$$

Therefore, it is extraction controlled when organic feeding is used. The extraction factors can be calculated by Eqs. (5.148) and (5.150):

$$E_M=\frac{1}{\sqrt{\beta}}=\frac{1}{\sqrt{2}}=0.707$$

$$E'_M=\frac{1-E_M f'_A}{f'_B}=\frac{1-0.707\times 0.5}{0.5}=1.293$$

The extraction and washing are:

$$S_o=\frac{E_M f'_B}{1-E_M}=\frac{0.707\times 0.414}{1-0.707}=1.206$$

$$W_a=S_o+f'_B=1.206+0.5=1.706$$

The intersection of the purity equilibrium line and purity operating line can be calculated by the approximation equation in Eqs. (5.113) and (5.119).

$$P^*_{A_{(a)}}\approx\frac{(\beta E_M-1)}{\beta-1}=\frac{2.0\times 0.707-1}{2.0-1}=0.414$$

$$P^{*}_{\mathrm{B(o)}} \approx \frac{\beta'/E'_{\mathrm{M}} - 1}{\beta' - 1} = \frac{2.0/1.293 - 1}{2.0 - 1} = 0.546$$

Since it is organic feeding, the purity of A and B in the organic phase in the feeding stage can be obtained according to Xu's assumption about feeding stage.

$$P_{\mathrm{A}_{n(\mathrm{o})}} = f_{\mathrm{A}} = 0.5, \; P_{\mathrm{B}_{n(\mathrm{o})}} = f_{\mathrm{B}} = 0.5$$

The purity of A and B in the aqueous phase in the feeding stage can be calculated by Eq. (5.91):

$$P_{\mathrm{B}_{n(\mathrm{a})}} = \frac{\beta P_{\mathrm{B(o)}}}{1 + (\beta - 1)P_{\mathrm{B(o)}}} = \frac{2.0 \times 0.5}{1 + 0.5} = 0.667$$

$$P_{\mathrm{A}_{n(\mathrm{a})}} = 1 - P_{\mathrm{B(a)}} = 1 - 0.667 = 0.333$$

The number of extraction stage can be calculated according to Eq. (5.116):

$$n = \log b/\log\beta E_{\mathrm{M}} + 2.303\log\frac{P^{*}_{\mathrm{A(a)}} - P_{\mathrm{A}_{1(\mathrm{a})}}}{P^{*}_{\mathrm{A(a)}} - P_{\mathrm{A}_{n(\mathrm{a})}}}$$

$$= \log 9999/\log(2.0 \times 0.707) + 2.303\log\frac{0.414 - 0.0001}{0.414 - 0.333} = 28.2$$

The number of scrubbing stage can be calculated according to Eq. (5.118):

$$m + 1 = \log a/\log\frac{\beta'}{E'_{\mathrm{M}}} + 2.303\log\frac{P^{*}_{\mathrm{B(o)}} - P_{\mathrm{B}_{n+m(\mathrm{o})}}}{P^{*}_{\mathrm{B(o)}} - P_{\mathrm{B}_{n(\mathrm{o})}}}$$

$$= \log 999/\log\frac{2.0}{1.293} + 2.303\log\frac{0.546 - 0.001}{0.546 - 0.5} = 18.3$$

Therefore, the numbers of stages of extraction and scrubbing can be determined as 29 and 18, totally 47 stages.

Similarly, the numbers of stages can be determined by using the extrema equations.

The flow ratio can be calculated based on the relationship between flow rates and concentrations. Material distribution and balance table can be constructed and used for the final overall process evaluation.

5.11 Circulating Solvent Extraction

In rare earth cascade solvent extraction, it is not uncommon that a very long time is required to reach steady state to produce targeted products for those rare earths with low separation factors but with high purity requirement. In production practice, large amount of low quality product can be produced in a long period. Therefore, circulating solvent extraction is always used to start the process so that the time to reach stead state can be reduced and the amount of low quality product can be minimized. Circulating process includes full circulation and partial circulation. It is common to start up a rare earth cascade solvent extraction process following the approach of:

Full Circulation → Partial Circulation → Normal Fractional Operation

5.11.1 *Full Circulation*

Full circulation solvent extraction is a cascade process with no feeding and no discharge. In full circulation solvent extraction, the extraction equals to washing. It meets the following conditions:

$$M_F = 0,\ f'_A = f'_B = 0,\ S_o = W_a$$
$$E_M = E'_M = 1,\ J_S \to \infty,\ J_W \to \infty$$

The material distribution and balance table of a full circulation solvent extraction process is shown in Table 5.9.

5.11.2 *Partial Circulation*

Partial circulation solvent extraction is a cascade process with limited feeding and only one end discharge. It includes scrubbing partial circulation and extraction partial circulation.

Table 5.9 Material distribution and balance table of full circulation process

	S_o ↓				$M_F=0$ ↓				$W_a=S_o$ ↓
Stage	1	…	i	…	n	…	j	…	$n+m$
$M_{(o)}$	S_o	…	S_o	…	S_o	…	S_o	…	$f'_A=0$
$M_{(a)}$	$f'_B=0$	…	W_a	…	W_a	…	W_a	…	W_a

Table 5.10 Material distribution and balance table of scrubbing partial circulation process

	S_o ↓				$M_F < 1$ ↓				$W_a = S_o$ ↓
Stage	1	...	i	...	n	...	j	...	$n+m$
$M_{(o)}$	S_o	...	S_o	...	S_o	...	S_o	...	$f'_A = 0$
$M_{(a)}$	$f'_B = M_F$	...	W_a	...	W_a	...	W_a	...	W_a

Table 5.11 Material distribution and balance table of extraction partial circulation process

	S_o ↓				$M_F < 1$ ↓				$W_a = S_o$ ↓
Stage	1	...	i	...	n	...	j	...	$n+m$
$M_{(o)}$	S_o	...	S_o	...	S_o	...	S_o	...	$f'_A = M_F$
$M_{(a)}$	$f'_B = 0$	...	W_a	...	W_a	...	W_a	...	W_a

Scrubbing Partial Circulation

In scrubbing partial circulation, product B is produced from the aqueous outlet (first stage of extraction). There is no product A produced from the organic outlet. The material distribution and balance of a scrubbing partial circulation process is shown in Table 5.10.

In scrubbing partial circulation, there are the following relations:

$$E_M = S_o/(W_a + M_F) < 1,\ \ E'_M = S_o/W_a = 1,\ \ J_S = S_o/M_F,\ \ J_W \to \infty$$

Extraction Partial Circulation

In extraction partial circulation, product A is produced from the organic outlet (the $n+m$ stage of scrubbing) while no product is produced from the aqueous outlet. The material distribution and balance of an extraction partial circulation is shown in Table 5.11.

The following relations exist in the extraction partial circulation process:

$$E_M = S_o/(W_a + M_F) = 1,\ \ E'_M = S_o/W_a > 1,\ \ J_S \to \infty,\ \ J_W = W_a/M_F$$

The reflux ratios in both full circulation process and partial circulation process are all higher than regular fractional process.

5.12 Process Design of Three-Outlet Process

In the previous sections of this chapter, the cascade process principles and design are discussed based on the two-outlet process for a system with only A and B. On the basis of the two-outlet process, Professor Chunhua Yan and coworkers at the Peking University developed the three-outlet process.

One single two-outlet cascade process can only produce one pure product and one mixed product from a feed containing more than two rare earth elements. The three-outlet cascade process is a new process that can produce two pure products and a concentrated mixed product in a single process.

This section discusses the principles of three-outlet process and its process design. In three-outlet process, rare earth elements are classified into three components: (1) A that is easily extractable and produced from the organic outlet (the $n+m$ stage of scrubbing), (2) C that is difficult to extract and produced from the aqueous outlet (the first stage of extraction), and (3) B that is a middle concentrated product and produced from the aqueous phase or organic phase of a middle stage between 1 and $n+m$.

5.12.1 Outlet Fraction, Purity, and Recovery

5.12.1.1 Basic Equations

Different with the two-outlet process, the three-outlet process has three products A, B, and C. The specifications include:

A Purity and Recovery: $P_{\mathrm{A}_{n+m(\mathrm{o})}}$ and Y_{A}
B Purity and Recovery: $P_{\mathrm{B}I(\mathrm{a})}$ or $P_{\mathrm{B}I(\mathrm{o})}$ and Y_{B}
C Purity and Recovery: $P_{\mathrm{C}1(\mathrm{a})}$ and Y_{C}

The feed composition can be expressed by the mole fractions of A, B, and C.

$$f_{\mathrm{A}}+f_{\mathrm{B}}+f_{\mathrm{C}}=1 \quad (5.172)$$

If the third outlet is in the aqueous phase, there are the following equations:

$$f_{\mathrm{A}}=A_{1(\mathrm{a})}+A_{I(\mathrm{a})}+A_{n+m(\mathrm{o})} \quad (5.173)$$

$$f_{\mathrm{B}}=B_{1(\mathrm{a})}+B_{I(\mathrm{a})}+B_{n+m(\mathrm{o})} \quad (5.174)$$

$$f_{\mathrm{C}}=C_{1(\mathrm{a})}+C_{I(\mathrm{a})}+C_{n+m(\mathrm{o})} \quad (5.175)$$

where $\mathrm{A}_{1(\mathrm{a})}$, $\mathrm{B}_{1(\mathrm{a})}$, $\mathrm{C}_{1(\mathrm{a})}$ are the mole flow rates of A, B, C in the first stage aqueous outlet, $\mathrm{A}_{i(\mathrm{a})}$, $\mathrm{B}_{i(\mathrm{a})}$, $\mathrm{C}_{i(\mathrm{a})}$ the mole flow rates in the third outlet or *I*th stage, $\mathrm{A}_{n+m(\mathrm{o})}$, $\mathrm{B}_{n+m(\mathrm{o})}$, $C_{n+m(\mathrm{o})}$ the mole flow rates in the last stage organic outlet.

The outlet fractions f'_A, f'_B, and f'_C can be obtained:

$$f'_A = A_{n+m(o)} + B_{n+m(o)} + C_{n+m(o)} \quad (5.176)$$

$$f'_B = A_{I(a)} + B_{I(a)} + C_{I(a)} \quad (5.177)$$

$$f'_C = A_{1(a)} + B_{1(a)} + C_{1(a)} \quad (5.178)$$

$$f'_A + f'_B + f'_C = 1 \quad (5.179)$$

According to the definition of purity and recovery, there are the following equations:

$$P_{A_{n+m(o)}} = \frac{A_{n+m(o)}}{f'_A} \quad (5.180)$$

$$P_{B_{I(a)}} = \frac{B_{I(a)}}{f'_B} \quad (5.181)$$

$$P_{C_{1(a)}} = \frac{C_{1(a)}}{f'_C} \quad (5.182)$$

$$Y_A = \frac{A_{n+m(o)}}{f_A} \quad (5.183)$$

$$Y_B = \frac{B_{I(a)}}{f_B} \quad (5.184)$$

$$Y_C = \frac{C_{1(a)}}{f_C} \quad (5.185)$$

Substituting the purity equations into the recovery equations, the recovery equations can be expressed by purity and outlet fraction:

$$Y_A = \frac{f'_A P_{A_{n+m(o)}}}{f_A} \quad (5.186)$$

$$Y_B = \frac{f'_B P_{B_{I(a)}}}{f_B} \quad (5.187)$$

$$Y_C = \frac{f'_C P_{C_{1(a)}}}{f_C} \quad (5.188)$$

Since A and C are normally high purity product, $A_{1(a)} \approx 0$ and $C_{n+m(o)} \approx 0$.

If the third outlet is in organic phase, similar equations exist.

$$f_{\mathrm{A}} = A_{1(\mathrm{a})} + A_{I(\mathrm{o})} + A_{n+m(\mathrm{o})} \tag{5.189}$$

$$f_{\mathrm{B}} = B_{1(\mathrm{a})} + B_{I(\mathrm{o})} + B_{n+m(\mathrm{o})} \tag{5.190}$$

$$f_{\mathrm{C}} = C_{1(\mathrm{a})} + C_{I(\mathrm{o})} + C_{n+m(\mathrm{o})} \tag{5.191}$$

$$f'_{\mathrm{B}} = A_{I(\mathrm{o})} + B_{I(\mathrm{o})} + C_{I(\mathrm{o})} \tag{5.192}$$

$$P_{\mathrm{B}_{I(\mathrm{o})}} = \frac{B_{I(\mathrm{o})}}{f'_{\mathrm{B}}} \tag{5.193}$$

$$Y_{\mathrm{B}} = \frac{f'_{\mathrm{B}} P_{\mathrm{B}_{I(\mathrm{o})}}}{f_{\mathrm{B}}} \tag{5.194}$$

5.12.1.2 Separation Target Specification

There are total six separation targets for the three-outlet cascade process. Three purity targets: $P_{\mathrm{A}_{n+m(\mathrm{o})}}$, $P_{\mathrm{B}_{I(\mathrm{a})}}$ or $P_{\mathrm{B}_{I(\mathrm{o})}}$, $P_{\mathrm{C}_{1(\mathrm{a})}}$ and three recovery targets: Y_{A}, Y_{B}, Y_{C}. They can be represented more generally by P_{λ} and Y_{λ} ($\lambda = \mathrm{A}$, B, or C). When $A_{1(\mathrm{a})} \approx 0$ and $C_{n+m(\mathrm{o})} \approx 0$, it is only need to specify four targets and the other two separation targets and outlet fractions can be calculated. There are total 15 different combinations of specifying separation targets.

Combination: $Y_{\lambda 1}$, $P_{\lambda 1}$, $Y_{\lambda 2}$, $P_{\lambda 2}$

1. Specifying $P_{\mathrm{A}_{n+m(\mathrm{o})}}$, Y_{A}, $P_{\mathrm{C}_{1(\mathrm{a})}}$, Y_{C}, we have the following:

$$f'_{\mathrm{A}} = f_{\mathrm{A}} Y_{\mathrm{A}} / P_{\mathrm{A}_{n+m(\mathrm{o})}}$$
$$f'_{\mathrm{C}} = f_{\mathrm{C}} Y_{\mathrm{C}} / P_{\mathrm{C}_{1(\mathrm{a})}}$$
$$f'_{\mathrm{B}} = 1 - f'_{\mathrm{A}} - f'_{\mathrm{C}}$$
$$A_{I(\mathrm{a})} = f_{\mathrm{A}} - A_{n+m(\mathrm{o})} = f_{\mathrm{A}}(1 - Y_{\mathrm{A}})$$
$$C_{I(\mathrm{a})} = f_{\mathrm{C}} - C_{1(\mathrm{a})} = f_{\mathrm{C}}(1 - Y_{\mathrm{C}})$$
$$B_{I(\mathrm{a})} = f'_{\mathrm{B}} - A_{I(\mathrm{a})} - C_{I(\mathrm{a})}$$
$$Y_{\mathrm{B}} = B_{I(\mathrm{a})} / f_{\mathrm{B}}$$
$$P_{\mathrm{B}_{I(\mathrm{a})}} = B_{I(\mathrm{a})} / f'_{\mathrm{B}}$$

2. If $P_{\mathrm{B}_{I(\mathrm{a})}}$, Y_{B}, $P_{\mathrm{C}_{1(\mathrm{a})}}$, Y_{C} or $P_{\mathrm{A}_{n+m(\mathrm{o})}}$, Y_{A}, $P_{\mathrm{B}_{I(\mathrm{a})}}$, Y_{B} are set, the others can be calculated in a similar way.

Combination: $Y_{\lambda 1}$, $P_{\lambda 1}$, $P_{\lambda 2}$, $Y_{\lambda 3}$

1. Specifying $P_{A_{n+m(o)}}$, Y_A, $P_{B_{I(a)}}$, Y_C, we have the following:

$$f'_A = f_A Y_A / P_{A_{n+m(o)}}$$
$$A_{I(a)} = f_A - A_{n+m(o)} = f_A(1 - Y_A)$$
$$C_{I(a)} = f_C - C_{1(a)} = f_C(1 - Y_C)$$
$$B_{I(a)} = \left(A_{I(a)} + C_{I(a)}\right) P_{B_{I(a)}} / \left(1 - P_{B_{I(a)}}\right)$$
$$f'_B = A_{I(a)} + B_{I(a)} + C_{I(a)}$$
$$f'_C = 1 - f'_A - f'_B$$
$$P_{C_{1(a)}} = f_C Y_C / f'_C$$
$$Y_B = B_{I(a)} / f_B$$

If Y_C, $P_{C_{1(a)}}$, Y_A, and $P_{B_{I(a)}}$ are set, the others like Y_B, $P_{A_{n+m(o)}}$, f'_A, f'_B, f'_C can be obtained in a similar way.

2. Specifying $P_{A_{n+m(o)}}$, Y_B, $P_{B_{I(a)}}$, Y_C, we have the following equations:

$$f'_B = f_B Y_B / P_{B_{I(a)}}$$
$$B_{I(a)} = f_B Y_B$$
$$C_{I(a)} = f_C(1 - Y_C)$$
$$A_{i(a)} = f'_B - B_{I(a)} - C_{I(a)}$$
$$A_{n+m(o)} = f_A - A_{i(a)}$$
$$Y_A = A_{n+m(o)} / f_A$$
$$f'_A = f_A Y_A / P_{A_{n+m(o)}}$$
$$f'_C = 1 - f'_A - f'_B$$
$$P_{C_{1(a)}} = Y_C f_C / f'_C$$

If $P_{C_{1(a)}}$, Y_B, $P_{B_{I(a)}}$, and Y_A are set, $P_{A_{n+m(o)}}$, similarly $Y_C f'_A, f'_B$, and f'_C can be obtained.

3. Specifying $P_{A_{n+m(o)}}$, Y_A, Y_B, $P_{C_{1(a)}}$, we have the following equations:

$$f'_A = f_A Y_A / P_{A_{n+m(o)}}$$
$$A_{n+m(o)} = f_A Y_A$$

$$B_{n+m(\mathrm{o})} = f'_{\mathrm{A}} - A_{n+m(\mathrm{o})}$$
$$B_{I(\mathrm{a})} = f_{\mathrm{B}} Y_{\mathrm{B}}$$
$$B_{1(\mathrm{a})} = f_{\mathrm{B}} - B_{I(\mathrm{a})} - B_{n+m(\mathrm{o})}$$
$$C_{1(\mathrm{a})} = f_{\mathrm{C}} Y_{\mathrm{C}}$$
$$f'_{\mathrm{C}} = B_{1(\mathrm{a})} + C_{1(\mathrm{a})}$$
$$f'_{\mathrm{B}} = 1 - f'_{\mathrm{A}} - f'_{\mathrm{C}}$$
$$P_{\mathrm{B}_{1(\mathrm{a})}} = B_{I(\mathrm{a})}/f'_{\mathrm{B}} = f_{\mathrm{B}} Y_{\mathrm{B}}/f'_{\mathrm{B}}$$

Similarly, if $P_{\mathrm{C}_{1(\mathrm{a})}}$, Y_{C}, Y_{B}, and $P_{\mathrm{A}_{n+m(\mathrm{o})}}$ are set, $P_{\mathrm{B}_{I(\mathrm{a})}}$, $Y_{\mathrm{A}}, f'_{\mathrm{A}}, f'_{\mathrm{B}}$, and f'_{C} can be calculated.

Combination: $P_{\mathrm{A}_{n+m(\mathrm{o})}}$, $P_{\mathrm{B}_{I(\mathrm{a})}}$, $P_{\mathrm{C}_{1(\mathrm{a})}}$, Y_{λ}

1. Specifying $P_{\mathrm{A}_{n+m(\mathrm{o})}}$, $P_{\mathrm{B}_{I(\mathrm{a})}}$, $P_{\mathrm{C}_{1(\mathrm{a})}}$, and Y_{A}, we have the following equations:

$$f'_{\mathrm{A}} = f_{\mathrm{A}} Y_{\mathrm{A}}/P_{\mathrm{A}_{n+m(\mathrm{o})}}$$
$$A_{i(\mathrm{a})} = f_{\mathrm{A}} - f'_{\mathrm{A}} P_{\mathrm{A}_{n+m(\mathrm{o})}}$$
$$B_{n+m(\mathrm{o})} = f'_{\mathrm{A}}\left(1 - P_{\mathrm{A}_{n+m(\mathrm{o})}}\right)$$
$$B_{I(\mathrm{a})} = f'_{\mathrm{B}} - A_{I(\mathrm{a})} - C_{I(\mathrm{a})} = B_{I(\mathrm{a})}/P_{\mathrm{B}_{I(\mathrm{a})}} - A_{I(\mathrm{a})} - C_{I(\mathrm{a})}$$
$$B_{i(\mathrm{a})} = f'_{\mathrm{B}} - A_{I(\mathrm{a})} - C_{I(\mathrm{a})} = \left(A_{I(\mathrm{a})} + C_{I(\mathrm{a})}\right) P_{\mathrm{B}_{I(\mathrm{a})}}/\left(1 - P_{\mathrm{B}_{I(\mathrm{a})}}\right)$$
$$B_{1(\mathrm{a})} = f'_{\mathrm{C}} - C_{1(\mathrm{a})} = C_{1(\mathrm{a})}/P_{\mathrm{C}_{1(\mathrm{a})}} - C_{1(\mathrm{a})}$$
$$C_{1(\mathrm{a})} = f_{\mathrm{C}} - C_{I(\mathrm{a})}$$
$$B_{1(\mathrm{a})} = \left(f_{\mathrm{C}} - C_{I(\mathrm{a})}\right)\left(1 - P_{\mathrm{C}_{1(\mathrm{a})}}\right)/P_{\mathrm{C}_{1(\mathrm{a})}}$$

Since $B_{1(\mathrm{a})} + B_{I(\mathrm{a})} = f_{\mathrm{B}} - B_{n+m(\mathrm{o})}$, $C_{\mathrm{I(a)}}$ can be calculated by the following equation by substituting $B_{1(\mathrm{a})}$ and $B_{\mathrm{I(a)}}$ into the B balance equation.

$$C_{i(\mathrm{a})} = \frac{f_{\mathrm{B}} - B_{n+m(\mathrm{o})} - f_{\mathrm{C}}\left(1 - P_{\mathrm{C}_{1(\mathrm{a})}}\right)/P_{\mathrm{C}_{1(\mathrm{a})}} - A_{I(\mathrm{a})} P_{\mathrm{B}_{I(\mathrm{a})}}/\left(1 - P_{\mathrm{B}_{I(\mathrm{a})}}\right)}{\left(1 - P_{\mathrm{C}_{1(\mathrm{a})}}\right)/P_{\mathrm{C}_{1(\mathrm{a})}} + P_{\mathrm{B}_{I(\mathrm{a})}}/\left(1 - P_{\mathrm{B}_{I(\mathrm{a})}}\right)}$$
$$f'_{\mathrm{B}} = A_{I(\mathrm{a})} + B_{I(\mathrm{a})} + C_{I(\mathrm{a})}$$
$$f'_{\mathrm{C}} = 1 - f'_{\mathrm{A}} - f'_{\mathrm{B}}$$

$$Y_B = B_{I(a)}/f_B$$

$$Y_C = C_{1(a)}/f_C$$

2. Following the similar approach, Y_A, Y_B, f_A', f_B', and f_C' can be calculated if $P_{A_{n+m(o)}}$, $P_{B_{I(a)}}$, $P_{C_{1(a)}}$, and Y_C are specified.
3. If $P_{A_{n+m(o)}}$, $P_{B_{I(a)}}$, $P_{C_{1(a)}}$, and Y_B are specified, following the similar approach, Y_A, Y_B, f_A', f_B', and f_C' cannot be solved. Therefore, another parameter must be specified.

Combination: Y_A, Y_B, Y_C and P_λ

1. Specifying Y_A, Y_B, Y_C, and $P_{A_{n+m(o)}}$, the others can be calculated by the following equations.

$$f_A' = f_A Y_A / P_{A_{n+m(o)}}$$

$$A_{I(a)} = f_A - f_A' P_{A_{n+m(o)}}$$

$$B_{n+m(o)} = f_A' \left(1 - P_{A_{n+m(o)}}\right)$$

$$C_{1(a)} = f_C Y_C$$

$$B_{I(a)} = f_B Y_B$$

$$B_{1(a)} = f_B - B_{I(a)} - B_{n+m(o)}$$

$$f_C' = B_{1(a)} + C_{1(a)}$$

$$f_B' = 1 - f_A' - f_C'$$

$$P_{B_{I(a)}} = B_{I(a)}/f_B'$$

$$P_{C_{1(a)}} = C_{1(a)}/f_C'$$

2. If Y_A, Y_B, Y_C, and $P_{C_{1(a)}}$ are specified, $P_{A_{n+m(o)}}$, $P_{B_{I(a)}}$, f_A', f_B', and f_C' can be calculated.
3. If Y_A, Y_B, Y_C, and $P_{B_{I(a)}}$ are specified, the others cannot be solved. Another parameter must be specified so the others can be calculated.

From above equations of the three-outlet process, it can be seen that only four separation specifications are independent in most of cases if the approximation is introduced. For accurate calculation, at least five specifications must be determined first.

5.12.2 *Amount of Extraction* S_o

In the three-outlet rare earth separation process, the required amount of extraction S_o is different with the different position and different phase of the third outlet. The calculation method of the amount of extraction S_o in three-outlet processes is developed by Professor Yan and coworkers based on the characteristics of the constant mixing ratio system.

5.12.2.1 Material Balance of Three-Outlet System

The material distribution and balance of the three-outlet system have eight different situations depending on the feed, the phase of the third outlet, and the position of the third outlet. The material distribution and balance of the extraction three-outlet process are shown in Tables 5.12, 5.13, 5.14, and 5.15. For the scrubbing three-outlet process, material distribution and balance are shown in Tables 5.16, 5.17, 5.18, and 5.19. Although there are eight different situations, the four with the aqueous third outlet are often used in industrial practice.

Table 5.12 Aqueous feeding and aqueous third outlet in extraction

	S_o ↓				$M_{F(a)}=1$ ↓		$W_a=S_o-f'_A$ ↓
Stage	1	$2\cdots I-1$	I	$I+1\cdots n-1$	n	$n+1\cdots n+m-1$	$n+m$
$M_{(o)}$	S_o	S_o	S_o	S_o	S_o	S_o	f'_A
$M_{(a)}$	f'_C	$S_o+f'_C$	W_a+1	W_a+1	W_a+1	W_a	W_a
			↓ f'_B				

Table 5.13 Aqueous feeding and organic third outlet in extraction

	S_o ↓		f'_B ↑		$M_{F(a)}=1$ ↓		$W_a=S_o-f'_A-f'_B$ ↓
Stage	1	$2\cdots I-1$	I	$I+1\cdots n-1$	n	$n+1\cdots n+m-1$	$n+m$
$M_{(o)}$	S_o	S_o	$S_o-f'_B$	$S_o-f'_B$	$S_o-f'_B$	$S_o-f'_B$	f'_A
$M_{(a)}$	f'_C	W_a+1	W_a+1	W_a+1	W_a+1	W_a	W_a

Table 5.14 Organic feeding and extraction aqueous third outlet in extraction

	S_o ↓				$M_{F(o)}=1$ ↓		$W_a=S_o+1-f'_A$ ↓
Stage	1	$2\cdots I-1$	I	$I+1\cdots n-1$	n	$n+1\cdots n+m-1$	$n+m$
$M_{(o)}$	S_o	S_o	S_o	S_o	S_o+1	S_o+1	f'_A
$M_{(a)}$	f'_C	$S_o+f'_C$	W_a	W_a	W_a	W_a	W_a
			↓ f'_B				

Table 5.15 Organic feeding and extraction organic third outlet in extraction

	S_o ↓		f'_B ↑		$M_{F(o)}=1$ ↓		$W_a=S_o+f'_C$ ↓
Stage	1	$2\cdots I-1$	I	$I+1\cdots n-1$	n	$n+1\cdots n+m-1$	$n+m$
$M_{(o)}$	S_o	S_o	$S_o-f'_B$	$S_o-f'_B$	$S_o+1-f'_B$	$S_o+1-f'_B$	f'_A
$M_{(a)}$	f'_C	W_a	W_a	W_a	W_a	W_a	W_a

Table 5.16 Aqueous feeding and aqueous third outlet in scrubbing

	S_o ↓		$M_{F(a)}=1$ ↓				$W_a=S_o-f'_A$ ↓
Stage	1	$2\cdots n-1$	n	$n+1\cdots I-1$	I	$I+1\cdots n+m-1$	$n+m$
$M_{(o)}$	S_o	S_o	S_o	S_o	S_o	S_o	f'_A
$M_{(a)}$	f'_C	$S_o+f'_C$	$S_o+f'_C-1$	$S_o+f'_C-1$	W_a	W_a	W_a
					↓ f'_B		

Table 5.17 Aqueous feeding and organic third outlet in scrubbing

	S_o ↓		$M_{F(a)}=1$ ↓		f'_B ↑		$W_a=S_o-f'_A-f'_B$ ↓
Stage	1	$2\cdots n-1$	n	$n+1\cdots I-1$	I	$I+1\cdots n+m-1$	$n+m$
$M_{(o)}$	S_o	S_o	S_o	S_o	S_o	$S_o-f'_B$	f'_A
$M_{(a)}$	f'_C	W_a+1	W_a	W_a	W_a	W_a	W_a

Table 5.18 Organic feeding and aqueous third outlet in scrubbing

	S_o ↓		$M_{F(o)}=1$ ↓				$W_a=S_o+1-f'_A$ ↓
Stage	1	$2\cdots n-1$	n	$n+1\cdots I-1$	I	$I+1\cdots n+m-1$	$n+m$
$M_{(o)}$	S_o	S_o	S_o	S_o+1	S_o+1	S_o+1	f'_A
$M_{(a)}$	f'_C	$S_o+f'_C$	$S_o+f'_C$	$S_o+f'_C$	W_a	W_a	W_a
					↓ f'_B		

Table 5.19 Organic feeding and organic third outlet in scrubbing

	S_o ↓		$M_{F(o)}=1$ ↓		f'_B ↑		$W_a=S_o+f'_C$ ↓
Stage	1	$2\cdots n-1$	n	$n+1\cdots I-1$	I	$I+1\cdots n+m-1$	$n+m$
$M_{(o)}$	S_o	S_o	S_o	S_o+1	S_o+1	$S_o+1-f'_B$	f'_A
$M_{(a)}$	f'_C	W_a	W_a	W_a	W_a	W_a	W_a

5.12.2.2 Minimum Amount of Extraction Rule

In order to achieve the three-outlet separation, the targeted purity of the three products must be met. Therefore, the follow relations exist:

$$P_{C_{i(a)}} > P_{C_{i+1(a)}} \; (i = 1, 2, \cdots, n+m) \quad (5.195)$$

$$P_{A_{j(o)}} < P_{A_{j+1(o)}} \; (j = 1, 2, \cdots, n+m) \quad (5.196)$$

In order to meet above conditions, the system should have a minimum amount of extraction. If the third outlet is in extraction section, the minimum amount of extraction can be denoted as $(S_{min})_o$ when feed is organic and $(S_{min})_a$ when feed is aqueous. If the third outlet is in scrubbing section, the minimum amount of extraction can be denoted as $(S'_{min})_o$ when feed is organic and $(S'_{min})_a$ when feed is aqueous.

According to the separator factors, separation targets, and feeding, the minimum amount of extraction can be determined based on extraction equilibrium and material balance.

$$(S_{\min})_{\mathrm{a}} = f'_{\mathrm{C}}\left(\frac{\beta_{\mathrm{AC}} \cdot P_{\mathrm{C}_{I(\mathrm{a})}} \cdot \left(1 - P_{\mathrm{C}_{I(\mathrm{a})}}\right) - \beta_{\mathrm{AB}} \cdot P_{\mathrm{B}_{I(\mathrm{a})}} \cdot P_{\mathrm{C}_{I(\mathrm{a})}}}{P_{\mathrm{C}_{I(\mathrm{a})}} \cdot \left[\beta_{\mathrm{AC}} \cdot \left(1 - P_{\mathrm{C}_{I(\mathrm{a})}}\right) - P_{\mathrm{A}_{I(\mathrm{a})}} - \beta_{\mathrm{AB}} \cdot P_{\mathrm{B}_{I(\mathrm{a})}}\right]} - 1\right) \tag{5.197}$$

$$(S_{\min})_{\mathrm{a}} = (S_{\min})_{\mathrm{o}} \tag{5.198}$$

$$\left(S'_{\min}\right)_{\mathrm{a}} = \frac{f'_{\mathrm{A}}\left(P_{\mathrm{A}_{n+m(\mathrm{o})}}/P_{\mathrm{A}_{I(\mathrm{a})}} - 1\right) \cdot \left(\beta_{\mathrm{AC}} \cdot P_{\mathrm{A}_{I(\mathrm{a})}} + \beta_{\mathrm{BC}} \cdot P_{\mathrm{B}_{I(\mathrm{a})}} + P_{\mathrm{C}_{I(\mathrm{a})}}\right)}{\beta_{\mathrm{AC}} \cdot \left(1 - P_{\mathrm{A}_{I(\mathrm{a})}}\right) - P_{\mathrm{C}_{I(\mathrm{a})}} - \beta_{\mathrm{BC}} \cdot P_{\mathrm{C}_{I(\mathrm{a})}}} \tag{5.199}$$

$$\left(S'_{\min}\right)_{\mathrm{a}} = \left(S'_{\min}\right)_{\mathrm{o}} \tag{5.200}$$

The above minimum amount of extraction is a function of separation targets, feed composition, and separation factor. According the minimum amount of extraction, the optimum location and phase of the third outlet can be determined. If $(S_{\min})_{\mathrm{o}} < (S'_{\min})_{\mathrm{o}}$, the third outlet should be opened in extraction section otherwise it should be in the scrubbing section. For the organic third outlet, its location can be determined similarly. This is the "Minimum amount of extraction" rule for the determination of the third outlet location.

It is known that the separation results are very poor at the conditions of minimum amount of extraction. A large number of stages are required to meet the separation targets. Therefore, the amount of extraction S_{o} used in the actual process design should be larger than the minimum amount of extraction in order to optimize the process design.

5.12.3 *The Number of Three-Outlet Process Stages*

A three-outlet rare earth cascade solvent extraction process can be considered as the combination of three countercurrent solvent extraction processes. Therefore, the number of stages of a three-outlet process can be simplified to the number of stages of three countercurrent processes. In the following section, the discussion is based on the location of the third outlet.

5.12.3.1 The Third Outlet Located in Extraction

Figure 5.8 shows the basic separation process of the three-outlet processes with the third outlet in extraction. Normally, it is required to produce two high purity products in the two end-outlets with high recovery of A in the organic outlet. Therefore, the A content in the third outlet is low.

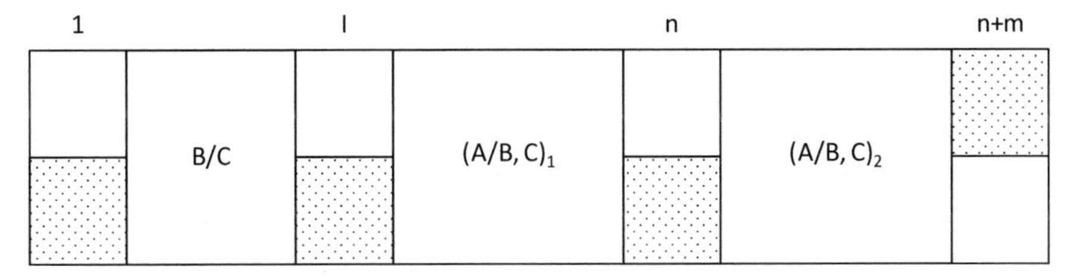

Fig. 5.8 Representation of a three-outlet process with the third outlet in extraction

The three-outlet process can be divided to three sections:

1. $1 \rightarrow \mathrm{I}$ extraction section where B/C is the major separation. I stage can be considered as the feeding stage of the B/C separation system. Therefore, the concentrating factor of C can be calculated by the following equation:

$$c = \frac{P_{\mathrm{C}_{1(\mathrm{a})}} / \left(1 - P_{\mathrm{C}_{1(\mathrm{a})}}\right)}{P_{\mathrm{C}_{I(\mathrm{a})}} / \left(1 - P_{\mathrm{C}_{I(\mathrm{a})}}\right)} \tag{5.201}$$

The extraction factor is:

$$E_{\mathrm{M1}} = \frac{S_{\mathrm{o}}}{W_{\mathrm{a}} + 1 - f_{\mathrm{B}}'} \tag{5.202}$$

2. $I + 1 \rightarrow n$ extraction section where $A/(B + C)$ is the major separation. n stage can be considered as the feeding stage. $(B + C)_{I(\mathrm{a})}$ can be considered as the aqueous outlet product. Therefore, the concentrating factor of $(B + C)$ can be calculated by the following equation:

$$bc = \frac{\left(P_{\mathrm{B}_{I(\mathrm{a})}} + P_{\mathrm{B}_{I(\mathrm{a})}}\right) / P_{\mathrm{A}_{I(\mathrm{a})}}}{(f_{\mathrm{B}} + f_{\mathrm{C}}) / f_{\mathrm{A}}} \tag{5.203}$$

The extraction factor is:

$$E_{\mathrm{M2}} = \frac{S_{\mathrm{o}}}{W_{\mathrm{a}} + 1} \tag{5.204}$$

3. $n + 1 \rightarrow n + m$ scrubbing section where $A/(B + C)$ is the major separation. In comparison with the $I + 1 \rightarrow n$ section, C is very low in this section. Therefore, it also can be considered as A/B separation. n can be considered as the feeding stage and $\mathrm{A}_{n+m(\mathrm{o})}$ is considered as the outlet product of scrubbing. Therefore, the concentrating factor of A is:

$$a = \frac{P_{\mathrm{A}_{n+m(\mathrm{o})}} / \left(1 - P_{\mathrm{A}_{n+m(\mathrm{o})}}\right)}{f_{\mathrm{A}} / (f_{\mathrm{B}} + f_{\mathrm{C}})} \tag{5.205}$$

The extraction factor is:

$$E'_{\mathrm{M}} = \frac{S_{\mathrm{o}}}{W_{\mathrm{a}}} \tag{5.206}$$

The n and m equations summarized in Table 5.6 can be used to calculate the number of stages of each section using the concentrating factors and extraction factors obtained.

5.12.3.2 The Third Outlet Located in Scrubbing

The basic separation process when the third outlet is in scrubbing can be represented by Fig. 5.9. It can be divided into three sections:

1. $1 \rightarrow n$ extraction section where B/C is the major separation since A is low in this section. n stage can be considered as the feeding stage. $C_{1(a)}$ is the aqueous outlet product. Therefore, the concentrating factor of C can be calculated by the following equation:

$$c = \frac{P_{\mathrm{C}_{1(\mathrm{a})}} / \left(1 - P_{\mathrm{C}_{1(\mathrm{a})}}\right)}{f_{\mathrm{C}} / (f_{\mathrm{A}} + f_{\mathrm{B}})} \tag{5.207}$$

The extraction factor is:

$$E_{\mathrm{M1}} = \frac{S_{\mathrm{o}}}{W_{\mathrm{a}} + 1 - f'_{\mathrm{B}}} \tag{5.208}$$

2. $n+1 \rightarrow I$ scrubbing section where (A + B)/C is the major separation. n stage can be considered as the feeding stage. $(\mathrm{A}+\mathrm{B})_{I(\mathrm{o})}$ can be considered as product of

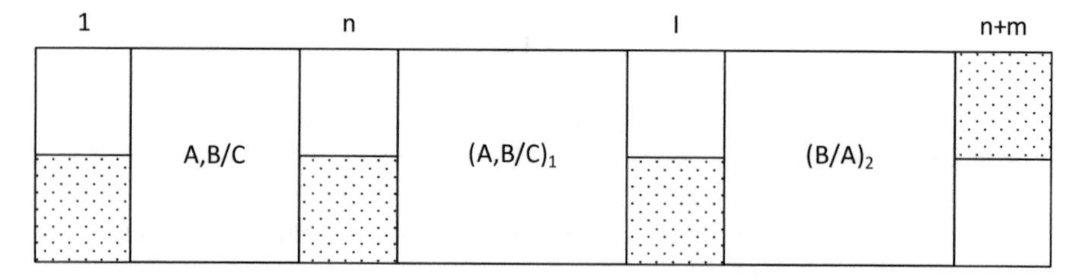

Fig. 5.9 Representation of a three-outlet process with the third outlet in scrubbing

scrubbing. Therefore, the concentrating factor of (A + B) can be calculated by the following equation:

$$ab = \frac{\left(P_{A_{I(o)}} + P_{B_{I(o)}}\right)/P_{C_{I(o)}}}{(f_A + f_B)/f_C} \tag{5.209}$$

where $P_{A_{I(o)}}$, $P_{B_{I(o)}}$, and $P_{C_{I(o)}}$ can be obtained through their equilibrium relationship with $P_{A_{I(a)}}$, $P_{B_{I(a)}}$, and $P_{C_{I(a)}}$.

The extraction factor is:

$$E'_{M1} = \frac{S_o}{W_a - f'_B} \tag{5.210}$$

3. $I + 1 \rightarrow n + m$ scrubbing section where A/B is the major separation since C is low. Stage I can be considered as the feeding stage and $P_{A_{n+m(o)}}$ as the outlet product. Therefore, the concentrating factor of A is:

$$a = \frac{P_{A_{n+m(o)}}/\left(1 - P_{A_{n+m(o)}}\right)}{P_{A_{I(a)}}/\left(P_{B_{I(a)}} + P_{C_{I(a)}}\right)} \tag{5.211}$$

The extraction factor is:

$$E'_{M2} = \frac{S_o}{W_a} \tag{5.212}$$

The n and m equations summarized in Table 5.6 can be used to calculate the number of stages of each section using the concentrating factors and extraction factors obtained.

5.12.4 Example of Three-Outlet Process Design

Example 1 The separation of Sm/Eu/Gd is to be done in an acidic extraction system with aqueous feeding. The feed composition is Gd/Eu/Sm = 0.39/0.10/0.51. The extraction system has the separation factors of $\beta_{Gd/Eu} = 1.40$ and $\beta_{Eu/Sm} = 2.00$.

Step 1—Separation Target Specification
Gd_2O_3 purity $P_{A_{n+m(o)}} = 0.999$, Sm_2O_3 purity $P_{C_{1(a)}} = 0.9999$, Sm_2O_3 recovery $Y_C = 0.999$, Eu_2O_3 concentrate purity $P_{B_{1(a)}} \geq 0.40$.

Step 2—Process Parameters Determination

$$f'_{\rm C} = \frac{f_{\rm C} Y_{\rm C}}{P_{{\rm C}_{1(a)}}} = \frac{0.51 \times 0.999}{0.9999} = 0.509541$$

$$C_{1(a)} = f_{\rm C} Y_{\rm C} = 0.51 \times 0.999 = 0.509490$$

$$B_{1(a)} = f'_{\rm C} - C_{1(a)} = 0.509541 - 0.509490 = 5.09541 \times 10^{-5}$$

$$C_{I(a)} = f_{\rm C} - C_{1(a)} = 0.51 - 0.509490 = 5.1 \times 10^{-5}$$

$$B_{I(a)} = \frac{\left[(1 - f'_{\rm C}) P_{{\rm A}_{n+m(o)}} - C_{I(a)} - f_{\rm A}\right]}{\left[1 - \frac{1 - P_{{\rm A}_{n+m(o)}}}{P_{{\rm B}_{I(a)}}}\right]}$$

$$= \frac{(1 - 0.509541) \times 0.999 - 5.1 \times 10^{-4} - 0.39}{1 - \frac{1 - 0.999}{0.40}} = 0.099708$$

$$f'_{\rm B} = \frac{B_{I(a)}}{P_{{\rm B}_{I(a)}}} = \frac{0.099708}{0.40} = 0.249270$$

$$A_{I(a)} = f'_{\rm B} - B_{I(a)} - C_{I(a)} = 0.249270 - 0.099708 - 5.1 \times 10^{-5} = 0.149052$$

$$f'_{\rm A} = 1 - f'_{\rm B} - f'_{\rm C} = 1 - 0.249270 - 0.509541 = 0.241189$$

$$A_{n+m(o)} = f_{\rm A} - A_{I(a)} = 0.39 - 0.149052 = 0.240948$$

$$B_{n+m(o)} = f'_{\rm A} - A_{n+m(o)} = 0.241189 - 0.240948 = 2.41 \times 10^{-4}$$

$$Y_{\rm B} = B_{I(a)}/f_{\rm B} = 0.099708/0.10 = 0.997080$$

$$Y_{\rm A} = A_{n+m(o)}/f_{\rm A} = 0.240948/0.39 = 0.617815$$

$$P_{{\rm A}_{I(a)}} = \frac{A_{I(a)}}{A_{I(a)} + B_{I(a)} + C_{I(a)}} = \frac{0.149052}{0.149052 + 0.099708 + 5.1 \times 10^{-5}} = 0.599057$$

$$P_{{\rm B}_{I(a)}} = \frac{B_{I(a)}}{A_{I(a)} + B_{I(a)} + C_{I(a)}} = \frac{0.099708}{0.149052 + 0.099708 + 5.1 \times 10^{-5}} = 0.400738$$

$$P_{{\rm C}_{I(a)}} = \frac{C_{I(a)}}{A_{I(a)} + B_{I(a)} + C_{I(a)}} = \frac{5.1 \times 10^{-5}}{0.149052 + 0.099708 + 5.1 \times 10^{-5}} = 2.05 \times 10^{-4}$$

Step 3—Minimum Amount of Extraction

The minimum amounts of extraction in extraction and scrubbing are calculated by substituting the values of $P_{{\rm A}_{I(a)}}$, $P_{{\rm B}_{I(a)}}$, $P_{{\rm C}_{I(a)}}$, $f'_{\rm C}$, $\beta_{\rm A/C}$, $\beta_{\rm A/B}$ into Eqs. (5.197) and (5.199).

$$(S_{\min})_{\rm a} = f'_{\rm C}\left(\frac{\beta_{\rm AC} \cdot P_{{\rm C}_{I(a)}} \cdot \left(1 - P_{{\rm C}_{I(a)}}\right) - \beta_{\rm AB} \cdot P_{{\rm B}_{I(a)}} \cdot P_{{\rm C}_{I(a)}}}{P_{{\rm C}_{I(a)}} \cdot \left[\beta_{\rm AC} \cdot \left(1 - P_{{\rm C}_{I(a)}}\right) - P_{{\rm A}_{I(a)}} - \beta_{\rm AB} \cdot P_{{\rm B}_{I(a)}}\right]} - 1\right) = 424.732$$

Table 5.20 Number of stages at different amounts of extraction

So	Number of stages					
	I	n	m	$n+m$	$I-n$	$S_o \times (n+m)$
1.4000	39	22	63	85	17	119
1.5000	36	21	55	76	15	114
1.7500	32	19	47	66	13	115
2.0000	30	18	42	60	12	120
2.2500	29	18	39	57	11	128
2.5000	28	17	38	55	11	138

$$\left(S'_{\min}\right)_a = \frac{f'_A\left(\frac{p_{A_{n+m(o)}}}{p_{A_{I(a)}}} - 1\right)\cdot\left(\beta_{AC}\cdot P_{A_{I(a)}} + \beta_{BC}\cdot P_{B_{I(a)}} + P_{C_{I(a)}}\right)}{\beta_{AC}\cdot\left(1 - P_{A_{I(a)}}\right) - P_{C_{I(a)}} - \beta_{BC}\cdot P_{C_{I(a)}}} = 1.23757$$

According to the minimum amount of extraction rule, the third outlet should be opened in scrubbing. The actual amount of extraction should be higher than the minimum extraction.

Step 4—Number of Stages
To calculate the number of stages, the process can be divided to three countercurrent solvent extraction processes. Following the procedures in Sect. 5.12.3, the number of stages of each section can be obtained by setting the amount of extraction to calculate the extraction factor in each section. For example: If $S_o = 2.0000$, $W_a = 1.7588$, $n = 18$, $m = 42$, $I = 30$. Setting S_o at different values higher than the minimum amount of extraction, a series of number of stages of each section can be obtained as shown in Table 5.20.

The separation process table can be constructed based on the results for further evaluation. For example, Table 5.21 is the separation process table at $S_o = 2.0000$, $W_a = 1.7588$, $n = 18$, $m = 42$, and $I = 30$.

Step 5—Overall Process Evaluation
The purpose of the cascade solvent process design is to determine the optimum process that requires the balance between capital investment and operating cost. For a given feed with fixed composition, the unit production cost depends on the normalized amount of extraction S_o. If mixer-settlers are used, the major fixed investment depends on the amount of organic holding and rare earth holding in extraction cells. When mixing and settling residence times are determined, the product of extraction amount S_o and the total number of stages $(n+m)$ can be used as the indicator of the fixed investment. The curve $S_o \times (n+m) - S_o$ as shown in Figure 5.10 is normally used to determine the actual process design. The minimum fixed investment is close to the bottom of this U shape curve. Since there is a minimum amount of extraction, the number of stages will dramatically increase when S_o is less than the minimum amount. When S_o is too high, the amounts of organic and rare earth holdings in cells are also too high. Therefore, the actual process parameters are normally determined at slightly right side of the cure bottom.

Table 5.21 Example of separation process table

	$S_o = 2.0000$ ↓		$M_{F(a)} = 1.0000$ ↓				$f'_A = 0.2412$ ↑
Stage	1	⋯	18	⋯	30	⋯	60
$M_{(o)}$	S_o	S_o	S_o	S_o	S_o	S_o	f'_A
$M_{(a)}$	f'_C	$S_o + f'_C$	$S_o + f'_C$	$W_a - f'_B$	W_a	W_a	W_a
	↓ $f'_C = 0.5095$				↓ $f'_B = 0.2493$		↑ $W_a = S_o - f'_A = 1.7588$

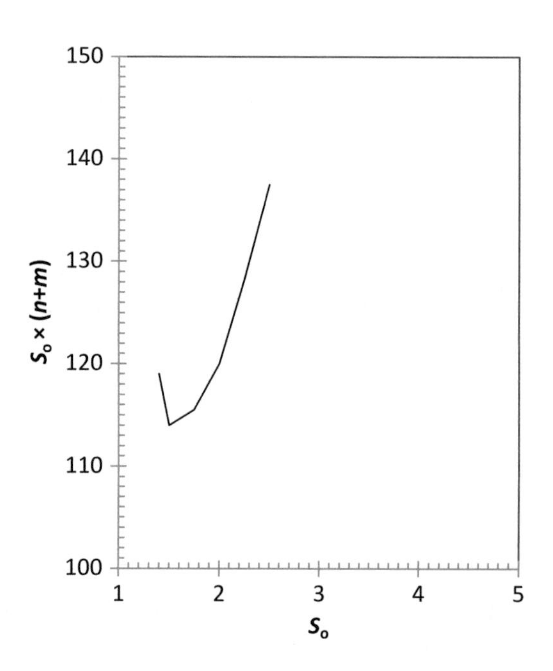

Fig. 5.10 Cost evaluation curve of process design

In this example at $S_o = 2.0000$, the composition of outlet products at steady state are calculated and shown in Table 5.22.

5.12.5 Three-Outlet Process Features

In comparison with the two-outlet process, the three-outlet process can produce one extra pure product and one concentrated product for the multiple rare earth separation. The third outlet concentrated product can be light rare earths like Pr_6O_{11} and

Table 5.22 Composition of outlet products at steady state

Component	Stage		
	1 (aqueous phase)	30 (aqueous phase)	60 (organic phase)
A: Gd_2O_3	5.436×10^{-6}	0.5979	0.9991
B: Eu_2O_3	9.280×10^{-5}	0.4000	8.925×10^{-4}
C: Sm_2O_3	0.9999	2.046×10^{-3}	3.383×10^{-6}

CeO_2 or middle heavy rare earths like Eu_2O_3 and Tb_4O_7. The third outlet product always has high recovery. Therefore, it is particularly suitable for concentrating low content but high value rare earths like Eu_2O_3 and Tb_4O_7 during multiple-rare earth element separation.

Due to the concentration increase and consequent volume reduction of the third outlet product, the following processing cost can be reduced significantly. It also increases the process flexibility. If required, the third outlet can be closed and the three-outlet process can be easily converted to two-outlet process.

It has been found that B concentration in stage I is normally lower than its maximum value. The maximum B concentration in organic phase is behind one stage than the maximum B concentration in aqueous phase.

Due to the open of a third outlet, the distribution and concentration profile of other components are changed. It is difficult to reach the optimum separation by only experience. Therefore, the continuous research and development of the three-outlet process principle and design are important.

5.13 Fuzzy Linkage Extraction

5.13.1 Introduction

Due to the complexity of the three-outlet process, fuzzy linkage extraction was developed by Jiankang Hu and coworkers (Deng et al. 2012). As shown in Fig. 5.11, the schematic of fuzzy linkage extraction process. A feed with A, B, C can be separated to individual components or multiple components. For example for the rare earth feed that contains LaCePrNd/SmEuGdTbDy/HoYErTmYbLu, A includes Ho, Y, Er, Tm, Yb, and Lu; B includes Sm, Eu, Gd, Tb, and Dy; C includes La, Ce, Pr, and Nd.

One fuzzy linkage extraction process (A/B/C) is called one separation module. The A/C separation is the first layer separation of the A/B/C module. The B/C separation and A/B separation are the second layer separation of the A/B/C module.

A/C separation, the first layer separation, utilizes the large effective separation factor between A and C. The separation target is to realize complete separation between A and C. That means the A concentrate does not contain C and the C concentrate does not contain A. The first layer separation does not consider the

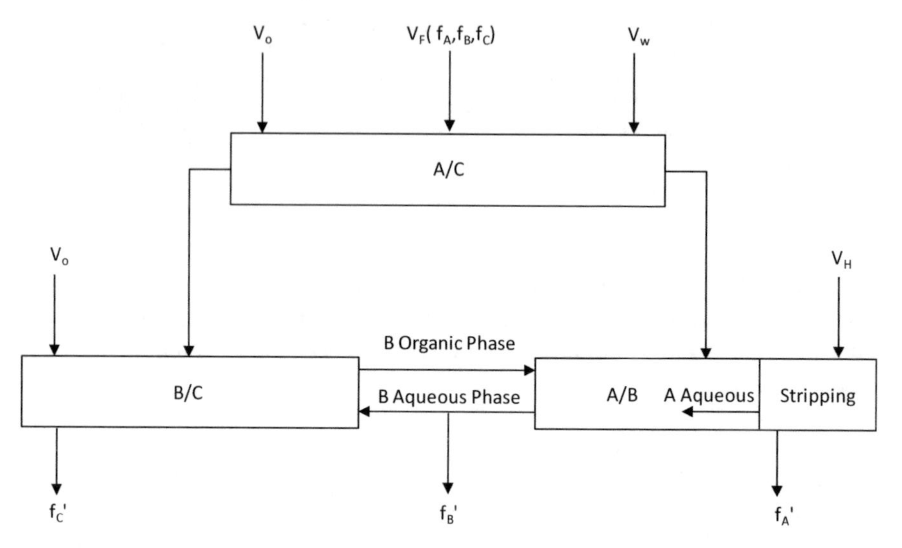

Fig. 5.11 Schematic of fuzzy linkage extraction process

distribution of the middle component B. B can be in A concentrate or C concentrate. This is why the process is called fuzzy separation. Fuzzy separation is actually a preliminary separation.

B/C separation and A/B separation, the second layers of separation, are the final separation of B/C and A/B on the basis of A/C preliminary separation. The aqueous outlet of A/C separation is directly fed to B/C separation and the organic outlet of A/C separation is directly used by A/B separation. This is why the process is also called linkage separation.

In conventional separation process, the scrubbing section and stripping separation are separated. The scrubbing section uses washing solution (V_w) and the stripping section uses stripping solution (V_H). The stripping solution normally contains high concentration of acid and will needs treatment before next stage of further separation. Therefore, reagent consumption is high. However, in fuzzy linkage separation process, the scrubbing solution and stripping solution are the same. Only stripping solution is added strip the rare earth out of the organic solution. Partial stripping solution that contains A will enter the scrubbing section of the A/B separation as washing solution. Therefore, reagent consumption can be reduced significantly.

5.13.2 Features of Fuzzy Linkage Separation Process

1. Xu's cascade rare earth solvent extraction principles are applicable to the fuzzy linkage separation process. Only flow configurations are different from case to case.

2. Production practice shows that reagent consumption can be reduced significantly to 30 % in comparison with the conventional separation processes. A fuzzy linkage process has lower amount of extraction (S_o) and lower amount of washing (W_a) which dictate the operating cost of a cascade rare earth separation process.
3. Phase contacts are increased in fuzzy linkage processes. For example in the second layer A/B separation, the washing solution containing A will contact with the organic solution containing B to have exchanges:

$$B_{(o)} + A_{(a)} = A_{(o)} + B_{(a)}$$

   Similar phase contacts and exchanges occur in the second layer B/C separation:

$$C_{(o)} + B_{(a)} = B_{(o)} + C_{(a)}$$

   Increased phase contacts will increase the separation efficiency.
4. Solvent extraction cells have smaller size due to the direct aqueous feed and organic feed from the first layer A/C separation to the second layer B/C and A/B separations. The products of the first layer separation have higher concentration than the feed and consequently smaller volume to handle for the remaining processes.
5. Fuzzy linkage process can be used to simplify the three-outlet processes. A three-outlet process can produce two pure end products and one middle concentrated product. But a fuzzy linkage separation module can produce three high purity products with high recoveries. Also, the fuzzy linkage process eliminates the impact of the third outlet that changes the rare earth distribution in the process. Therefore, the operation is simplified. In comparison with three-outlet processes, the fuzzy linkage process is simple and stable.

References

Xu, G., & Yuan, C. (1987). *Solvent extraction of rare earths* (pp. 392–479). Beijing: China Science Press.
Xu, G. (1995). *Rare earths I* (pp. 626–727). Beijing: China Metallurgy Industry Press.
Deng, Z., Xu, T., Hu, J., & Yang, F. (2012). Interpretations of fuzzy linkage extraction technology. *Nonferrous Metals Science and Engineering, 13*(1), 10–12.

Chapter 6
Rare Earth Solvent Extraction Equipment

6.1 Classification of Solvent Extraction Equipment

According to the structure and working principles, industrial solvent extraction equipment can be classified into three categories: (1) stage-type extractors, (2) column-type extractors, and (3) centrifugal extractors. In the following sections each category of extractor is briefly discussed.

6.1.1 Stage-Type Extractor

In stage-type extractors, the organic phase and aqueous phase mix together to reach equilibrium. The two phases are incrementally separated when one flows counter-currently from the other. One of the most common stage-type extractors is mixer-settler. A schematic representation of a mixer-settler is shown in Fig. 6.1. It consists of a small mixing chamber followed by a large gravity settling chamber. Each mixer-settler provides one stage of extraction. The aqueous phase and organic phase are drawn into the mixing chamber where they are mixed by an impeller. Extraction occurs during the mixing of the two phases. The extraction normally takes a few seconds to a few minutes to reach steady state. The mixing solution flows into the settler where the two phases are allowed to separate by gravity due to their density difference. Normally the settling takes longer than the mixing. Therefore, the settler chamber is larger than the mixing chamber. In rare earth separation a mixer/settler ratio of 1/2.5 is commonly used.

Mixer-settlers are large and bulky with high stage efficiency. They can be used in batch mode or in continuous mode by staging multiple units. Mixer-settlers can form any number of stages with good phase contacting. They can handle high-viscosity solvents. However, mixer-settlers take a large floor area with high capital

J. Zhang et al., *Separation Hydrometallurgy of Rare Earth Elements*,
DOI 10.1007/978-3-319-28235-0_6

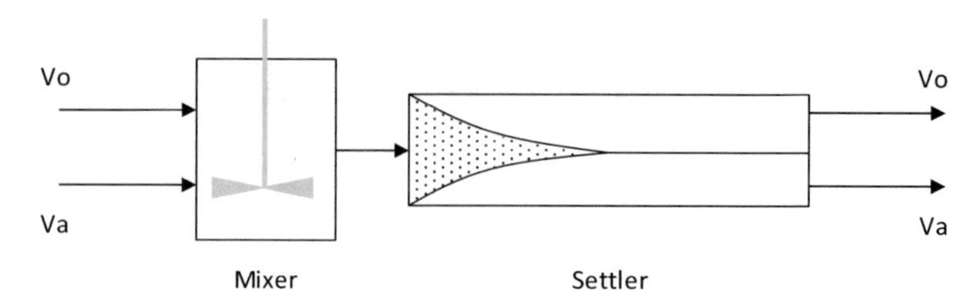

Fig. 6.1 Representation of a mixer-settler. V_o: organic flow, V_a: aqueous flow

cost and operation cost. Mixer-settlers are used when a long settling time is required and the solutions are readily separated by gravity.

The primary function of extractors is to mix two phases and subsequently to separate the two phases. To achieve the separation of solutes, the right amount of mixing is important. Sufficient mixing can create a large interfacial area to ensure good mass transfer and stage efficiency. Less mixing will cause the formation of large droplets and the reduction of the interfacial area. Overmixing can minimize mass transfer resistance but can also cause the formation of very small droplets or emulsions which are difficult to settle.

6.1.2 *Column-Type Extractors*

Common column-type extractors include packed tower, pulse column, rotating disc contactors, tray column, etc. Column-type extractors are also called differential/continuous contact extractors due to the continuous concentration gradient along the height of the column. The driving force of solute transfer from one phase to the other exists during the entire countercurrent contacting process. In comparison with the stage type of extractors, column-type extractors maximize the utilization of the mass transfer driving force. The stage retention time or the amount of materials to complete a theoretical stage is only about 30 % of that in a mixer-settler when the processing capacity is the same. Column-type extractors are of high efficiency with small footage. However, they need high headroom and are difficult to scale up to handle high flow rates and high phase ratios. It is less efficient to separate those phases with a small density difference.

6.1.2.1 Packed Tower

As shown in Fig. 6.2, a packed tower is composed of three parts: (1) tower head, (2) tower body, and (3) tower bottom. The phase separation occurs in the tower

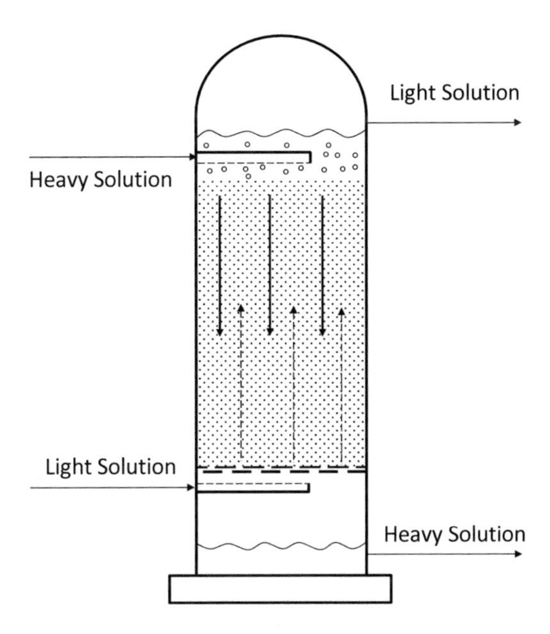

Fig. 6.2 Representation of a packed tower

head and tower bottom while phase mixing occurs in the void space of the packing in the tower body. The packing is rested on support plates.

The heavy solution is fed from the top while the light solution is fed from the bottom. The heavy solution fills in a large portion of the void space of the packing as the continuous phase and flows downwards. The light solution is dispersed in the remainder of the void space and rises upwards. The packing in the tower body provides large interfacial areas for phase contacting which causes the drops to coalesce and reform. Mass transfer efficiency in packed towers is 2–3 times higher than that in spray towers which are similar to the packed tower but the tower body is empty without packing. It is suitable for the separation of clear solutions with low viscosity.

In order to provide large interfacial areas, the packing materials should be able to be wetted by the continuous phase. The phase with lower viscosity is normally chosen as the continuous phase. The phase with higher flow rate is chosen as the dispersed phase to create more interfacial area and turbulence. In general, aqueous phase can wet metallic surfaces and organic phases can wet nonmetallic surfaces.

6.1.2.2 Pulse Column

Packed towers have no moving parts and are simple to operate but they are not very efficient for separation. Very tall columns are required to achieve good separation. To reduce the height of the column, trays or perforated plates are used and a mechanical force is applied to reduce the droplet size of the dispersed phase.

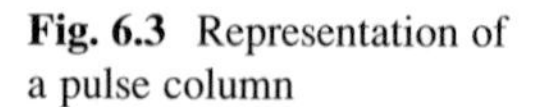

Fig. 6.3 Representation of a pulse column

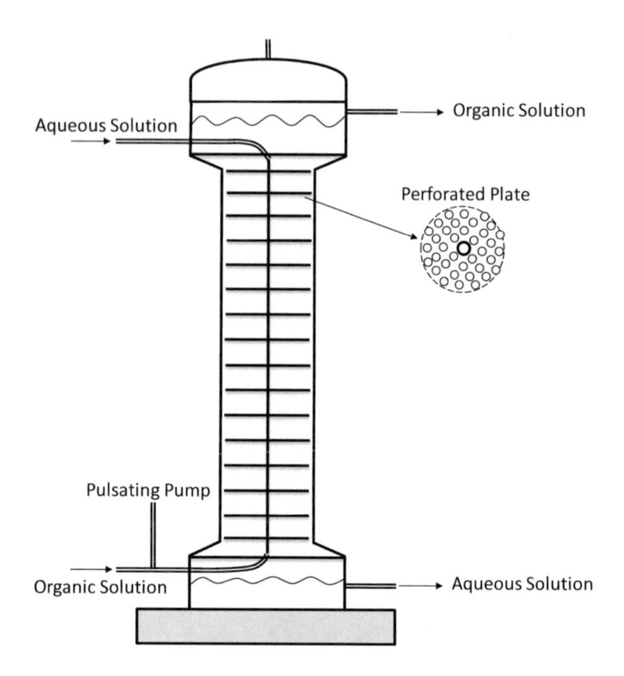

A pulse column typically requires less than a third of the number of theoretical stages as compared to a non-pulsating column.

Figure 6.3 shows a schematic representation of a pulse column. The same as the packed column, solutions are fed to the column continuously and flow countercurrently. The difference is that mechanical energy is applied to pulse the solutions in the column up and down. A reciprocating pump is commonly used to pulse the entire content of the column at frequent intervals. A pulse amplitude of 5–25 mm is commonly used at the frequency of 100–200 cycles per minute. The column is installed with perforated plates to promote droplet formation as the dispersed phase is pushed through the plates. The pulsating action causes the light solution to be dispersed into the heavy solution on the upward stroke and the heavy solution to jet into the light phase on the downward stroke. The droplet size of the dispersed phase is reduced and mass transfer is improved by the pulsating action.

6.1.2.3 Rotating Disc Contactor

A rotating disc contactor is an agitated countercurrent extractor. As shown in Fig. 6.4, the high-speed rotating discs provide agitation inside the compartments that are formed by stationary stator rings. The rotating discs are in the center of the column and are slightly smaller than the opening in the stationary stator rings. Typically, the disc diameter is 33–36 % of the column diameter. There are three flow patterns in the rotating disc contactor: (1) the countercurrent flows of the two phases due to the density difference, (2) the outward flow of the liquids dispersed

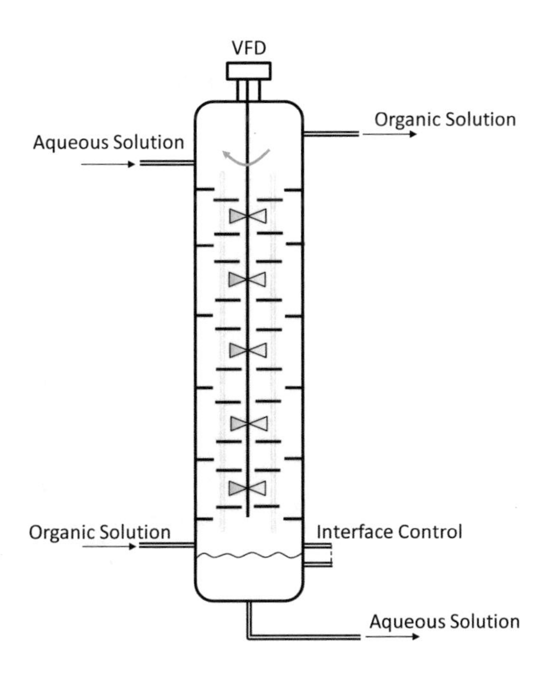

Fig. 6.4 Representation of a rotating disc contactor

and forced by the discs, and (3) the inward flow of the liquids due to the resistance of the wall and stator rings where quiet zones are formed and two phases are allowed to separate. Due to the flow patterns, the efficiency of the rotating disc contactor is much higher than packed columns. However, axial mixing is severe in large-diameter columns and at large phase ratios.

6.1.2.4 Tray Column

Figure 6.5 shows the general arrangement of a tray column. The organic solution is fed from the bottom passing through the perforation on the plate. The organic solution is dispersed as fine droplets rising through the continuous aqueous solution. The fine droplets coalesce into a layer under the plate and then are dispersed above the plate. The aqueous solution is fed from the top of the column and flows downward through the downcomers. If the organic solution is the continuous phase and the aqueous is the dispersed phase, the downcomers will become upcomers where the organic solution flows upwards. The aqueous solution will be dispersed and coalesced through the plates. The perforations in the plates are normally 3–6 mm in diameter and account for 15–25 % of the plates.

The advantages of the tray column are that the axial mixing is restricted in the space between trays and mass transfer is improved by the dispersion at each tray.

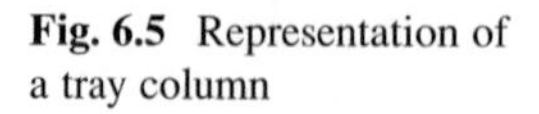

Fig. 6.5 Representation of a tray column

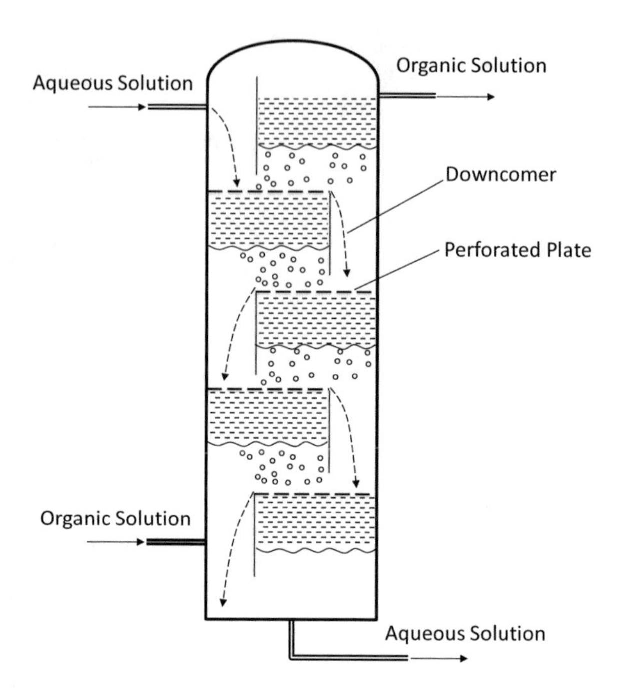

6.1.3 *Centrifugal Extractors*

Centrifugal extractors are high-speed rotary machines. Like mixer-settlers they are discrete-stage units providing one stage of extraction per unit. Multiple units can be linked together. Due to the centrifugal forces, the phase separation inside a centrifugal contractor is faster and more efficient. The retention time is very short and the process capacity therefore is large.

The advantages of the centrifugal extractors include high efficiency, high capacity, and low material holdup. They can be used to separate phases with density difference as low as 0.01 g/mL. They are suitable for the separation of phases with high viscosity, high tendency of emulsification, and unstable chemical properties. However, the capital cost and maintenance cost of centrifugal extractors are high (Fig. 6.6).

6.2 Selection Criteria of Extractors

An ideal extractor will have fast mass transfer, good phase separation, high throughput, simple structure, low cost, and easy maintenance. It is also flexible, reliable, and operable. However, there is no such extractor with all these merits. Each type of extractors has its advantages and disadvantages. In the selection of extractors, factors to be considered include the following:

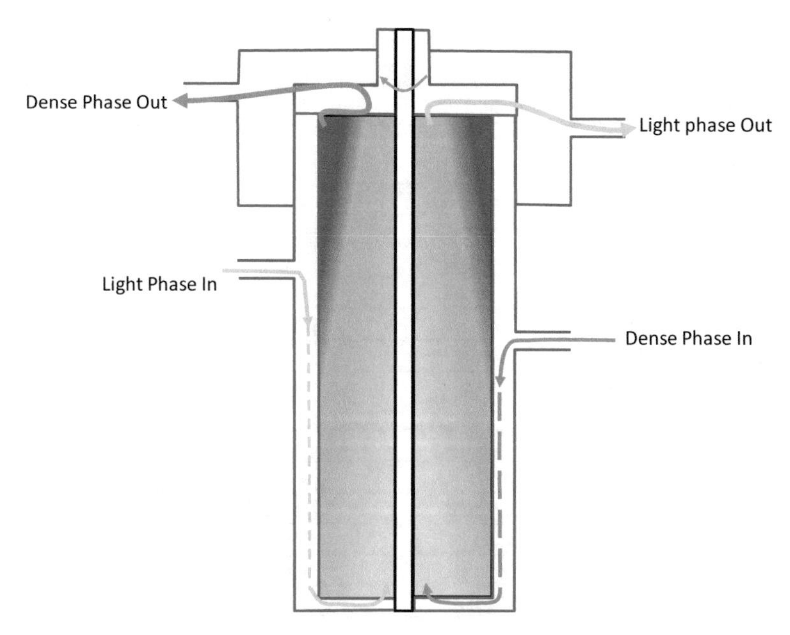

Fig. 6.6 Schematic representation of a centrifugal contractor

1. Extraction System
   The selection of the contactor device is essentially dependent on the physical properties of the extraction system. The physical properties like density and interfacial tension of the solvents have important impact on the phase mixing and separation. For solvents with small surface tension and high viscosity, external energy (forces) can be avoided in mixing to facilitate phase separation. For those with high surface tension, they are difficult to mix well but can separate easily. Mechanical energy is normally applied to get sufficient mixing.
   For the reaction-controlled extraction systems, longer residence time is normally required. Mixer-settlers are good choices for systems that require long residence time like 5 min. The extractors such as centrifugal extractors with very short residence time are not suitable. However, centrifugal extractors are good for the extraction systems with unstable chemical properties.
2. Extractor Capacity
   Mixer-settlers and columns without agitation have low process capacity in comparison with columns with agitation and centrifugal extractors.
3. Number of Stages
   Another important criterion in selecting a contactor device is the required number of theoretical stages. For the separation that requires a small number of stages, all type of extractors can be considered. When a high number of stages is required, high-efficiency columns or centrifugal extractors are preferred.

Table 6.1 Summary and comparison of different types of extractors

Parameters	Mixer-settlers	Columns without agitation	Columns with agitation	Centrifugal extractor
Separation efficiency	High	Poor	Average	High
Mass transfer	Average	High	Low	High
Residence time	Long	Average	Average	Short
Stage required	Low	Average	High	Average
Capacity	Low	Low	High	High
Capital cost	High	Low	Average	High
Operation cost	High	Low	Average	High
Maintenance cost	Average	Low	Average	High
Plant headroom	Low	High	High	Low
Floor area	Large	Low	Low	Moderate
Holdup volume	High	Average	Average	Low
Reliability	Good	Good	Good	Average
Flexibility	Good	Average	Average	Good
Complexity	Simple	Simple	Average	Complex
Quick restart	Quick/easy	Difficult	Difficult	Quick/easy

4. Critical Constraints
   It is common that the process or separation is restricted by actual operating conditions and plant limitation. For the process with very high phase ratio, mixer-settler or centrifugal contractors are normally considered to minimize axial mixing. If there is a plant height limit, column-type extractors will not be considered. For the plant limited by area, mixer-settlers are not a good choice. Other constraints may include material properties, health and safety, and maintenance requirement.
5. Project Economics
   Overall the project economics will dictate the process design and equipment selection. Some advantages of the extractors may be sacrificed and some disadvantages may be tolerated to make overall sound project economics.

Table 6.1 is the summary and comparison of the different type of extractors. In general, mix-settlers are suitable for the easy phase-separating systems that require a large number of theoretical stages. Columns without agitation are often used for easy phase separation with a low number of theoretical stages while agitated columns are suitable for the moderate phase separation with a large number of theoretical stages. Centrifugal extractors are often used for difficult phase separation with a low number of stages.

6.3 Basics of Extractor Design

6.3.1 Rate of Extraction (N)

Rate of extraction is the measurement of the mass transfer of a solute from one phase to another in unit time. The rate of extraction directly affects the extractor efficiency. The rate of extraction of a system is restricted by (1) the formation rate of the extractant-metal complexes, or (2) the mass transfer rate of the complexes from one phase to the other. For a simple molecule extraction system, the rate of extraction is determined by the mass transfer rate of extractant-metal complexes. According to mass transfer principles, the mass transfer rate of a solute from aqueous phase to organic phase can be expressed by the following equation:

$$N = K_{\mathrm{a}}F\left(x - x^{*}\right) \tag{6.1}$$

$$N = K_{\mathrm{o}}F\left(y^{*} - y\right) \tag{6.2}$$

where N is the rate of extraction (kg/h); F is the interfacial area (m^2); x is the solute concentration in the aqueous phase ($\mathrm{kg/m}^3$); y is the solute concentration in the organic phase ($\mathrm{kg/m}^3$); x^* is the aqueous equilibrium concentration of y ($\mathrm{kg/m}^3$); y^* is the organic equilibrium concentration of x ($\mathrm{kg/m}^3$); K_{a} is the mass transfer coefficient in aqueous phase (m/h); and K_{o} is the mass transfer coefficient in organic phase (m/h).

Equations 6.1 and 6.2 can be expressed by a general equation:

$$N = KF\Delta C \tag{6.3}$$

where K is the mass transfer coefficient (m/h) and ΔC is the concentration difference ($\mathrm{kg/m}^3$).

6.3.2 Height Equivalent of a Theoretical Stage

In the design of differential extractors, they are considered to consist of a series of theoretical extraction stages. A height equivalent of a theoretical stage (HETS) is a section of a tower or column height which equals to a theoretical extraction stage. The HETS is considered as an indicator of extraction efficiency of a differential contactor. Its value depends on the type of contactor, extraction system, and operating conditions and can be determined by testing or empirical equations.

The effective height of a differential contactor can be calculated by the following equation:

$$H = n \cdot (\mathrm{HETS}) \tag{6.4}$$

where H is the effective height of the contactor (m) and n the number of theoretical extraction stages.

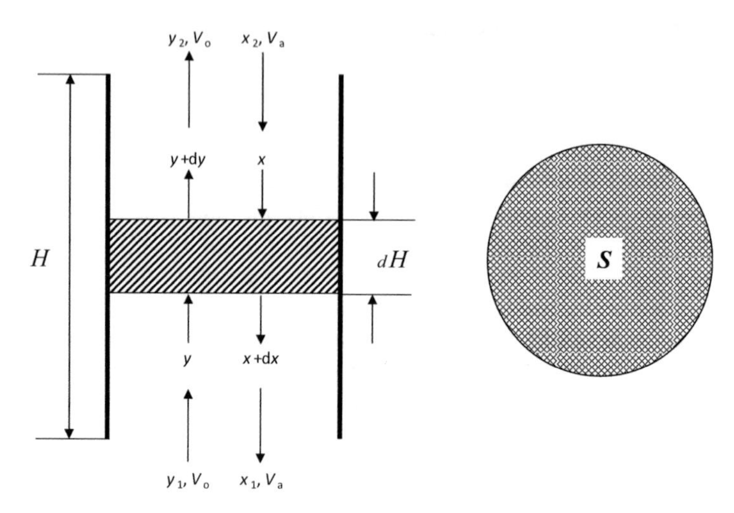

Fig. 6.7 Differential countercurrent extraction column

6.3.3 *Mass Transfer Unit*

In the design of differential extractors, a mass transfer unit approach is often used. A mass transfer unit is a physical piece of equipment that performs a specific function. However, there is no simple, clear-cut physical representation that can easily be made as an example of a physical mass transfer unit. Therefore, Fig. 6.7, a differential countercurrent extraction column, is used as an example to introduce the concept of mass transfer unit.

As shown in Fig. 6.7, the height of the column is H with cross-sectional area of S. The aqueous phase has the flow rate of V_a and solute concentration of x. The organic phase has the flow rate of V_o and solute concentration of y. In any section height of dH, the solute concentration changes in organic phase and aqueous phase are dy and dx, respectively. Assuming that the volume changes of the two phases are negligible, the rate of extraction in the section dH is

$$N = V_a \mathrm{d}x = V_o \mathrm{d}y \tag{6.5}$$

The interfacial area between the two phases in the section dH can be expressed by the following equation:

$$F = \alpha S \mathrm{d}H \tag{6.6}$$

where α is the interfacial area between the two phases in unit column volume and it is often referred to as specific interfacial area.

Substituting Eq. 6.6 into the extraction rate equation Eq. 6.1 gives

$$N = K_a \alpha S \mathrm{d}H \left(x - x^*\right) \tag{6.7}$$

Substituting Eq. 6.5 into Eq. 6.7 and integration gives

$$\int_0^H \mathrm{d}H = \int_{x_1}^{x_2} \frac{V_a}{K_a \alpha S} \frac{\mathrm{d}x}{x - x^*} \tag{6.8}$$

Assuming that $\frac{V_a}{K_a \alpha S}$ is constant in the whole column, the height of the column can be obtained in terms of the aqueous phase:

$$H = \frac{V_a}{K_a \alpha S} \int_{x_1}^{x_2} \frac{\mathrm{d}x}{x - x^*} \tag{6.9}$$

Following the same approach, the height of the column can be obtained in terms of the organic phase:

$$H = \frac{V_o}{K_o \alpha S} \int_{y_1}^{y_2} \frac{\mathrm{d}y}{y^* - y} \tag{6.10}$$

According to Eqs. 6.9 and 6.10, the column height is the product of two items: (1) $\frac{V_a}{K_a \alpha S}$ or $\frac{V_o}{K_o \alpha S}$ and (2) $\int_{x_1}^{x_2} \frac{\mathrm{d}x}{x - x^*}$ or $\int_{y_1}^{y_2} \frac{\mathrm{d}y}{y^* - y}$. The former includes phase flow rates, mass transfer coefficients, specific interfacial area, and cross-section area of the column. The former is defined as the height of the mass transfer unit (HTU) and is denoted by $(\mathrm{HTU})_a$ or $(\mathrm{HTU})_o$. The HTU is normally determined by testing and is the measurement of extraction efficiency. The latter relates to the solute concentration. It is defined as the number of mass transfer units and denoted by $(\mathrm{NTU})_a$ or $(\mathrm{NTU})_o$. Therefore, the height of the extraction column can be calculated by the following equation:

$$H = (\mathrm{HTU})_a \cdot (\mathrm{NTU})_a = (\mathrm{HTU})_o \cdot (\mathrm{NTU})_o \tag{6.11}$$

6.3.4 *Stage Efficiency*

In stage type of extractors, one actual physical stage is rarely 100 % of the efficient equilibrium stage or theoretical stage. Therefore, stage efficiency is used to quantify the actual separation in comparison with the theoretical separation. Overall stage efficiency is normally calculated by the ratio of the number of theoretical stages to the number of actual stages to reach the same separation results:

$$E_{\mathrm{o}} = \frac{\text{Number of Theroetical Stages}}{\text{Number of Actual Stages}} = \frac{n_{\mathrm{T}}}{n_{\mathrm{A}}} \tag{6.12}$$

where E_{o} denotes the overall stage efficiency, n_{T} the number of theoretical stages, and n_{A} the number of actual stages.

For each individual stage, stage efficiency can be calculated by either the solute concentration in organic phase or the solute concentration in aqueous phase:

$$E_y = \frac{y_t - y_0}{y_{\mathrm{e}} - y_0} \tag{6.13}$$

$$E_x = \frac{x_0 - x_t}{x_0 - x_{\mathrm{e}}} \tag{6.14}$$

where E_y and E_x are the individual stage efficiencies expressed by solute concentrations in organic phase and aqueous phase, respectively. y and x are the solute concentrations in organic phase and aqueous phase. The subscripts of solute concentration, 0, t, and e, indicate the solute concentration at the beginning, the solute concentration at time t, and the solute concentration at equilibrium, respectively.

6.3.5 *Axial Mixing*

The back mixing of either phase in the column type of extractors reduces the extraction efficiency. Many factors like column type, diameter, external forces, flow rate, phase ratio, and fluid properties can affect the axial mixing. The effects of axial mixing are considered by introducing an axial mixing factor (f_{m}) in the extraction rate equation:

$$N = KFf_{\mathrm{m}}\Delta C \tag{6.15}$$

The negative effects of axial mixing on extraction efficiency cannot be neglected for column type of extractors, especially for those large-size columns.

6.3.6 Dispersed Phase Holdup

The mass transfer between the organic and aqueous phases in column-type extractors depends on the interfacial area between the continuous and dispersed phases which is determined by the volume fraction or holdup of the dispersed phase as well as the mean droplet size. It is therefore important, at the design stage, to predict the dispersed liquid holdup for a given system, column geometry, and set of operating conditions. It is defined as the volume ratio of the dispersed phase to the column.

6.3.7 Equipment Efficiency

In the comparison of the efficiency of different extractors, using the rate of extraction (N) or the height of mass transfer unit (HTU) alone is not enough. It is reasonable to consider the mass transfer and flow rate together. Therefore, equipment efficiencies are defined as follows:

$$\varphi = \frac{V_{\mathrm{t}}/S}{(\mathrm{HETS})} = \frac{1}{\tau}(\text{Column type extractors}) \tag{6.16}$$

$$\varphi = \frac{V_{\mathrm{t}}}{V_{\mathrm{stage}}} = \frac{1}{\tau}(\text{Stage type extractors}) \tag{6.17}$$

where φ is the equipment efficiency, V_{t} is the maximum flow of the two phases, S is the cross-section area of the column, V_{stage} is the volume of a theoretical stage at the maximum flow, and HETS is the height equivalent of a theoretical stage at the maximum flow.

There are many factors that affect the equipment efficiency. They include equipment geometry, throughput, phase ratio, flow rate, concentration, temperature, fluid properties, and dispersed phase. For a given extraction system at determined operation conditions, the equipment geometry and throughput are the major factors that can be modified to improve equipment efficiency.

6.4 Design of Mixer-Settler

6.4.1 Number of Stages

Mixer-settler can be scaled up very well. The overall stage efficiency of a full-size mixer-settler is very close to the stage efficiency of a small-scale testing unit. The number of theoretical stages can be determined using Xu's cascade solvent extraction principles. The overall stage efficiency can be determined by small-scale testing. The number of actual stages is therefore determined by the number of theoretical stages and the overall stage efficiency.

6.4.2 Design of Mixer

In the design of mixer, it is important to consider phase mixing and mass transfer. The design of mixer also has a direct impact on settling. The parameters that have important impact on the mixer design include mixing residence time, agitation, and the impeller.

To ensure stage efficiency, the size of mixer should provide sufficient mixing or residence time. The residence time depends on the extractant, agitation, impeller diameter, etc. It is normally determined by bench testing or small-scale pilot testing. In rare earth solvent extraction, a typical mixing time is 5 min. As shown in Fig. 6.8 the cross section of mixers is normally square shape. The effective cell volume is determined by Eq. 6.18.

$$V_{\text{mixer}} = (Q_o + Q_a)\tau/60 = 1000x^2 z \tag{6.18}$$

$$r = \frac{y}{z},\ L = y + z = (1 + r)z \tag{6.19}$$

where V_{mixer} is the effective mixing volume (m^3), Q_o the volumetric flow rate of organic phase (m^3/h), Q_a the volumetric flow rate of aqueous phase (m^3/h), τ the residence time (min), x the length and width of the cell cross section (m), L the height of the cell (m), z the height of liquid surface (m), y the distance between the liquid surface and the top of the mixer (m), and r the ratio of the distance

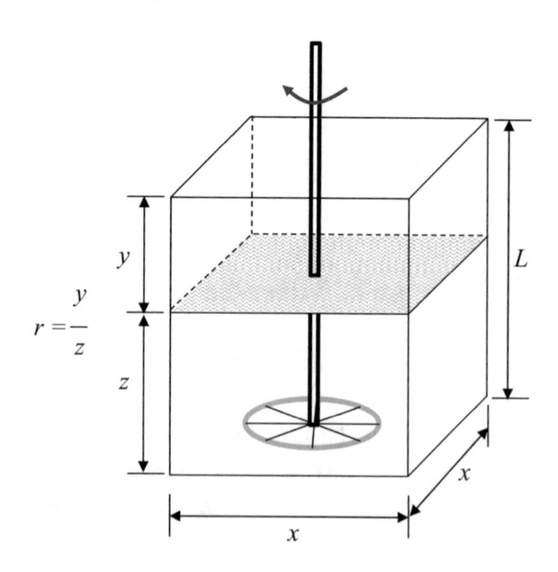

Fig. 6.8 Dimension of a mixer with square cross section

between the liquid and the top of the mixer to the height of the liquid. The ratio can be determined in small-scale testing.

To ensure good mixing and mass transfer, proper agitation is necessary. Higher speed agitation can improve mass transfer due to the reduction of droplet size of the dispersed phase and the increase of turbulence of the continuous phase. However, the mass transfer coefficient will decrease when agitation is too strong that decreases the internal circulation in small droplets and reduces droplet collision.

There are many different impeller shapes. But the turbo impeller is most common in mixer-settlers. A turbo impeller can generate large shear stress to the fluid and cause large circulation speed. It is often used to mix immiscible liquids. Open straight vane turbo impellers can generate the largest shear stress and dispersion. When the viscosity of the liquids is high, curved vane turbo impellers can be used to save energy. Currently, semi-closed turbo impeller with straight vanes is commonly used in large mixer-settlers.

6.4.3 Design of Settler

The scale-up of settler normally follows the principle of constant phase separation rate R_S ($m^3/(m^2\ h)$). The assumption is that the phase separation rate is constant for the same extraction system with the same mixing in the same type of settler. The area of settlers can be determined based on the phase separation rate which is normally determined by experiments and testing:

$$A_{\text{settler}} = (Q_o + Q_a)/R_S \tag{6.20}$$

where A_{settler} is the area of settler (m^2), Q_o the volumetric flow rate of organic phase (m^3/h), Q_a the volumetric flow rate of aqueous phase (m^3/h), and R_S the phase separation rate ($m^3/(m^2\ h)$).

After the area of the settler is determined, the ratio of length to width of the settler is to be determined as well. For industrial mixer-settlers, the length/width ratio is normally between 2 and 3.

It is common that the mixer and settler have the same width and height. Therefore, the design of the mixer and settler should be considered together.

6.5 Design of Packed Column

The scalability of column type of extractors is not very good in comparison with mixer-settlers due to many affecting factors such as HTU, HETS, and flooding rate. In the scale-up of a column extractor, it is important to consider the physical properties of the extraction system, operating conditions, mass transfer, equipment structure, etc.

6.5.1 Determination of the Dispersed Phase

For a given extraction system, the dispersed phase can be determined based on a few criteria: (1) the phase with higher volumetric flow rate, (2) the phase with higher viscosity, (3) the phase with weaker wetting ability for the packing materials, and (4) the phase that is inflammable.

6.5.2 Packing Material

The packing material should be easily wetted by the continuous phase. The wetting by the dispersed phase should be minimized. Generally, metal packing is easily wetted by the aqueous phase while stainless steel packing can be wetted by both organic and aqueous phases. Plastic and ceramic packing can be considered to avoid corrosion.

6.5.3 Column Diameter

The diameter of a packed column can be calculated using the following equation:

$$D = \sqrt{\frac{4Q_c}{\pi v_c}} \tag{6.21}$$

where D is the inner diameter of the column (m), Q_c the volumetric flow rate of the continuous phase (m^3/h), and v_c the operating line speed of the continuous phase (m/h).

According to Eq. 6.21, the column diameter is smaller at a higher operating line speed when the volumetric flow rate of the continuous phase is constant. However, the operating line speed is limited by the flooding capacity of the column. Flooding is the occurrence where one phase flows with the other with high operating line speed. Flooding speed is the operating line speed that starts to cause flooding and is the upper limit of allowed operating line speed. The actual operating line speed is always lower than the flooding speed in order to maintain steady operation. The following equation can be used to calculate the operating line speed:

$$v_c = m v_F \tag{6.22}$$

where m is the coefficient ($m < 1$) and v_F is the flooding speed.

The coefficient m is different for different types of contactors. For spray towers, packed columns, and pulse columns, m is approximately between 0.5 and 0.6. The flooding speed of a column can be estimated using empirical models.

The operating line speed can also be determined through testing. While steady operation is maintained, higher operating line speed should be used.

6.5.4 *Column Height*

According to phase equilibrium data obtained from testing and the process information about feed, solvent, and raffinate concentrations and flow rates, the number of theoretical stages can be determined by the HETS method. The height of the column can be determined by the number of theoretical stages and HETS according to Eq. 6.4.

If the mass transfer coefficients can be obtained and the effective mass transfer area is known, the HTU method can be used to calculate the column height using Eq. 6.11.

6.5.5 *Testing and Scale-Up*

In the design of column extractors, testing is often performed in small columns to confirm the estimated maximum throughput capacity, average drop size, and equipment efficiency. The tray distance, pulse amplitude, and flow rate of unit cross-sectional area are normally kept the same in the scale-up.

The HETS can be calculated by the following equation in the scale-up design:

$$\frac{(\mathrm{HETS})_2}{(\mathrm{HETS})_1} = \left(\frac{D_2}{D_1}\right)^{0.38} \tag{6.23}$$

where the subscripts of 1 and 2 denote the testing column and the full scale column, respectively. The power index 0.38 is not constant and will be adjusted based on the extraction system.

The pulse frequency (SMP) can be calculated by the following equation:

$$\frac{(\mathrm{SMP})_2}{(\mathrm{SMP})_1} = \left(\frac{D_1}{D_2}\right)^{0.14} \tag{6.24}$$

References

Ryon, A. D., Daley, F. L., & Lowrie, R. S. (1960). Design and scaleup of mixer-settlers for the DAPEX solvent extraction process. U.S. Atomic Energy Commission.

Tai, G., Bao, W., & He, P. (1995). Selection and application of extractors in separating rare earth elements. *Journal of Rare Earths, 16*(2), 65–69.

Xu, G., & Yuan, C. (1987). *Solvent extraction of rare earths* (pp. p480–521). China: China Science Press.

Printed in the United States
By Bookmasters